INTERNATIONAL TECHNOLOGICAL UNIVERSITY
This Book is Donated by:
Christina Bell
Date: 6/22/1999

32 Springer Series in Solid-State Sciences
Edited by Peter Fulde

Springer Series in Solid-State Sciences
Editors: M. Cardona P. Fulde H.-J. Queisser

Volume 40 **Semiconductor Physics** – An Introduction By K. Seeger

Volume 41 **The LMTO Method** By H.L. Skriver

Volume 42 **Crystal Optics with Spatial Dispersion and the Theory of Excitations**
By V.M. Agranovich and V.L. Ginzburg

Volume 43 **Resonant Nonlinear-Interactions of Light with Matter**
By V.S. Butylkin, A.E. Kaplan, Yu.G. Khronopulo, and E.I. Yakubovich

Volume 44 **Elastic Media with Microstructure II** Three-Dimensional Models
By I.A. Kunin

Volume 45 **Electronic Properties of Doped Semiconductors**
By B.I. Shklovsky and A.L. Efros

Volumes 1 – 39 are listed on the back inside cover

Robert M. White

Quantum Theory of Magnetism

Second Corrected and Updated Edition

With 113 Figures

Springer-Verlag Berlin Heidelberg New York 1983

Professor Dr. *Robert M. White*
Xerox Corporation, Palo Alto Research Center
Palo Alto, CA 94304, USA

Series Editors:
Professor Dr. Manuel Cardona
Professor Dr. Peter Fulde
Professor Dr. Hans-Joachim Queisser
Max-Planck-Institut für Festkörperforschung, Heisenbergstrasse 1
D-7000 Stuttgart 80, Fed. Rep. of Germany

Title of the 1. Edition: *Quantum Theory of Magnetism*
© by McGraw-Hill, Inc., New York 1970

ISBN 3-540-11462-9 Springer-Verlag Berlin Heidelberg New York
ISBN 0-387-11462-9 Springer-Verlag New York Heidelberg Berlin

Library of Congress Cataloging in Publication Data. White, Robert M., 1938-. Quantum theory of magnetism. (Springer series in solid-state sciences ; 32). Bibliography: p. Includes index. 1. Quantum theory. 2. Magnetism. 3. Magnetic susceptibility. I. Title. II. Series. QC754.2.Q34W48 1982 530.1'41 82-6014 AACR2

This work is subject to copyright. All rights are reserved, whether the whole or part of the material is concerned, specifically those of translation, reprinting, reuse of illustrations, broadcasting, reproduction by photocopying machine or similar means, and storage in data banks. Under § 54 of the German Copyright Law, where copies are made for other than private use, a fee is payable to "Verwertungsgesellschaft Wort", Munich.

© by Springer-Verlag Berlin Heidelberg 1983
Printed in Germany

The use of registered names, trademarks, etc. in this publication does not imply, even in the absence of a specific statement, that such names are exempt from the relevant protective laws and regulations and therefore free for general use.

Offset printing: Beltz Offsetdruck, 6944 Hemsbach/Bergstr. Bookbinding: J. Schäffer OHG, 6718 Grünstadt.

*To Sara,
Victoria, and Jonathan*

Preface to the Second Edition

Although it is one of the oldest physical phenomena studied, magnetism continues to be an active and challenging subject. This is due to the fact that magnetic phenomena represent a complex application of quantum mechanics, statistical physics, and electromagnetism. As new magnetic materials are synthesized and new experimental conditions realized, the very fundamentals of these subjects are expanded. Thus, the Kondo effect, like superconductivity, stimulated the development of many-body techniques; spin glasses with their competing interactions are leading to advances in statistical physics; and angle- and spin-resolved photoemission is probing details of transition-metal electronic states never before possible.

I have not tried to incorporate all the new developments in this subject since the first edition ten years ago. My purpose is still the same — to use linear response theory to establish a common conceptual basis for understanding a variety of magnetic phenomena. Many recent developments fit into this framework and have been included.

I would like to thank Professor Peter Fulde and Dr. Helmut Lotsch who encouraged me to undertake this version. Much of the work was done during six months I spent at the Max−Planck−Institut in Stuttgart as the result of an award from the Alexander von Humboldt Foundation. I benefited from many discussions with the members of Professor Fulde's *Abteilung,* particularly Dr. A.M. Oles. A number of other colleagues also deserve special thanks − Dr. Ralph Moon kindly read the chapter on neutron scattering and made helpful comments, and Professor Conyers Herring clarified various aspects of the derivation of the Onsager relation in Chapter 1, and Elizabeth Plowman cheerfully typed much of the new material. Finally, I want to thank my colleague Professor T. Geballe for his continued interest and support.

Palo Alto, December 1982 *R.M. White*

Contents

1. **The Magnetic Susceptibility**
   1.1 The Magnetic Moment 2
   1.2 The Magnetization 6
   1.3 The Generalized Susceptibility 11
       1.3.1 The Kramers–Kronig Relations 13
       1.3.2 The Fluctuation-Dissipation Theorem 15
       1.3.3 Onsager Relation 19
   1.4 Second Quantization 20

2. **The Magnetic Hamiltonian**
   2.1 The Dirac Equation 29
   2.2 Sources of Fields 31
       2.2.1 Uniform External Field 31
       2.2.2 The Electric Quadrupole Field 32
       2.2.3 The Magnetic Dipole (Hyperfine) Field 36
       2.2.4 Other Electrons on the Same Ion 38
       2.2.5 Crystalline Electric Fields 38
       2.2.6 Dipole–Dipole Interaction 44
       2.2.7 Direct Exchange 45
       2.2.8 Superexchange 52
   2.3 The Spin Hamiltonian 55
       2.3.1 Transition-Metal Ions 56
       2.3.2 Rare-Earth Ions 63
       2.3.3 Semiconductors 66

3. **The Static Susceptibility of Noninteracting Systems**
   3.1 Localized Moments 69
       3.1.1 Diamagnetism 72
       3.1.2 Paramagnetism of Transition-Metal Ions 73
       3.1.3 Paramagnetism of Rare-Earth Ions 76
   3.2 Metals 78
       3.2.1 Landau Diamagnetism 78
       3.2.2 The de Haas–van Alphen Effect 83
       3.2.3 Quantized Hall Conductance 86
       3.2.4 Pauli Paramagnetism 90
   3.3 Measurement of the Susceptibility 96

4. The Static Susceptibility of Interacting Systems
- 4.1 Localized Moments ... 101
  - 4.1.1 High Temperatures ... 103
  - 4.1.2 Low Temperatures ... 111
  - 4.1.3 Temperatures near T_c ... 115
  - 4.1.4 Topological Long-Range Order ... 121
- 4.2 Metals ... 123
  - 4.2.1 Fermi Liquid Theory ... 124
  - 4.2.2 The Stoner Model ... 130
  - 4.2.3 The Hubbard Model ... 139

5. The Dynamic Susceptibility of Weakly Interacting Systems
- 5.1 Localized Moments ... 143
  - 5.1.1 The Bloch Equations ... 147
  - 5.1.2 Resonance Line Shape ... 150
  - 5.1.3 Measurement of T_1 ... 158
  - 5.1.4 Calculation of T_1 ... 161
- 5.2 Metals ... 165
  - 5.2.1 Paramagnons ... 167
  - 5.2.2 Fermi Liquid Theory ... 169
- 5.3 Faraday Effect ... 181

6. The Dynamic Susceptibility of Strongly Interacting Systems
- 6.1 Broken Symmetry ... 183
- 6.2 Insulators ... 184
  - 6.2.1 Spin–Wave Theory ... 186
  - 6.2.2 Magnetostatic Modes ... 193
  - 6.2.3 Solitons ... 196
  - 6.2.4 Thermal Magnon Effects ... 197
  - 6.2.5 Parametric Excitation ... 201
  - 6.2.6 Optical Processes ... 203
  - 6.2.7 High Temperatures ... 206
- 6.3 Metals ... 209

7. Magnetic Impurities
- 7.1 Local Modes ... 214
- 7.2 Local Moments in Metals ... 224
  - 7.2.1 Anderson's Theory of Moment Formation ... 227
- 7.3 The Kondo Effect ... 232
- 7.4 Random Exchange ... 239
  - 7.4.1 The RKKY Interaction ... 239
  - 7.4.2 Spin Glasses ... 242
  - 7.4.3 Mictomagnetism ... 245

8. Neutron Scattering

8.1 Neutron Scattering Cross Section 248
8.2 Nuclear Scattering 249
    8.2.1 Bragg Scattering 252
    8.2.2 Scattering by Phonons 253
8.3 Magnetic Scattering 256
    8.3.1 Bragg Scattering 261
    8.3.2 Diffuse Scattering 266

References 271

Subject Index 279

1. The Magnetic Susceptibility

Any system may be characterized by its response to external stimuli. For example, in electronics the proverbial "black box" is characterized by its measured output voltage when an input current is applied. This *transfer impedance*, as it is called, provides all the information necessary to understand the operation of the black box. If we know what is in the black box—for example, the detailed arrangement of resistors, diodes, etc.—then we can predict, through analysis, what the transfer impedance will be.

Similarly, a system of charges and currents, such as a crystal, may be characterized by a *response function*. In this text we shall be concerned mainly with the response of such a system to a magnetic field. In this case the "output" is the magnetization and the response function is the *magnetic susceptibility*. A complete analysis of the magnetic susceptibility is virtually impossible since the system consists of about 10^{23} particles. Therefore we usually look to a measured susceptibility for clues to the important mechanisms active in the system and then use these to analyze the system. In order to carry out such a program, we must know what possible mechanisms exist and what effect they have on the susceptibility.

Determination of the susceptibility entails evaluation of the magnetization produced by an applied magnetic field. In general, this applied field may depend on space and time. The resulting magnetization will also vary in space and time. If the spatial dependence of the applied field is characterized by a wave vector q and its time dependence is characterized by a frequency ω, and if we restrict ourselves for the time being to the magnetization with this wave vector and frequency, we obtain the *susceptibility* $\chi(q, \omega)$. As we shall see shortly, the magnetization is the average magnetic moment. The magnetic moment itself is a well-defined quantity. The problem, however, is the computation of its average value. In order to compute this average it is necessary to know the probabilities of the system being in its various configurations. This information is contained in the *distribution function* associated with the system.

We shall see in this chapter that the distribution function depends on the total energy, or *Hamiltonian*, of the system. Therefore the first step in understanding magnetic properties is the identification of those interactions relevant to magnetism. In Chap. 2 the origin of these interactions is discussed, and they are expressed in a form which facilitates their application in later chapters. The reader is asked to keep in mind that Chaps. 1, 2 both constitute background material for the theoretical development which begins in Chap. 3. The motivation for the material in these first two chapters should become clear as this theory unfolds.

In the absence of time-dependent fields we may assume that the system is in thermal equilibrium. In this case the distribution function is easily obtained. In Chap. 3 this is used to compute the response of noninteracting moments to a static field. This computation leads to the susceptibility $\chi(\boldsymbol{q}, 0)$. In Chap. 4 the response $\chi(\boldsymbol{q}, 0)$ of an interacting system of moments to a static field is investigated in the random-phase approximation.

In the presence of time-dependent fields the distribution function must be obtained from its equation of motion. In the case of localized moments this consists of solving the *Bloch equations*. For itinerant moments the distribution function is obtained from a *Boltzmann equation*. In Chap. 5 these equations are solved for weakly interacting systems to obtain the generalized susceptibility $\chi(\boldsymbol{q}, \omega)$. Finally, in Chap. 6 the generalized susceptibility associated with strongly interacting systems is investigated. This function is of particular interest because its singularities determine the magnetic-excitation spectrum of the system.

The next few sections introduce the basic quantities with which we shall be concerned throughout this text. Since these quantities may be defined in various ways, the reader may find it informative to compare other approaches (especially the classic work [1.1]).

1.1 The Magnetic Moment

Let us begin by discussing the magnetic moment. To see why this particular object is of interest let us consider the classical description of a system of charges and currents. Such a system is governed by Maxwell's equations. The appropriate forms of these equations in a medium are the so-called *macroscopic Maxwell equations*, which are obtained from the microscopic equations by averaging over a large number of particles. The microscopic equation in which we shall be particularly interested is the one representing Ampere's law, which has the differential form

$$\nabla \times \boldsymbol{h} = \frac{4\pi}{c}\boldsymbol{j} + \frac{1}{c}\frac{\partial \boldsymbol{e}}{\partial t}. \tag{1.1}$$

We define the average fields

$$\begin{aligned}\langle \boldsymbol{h} \rangle &\equiv \boldsymbol{B}, \\ \langle \boldsymbol{e} \rangle &\equiv \boldsymbol{E}.\end{aligned} \tag{1.2}$$

Here $\langle \ldots \rangle$ is a spatial average over a region which is small compared with the size of the sample, yet large enough to contain many atomic systems.

When we write $B(r)$ or $E(r)$, the coordinate r refers to the center of the region over which the average is taken. Thus the first equation of (1.2) might have been written as $B(r) = \langle h \rangle_r$. In this description it is assumed that any spatial variations are large in comparison with interatomic spacings. The actual details of the averaging will be discussed in Sect. 1.2. With this notation (1.1) becomes

$$\nabla \times B = \frac{4\pi}{c} \langle j \rangle + \frac{1}{c} \frac{\partial E}{\partial t} . \tag{1.3}$$

The objective now is to calculate the average current density. To do this we separate the total current density into two parts, that associated with conduction electrons and that localized at an ionic site. The average value of the conduction-electron current density is the *free current density* j_{free}.

The ionic current density may be further separated into two contributions. First of all, the ion may possess an electric-dipole moment which is characterized by a *dipole charge density* ρ_{dip}. If this charge density is time dependent, there is a *polarization current density* j_{pol} which satisfies the continuity equation

$$\nabla \cdot j_{\text{pol}} = -\frac{\partial \rho_{\text{dip}}}{\partial t} . \tag{1.4}$$

Taking the average of this equation and assuming that the average commutes with the time and space derivatives, we obtain

$$\langle \sum_{\text{ions}} j_{\text{pol}} \rangle = \frac{\partial P}{\partial t}, \tag{1.5}$$

where the sum is over those ions within the averaging volume and P is the electric polarization defined by

$$\langle \sum_{\text{ions}} \rho_{\text{dip}} \rangle = -\nabla \cdot P .$$

The second contribution to the ionic current density arises from the internal motion of the ionic electrons. Since this current density j_{mag} is stationary, $\nabla \cdot j_{\text{mag}} = 0$. This is the current density responsible for the *magnetic moment* m of the ion. If the center of mass of the ion is at R, the magnetic moment is defined as

$$\boxed{m = \frac{1}{2c} \int dr(r - R) \times j_{\text{mag}}} . \tag{1.6}$$

A convenient representation for j_{mag} which has zero divergence and satisfies (1.6) is

$$j_{\text{mag}} = -cm \times \nabla f(|r - R|), \tag{1.7}$$

where $f(|\boldsymbol{r} - \boldsymbol{R}|)$ is an arbitrary smoothly varying function centered at $\boldsymbol{R}$ which goes to 0 at the ionic radius and is normalized to 1. Then

$$\langle \sum_{\text{ions}} \boldsymbol{j}_{\text{mag}} \rangle = c \langle \sum_{\text{ions}} \nabla f(|\boldsymbol{r} - \boldsymbol{R}|) \times \boldsymbol{m} \rangle = c \nabla \times \langle \sum_{\text{ions}} f(|\boldsymbol{r} - \boldsymbol{R}|) \boldsymbol{m} \rangle . \tag{1.8}$$

The last average in (1.8) is the magnetization $\boldsymbol{M}$, defined by

$$\boldsymbol{M} \equiv \langle \sum_{\text{ions}} f(|\boldsymbol{r} - \boldsymbol{R}|) \boldsymbol{m} \rangle . \tag{1.9}$$

Combining these results, we may now write (1.3) as

$$\nabla \times \boldsymbol{B} = \frac{4\pi}{c} \boldsymbol{j}_{\text{free}} + \frac{4\pi}{c} \frac{\partial \boldsymbol{P}}{\partial t} + 4\pi \nabla \times \boldsymbol{M} + \frac{1}{c} \frac{\partial \boldsymbol{E}}{\partial t} . \tag{1.10}$$

Defining

$$\boldsymbol{H} = \boldsymbol{B} - 4\pi \boldsymbol{M} \tag{1.11}$$

and

$$\boldsymbol{D} = \boldsymbol{E} + 4\pi \boldsymbol{P} , \tag{1.12}$$

we have the familiar result

$$\nabla \times \boldsymbol{H} = \frac{4\pi}{c} \boldsymbol{j}_{\text{free}} + \frac{1}{c} \frac{\partial \boldsymbol{D}}{\partial t} . \tag{1.13}$$

Thus we see that the magnetization which appears in the macroscopic Maxwell's equations is the average of the ionic magnetic moment given by (1.6).

As an example of the use of definition (1.6), let us neglect the possibility of nuclear currents and consider only the electron currents within the ion. Then

$$\boldsymbol{j}_{\text{mag}}(\boldsymbol{r}) = \sum_{\alpha} e \boldsymbol{v}_\alpha \delta(\boldsymbol{r} - \boldsymbol{r}_\alpha) , \tag{1.14}$$

where e is the charge on the electron, which is $-|e|$, and $\boldsymbol{v}_\alpha$ is the velocity of the αth electron. From (1.6) we find for the total magnetic moment of the ion

$$\boldsymbol{m} = \frac{e}{2c} \sum_{\alpha} \boldsymbol{r}_\alpha \times \boldsymbol{v}_\alpha . \tag{1.15}$$

Recalling that the orbital angular momentum of an electron is

$$\boldsymbol{l}_\alpha = \boldsymbol{r}_\alpha \times m \boldsymbol{v}_\alpha , \tag{1.16}$$

we have

$$m = \sum_\alpha \frac{e}{2mc} l_\alpha \,. \tag{1.17}$$

Since $e = -|e|$, we see that the orbital magnetic moment of an electron is in the opposite direction to its angular momentum.

We shall find it convenient to adopt a more general definition of the magnetic moment than that given by (1.6). This definition is based on the energy of the magnetic system, which is often a confusing subject (magnetic energy is discussed in [1.2]). The source of this confusion lies in the definition of the magnetic system. Let us define our magnetic system as characterized by an ionic magnetic current density j_{mag}. This clearly excludes the free currents which are the source of an external field H in which our magnetic ion is to be located. We now want to know the change in energy of this magnetic system when the field H is applied. This energy difference results from the work done on the field by the magnetic currents in their effort to maintain themselves in the presence of the external field. Since the magnetic field itself does no work on moving charges, we must use the induced electric field which is present while the external magnetic field is turned on. This is given by

$$\nabla \times E = -\frac{1}{c}\frac{\partial H}{\partial t}. \tag{1.18}$$

The work done on the field E by the magnetic currents in a time δt is

$$W = -\int j_{\text{mag}} \cdot E \, dr \, \delta t \,. \tag{1.19}$$

Making use of the representation (1.7) for j_{mag}, integrating by parts, and then using (1.18), we obtain

$$W = \int f(|r - R|) m \cdot \delta H \, dr \,. \tag{1.20}$$

If the field H is uniform over the ionic dimension, δH may be taken outside the integral. Since m is just a constant vector and $f(|r - R|)$ is normalized to unity,

$$W = m \cdot \delta H \,. \tag{1.21}$$

The resulting change in the energy of the magnetic system is $\delta E = -W$. Thus we may write

$$\boxed{m = -\frac{\partial E}{\partial H}} \,. \tag{1.22}$$

In the case we have been discussing our magnetic system was an ion. However, (1.22) applies to any system in which the ionic moment m is replaced by the total moment MV, where M is the magnetization and V is the volume of the system. As an example of the application of this definition, consider the ionic system of electrons which gave rise to the current density of (1.14). In the presence of a uniform field H, which may be obtained from a vector potential A by $H = \nabla \times A$, the energy of such a system is

$$E = \sum_\alpha \tfrac{1}{2}mv_\alpha^2 + \sum_\alpha e\phi_\alpha , \tag{1.23}$$

where ϕ_α is the ionic potential. We see that the magnetic field H does not appear explicitly in this energy. However, the velocity is, in fact, a function of the field. In general, the task of finding the field dependence of the total energy of the system is a difficult one. However, in this case the actual field dependence is revealed by expressing the energy in terms of the canonical coordinates of the system. The reason for this is that in a slowly varying uniform field the canonical momentum does not change. When expressed in canonical coordinates, the energy is the same as the Hamiltonian function. For this reason the *Hamiltonian* $\mathcal{H}$ is often used in place of the energy E in definition (1.22).

In Chap. 2 we shall find that the Hamiltonian is

$$\mathcal{H} = \sum_\alpha \frac{1}{2m}\left(p_\alpha - \frac{e}{c}A\right)^2 + \sum_\alpha e\phi_\alpha . \tag{1.24}$$

With the gauge $A = \tfrac{1}{2} H \times r$ this becomes

$$\mathcal{H} = \sum_\alpha \frac{p_\alpha^2}{2m} - \sum_\alpha \frac{e}{2mc}(r_\alpha \times p_\alpha)\cdot H + \frac{e^2}{8mc^2}\sum_\alpha (H \times r_\alpha)^2 + \sum_\alpha e\phi_\alpha . \tag{1.25}$$

Differentiating with respect to H and using the fact that the canonical momentum is given by $p_\alpha = mv_\alpha + (e/c)A$, we obtain (1.15).

1.2 The Magnetization

The magnetization is obtained by averaging the ionic moments over a region of space which is large enough to give such an average a meaning but smaller than spatial variations in the system. In order to perform this average we must know the ionic-current distributions. In general they are not known. In fact, herein lies the principle difficulty in the theory of magnetism. In any real system the motion of charge in one region is governed by the charge and currents throughout the system. Thus we have a many-body problem. There have been two historical approaches to this problem, the method of localized moments and the method of itinerant moments. The choice between these two methods depends on the nature of the material and in many cases is a difficult one to make.

In certain cases the relevant current distributions are localized within a lattice cell. In such cases the ionic magnetic moment is relatively unambiguous. The interaction with external charge and current distributions is then expressed in terms of this moment. This approach leads to the *spin Hamiltonian*, which has proved extremely useful. In other cases we begin by assuming that the current distributions are those associated with free electrons. Thus, although these electrons may extend throughout the lattice, the fact that they may be considered as a "gas" provides a certain simplification. With these two types of current distributions—that corresponding to very localized electrons and that corresponding to itinerant electrons—we can proceed to determine the average moment. The computation of averages necessarily relies on the techniques of statistical mechanics. Since this is not the place to develop such techniques, they will be introduced in a rather ad hoc manner. A more thorough derivation may be found in texts on statistical physics [1.3].

If the system possesses translational invariance, then the statistical average over numerous unit cells of the crystal is equivalent to the time average over one cell. This average is determined by the probability that the system will have some particular current distribution. For example, in the case of a magnetic insulator the average over a cell is an average of the magnetic ion. If this ion consists of h electrons, then, classically, the state of the ion is characterized by the $6h$ coordinates and momenta, $(q_1, \ldots, q_{3h}, p_1, \ldots, p_{3h})$. The magnetization is obtained by multiplying the magnetic moment, which is a function of all the coordinates and momenta, by the probability that the system is in the state $(q_1, \ldots, p_1, \ldots)$ and then integrating over all the coordinates and momenta. This probability is determined by the ion's environment, which defines a temperature T. For the most part this will be the temperature of the lattice in which the ions are located. The equilibrium probability function is then just the Boltmann distribution function $\exp(-\beta \mathcal{H}_{\text{ion}})$, where $\mathcal{H}_{\text{ion}}$ is the classical Hamiltonian for the ion and $\beta = 1/k_B T$. Therefore the equilibrium magnetization associated with N/V ions per unit volume is

$$M = \frac{N}{V} \langle m \rangle = \frac{N}{V} \frac{\int \ldots \int m \exp(-\beta \mathcal{H}_{\text{ion}}) dq_1 \ldots dp_1 \ldots}{\int \ldots \int \exp(-\beta \mathcal{H}_{\text{ion}}) dq_1 \ldots dp_1 \ldots}. \quad (1.26)$$

It is interesting that this classical averaging procedure leads to the conclusion that there can be no magnetism in thermodynamic equilibrium. The reason for this is that the integrals over the momenta in (1.26) run from $-\infty$ to $+\infty$. Therefore adding a vector potential may shift the momentum origin, but it will not affect the limits of integration. Since the vector potential always enters the integrand as an addition to the momentum, it may be transformed away. For example, consider the partition function for an electron in a magnetic field

$$Z = \int_{-\infty}^{\infty} dx\, dy\, dz \int_{-\infty}^{\infty} dp_x dp_y dp_z \exp\{-\beta[p_x - (e/c)A_x]^2/2m + \ldots\}. \quad (1.27)$$

We introduce

$$u = p_x - \frac{e}{c} A_x, \ldots, \qquad (1.28)$$

where, in general, A may be a function of r. Then

$$Z = V \int_{-\infty}^{\infty} du\, dv\, dw \exp - \beta(u^2 + v^2 + w^2)/2m \qquad (1.29)$$

which is independent of A. Therefore the derivative of Z with respect to the field H, which can be shown to be proportional to the magnetization, is 0. This result, known as *Miss van Leeuwen's theorem*, forces us to consider the discreteness of the eigenvalues of the system and hence its quantum-mechanical nature. This interesting result has the following physical interpretation (see the discussion in [Ref. 1.1, Sect. 26]). In the presence of a magnetic field the electrons move in circular orbits in the plane perpendicular to the field. Those electrons that complete such orbits contribute a diamagnetic moment. However, those electrons which strike the boundary have their orbits interrupted with the result that they creep around the boundary giving rise to a paramagnetic moment. It turns out that this paramagnetic moment just cancels the diamagnetic moment. Furthermore, it is independent of the size and nature of the boundary.

Quantum mechanically, the magnetic system is described by a Hamiltonian $\mathcal{H}$ which has eigenfunctions ψ with eigenvalues E. The total magnetic moment of the system when it is in the state ψ is, according to (1.22).

$$MV = -\frac{\partial E}{\partial H}. \qquad (1.30)$$

This may be written in a more useful form. First we differentiate the eigenvalue relation

$$(\mathcal{H} - E)\psi = 0 \qquad (1.31)$$

with respect to H to obtain

$$\left(\frac{\partial \mathcal{H}}{\partial H} - \frac{\partial E}{\partial H}\right)\psi = -(\mathcal{H} - E)\frac{\partial \psi}{\partial H}. \qquad (1.32)$$

Forming the scalar product with ψ and using the hermiticity of $\mathcal{H}$, we find

$$\left\langle \psi \left| \frac{\partial \mathcal{H}}{\partial H} \right| \psi \right\rangle - \frac{\partial E}{\partial H} = -\left\langle \psi \left| (\mathcal{H} - E) \frac{\partial \psi}{\partial H} \right.\right\rangle = -\left\langle \frac{\partial \psi}{\partial H} \left| (\mathcal{H} - E) \right| \psi \right\rangle^* = 0. \qquad (1.33)$$

Therefore

1.2 The Magnetization

$$MV = -\left\langle \psi \left| \frac{\partial \mathcal{H}}{\partial H} \right| \psi \right\rangle. \qquad (1.34)$$

This leads us to define a *magnetic-moment operator*

$$\boxed{\mathcal{M} = -\frac{\partial \mathcal{H}}{\partial H}}. \qquad (1.35)$$

Hereafter $\mathcal{M}$ will be understood to be an operator quantity.

Since the derivation above was independent of the detailed form of the Hamiltonian and its eigenfunctions, the result (1.35) is quite general. For example, the magnetic-moment operator for a particle governed by the nonrelativistic Schrödinger Hamiltonian (1.25) is

$$\mathcal{M}_z = \frac{e}{2mc}(xp_y - yp_x) - \frac{He^2}{4mc^2}(x^2 + y^2) = \frac{e}{2c}(x\dot{y} - y\dot{x}). \qquad (1.36)$$

For a relativistic electron governed by the Dirac equation (which will be discussed briefly in Chap. 2) the magnetic moment becomes

$$\mathcal{M}_z = -\frac{e}{2}(\alpha_x y - \alpha_y x). \qquad (1.37)$$

Since the α's are 4×4 matrices acting on negative as well as positive energy states, the physical meaning of the operator is not clear. However, if we apply a transformation to this operator which separates the positive and negative energy states, then we find that an intrinsic spin contribution to the magnetic moment emerges automatically.

To find the magnetization we must find the average of this magnetic-moment operator. If we knew the wave function ψ, this would be a straight-forward matter. However, the fact that we are describing the system at a temperature T implies that the system is in equilibrium with some temperature bath. Therefore the wave function ψ would have to include these extra degrees of freedom. Rather than do this, we use a wave function ϕ which is an eigenfunction of the magnetic system alone, but we allow for the fact that as time progresses this wave function changes in some fashion because it is not a true eigenfunction. Thus to find the thermal average of $\mathcal{M}$ we must find the expectation value of this operator in some state and then average over the states through which the system evolves in time. To facilitate this process let us expand the wave function $\phi(r_1, r_2, \ldots, r_N, t)$ in products of singleparticle wave functions. If we write such a product as φ_k, then

$$\phi(r_1, r_2, \ldots, r_N, t) = \sum_k c_k(t)\varphi_k(r_1, r_2, \ldots, r_N) \qquad (1.38)$$

where the $c_k(t)$ are time-dependent expansion coefficients.

At some later time t' the system will be characterized by $\phi(t')$. The probability that the system will be in a given state is proportional to the number of times it passes through that state in time. However, instead of thinking of the system at different times, we can just as well think of $\mathcal{N}$ identical systems at the same time. Thus, instead of averaging over time, we can average over this *ensemble*. Suppose the wave function for the nth system is

$$\phi^n(t) = \sum_k c_k^n(t) \varphi_k . \tag{1.39}$$

Then the average magnetic moment for this system is

$$\mathcal{M}^n = \int \phi^n(t)^* \mathcal{M} \phi^n(t) d\tau \tag{1.40}$$

where $d\tau \equiv dr_1 dr_2 \ldots dr_N$. The ensemble average is

$$\langle \mathcal{M} \rangle = \frac{1}{\mathcal{N}} \sum_n \mathcal{M}^n . \tag{1.41}$$

Using (1.39, 40), we may write this as

$$\langle \mathcal{M} \rangle = \sum_{k,k'} \rho_{k'k} \mathcal{M}_{kk'} = \text{tr}\{\rho \mathcal{M}\} , \tag{1.42}$$

where

$$\mathcal{M}_{kk'} = \int \varphi_k^* \mathcal{M} \varphi_{k'} d\tau$$

and

$$\rho_{k'k} \equiv \frac{1}{\mathcal{N}} \sum_n c_k^n(t)^* c_{k'}^n(t) \tag{1.43}$$

defines the *density matrix*.

This matrix may also be defined by its equation of motion. From the Schrödinger equations

$$i\hbar \frac{\partial \phi^n(t)^*}{\partial t} = - \mathcal{H} \phi^n(t)^* , \tag{1.44}$$

$$i\hbar \frac{\partial \phi^n(t)}{\partial t} = \mathcal{H} \phi^n(t) , \tag{1.45}$$

we obtain the equations for the expansion coefficients. Using (1.43), we find that

$$i\hbar \frac{\partial \rho_{k'k}}{\partial t} = - \sum_{k''} \rho_{k'k''} \mathcal{H}_{k''k} + \sum_{k''} \mathcal{H}_{k'k''} \rho_{k''k} \tag{1.46}$$

or

$$i\hbar \frac{\partial \rho}{\partial t} = [\mathscr{H}, \rho] \ .\tag{1.47}$$

This is often a more convenient approach to the density matrix, for, as we shall see below, when perturbation theory applies, (1.47) may be solved iteratively.

Notice that (1.42) gives the average of the magnetic moment over the entire system. If we are interested in the magnetization at point r, this behavior can be projected out as

$$\mathscr{M}(r) = \tfrac{1}{2} \sum_\alpha [\boldsymbol{\mu}_\alpha \delta(r - r_\alpha) + \delta(r - r_\alpha)\boldsymbol{\mu}_\alpha] \ .\tag{1.48}$$

Since the delta functions have dimensions of a reciprocal volume, $\mathscr{M}(r)$ is the magnetic-moment operator per unit volume. Here $\boldsymbol{\mu}_\alpha$ is the magnetic-moment operator associated with the αth electron. Notice that since this is a function of r_α and p_α, we must form the symmetric product. We may therefore generalize (1.42) by writing

$$M(r) = \mathrm{Tr}\{\rho \mathscr{M}(r)\} \ .\tag{1.49}$$

1.3 The Generalized Susceptibility

When we speak of a *susceptibility*, we are usually referring to a medium in which the response is proportional in some sense to the excitation. If the medium is linear, the response is directly proportional to the excitation. If the medium is nonlinear, the proportionality involves higher powers of the excitation. However, if the excitation is very small, the response will be given to a good approximation by the linear susceptibility. Since time- and space-varying magnetic fields are generally quite small, a linear response theory is usually adequate. Nonlinear effects become important in dealing with hysteresis phenomena or high-power absorption in magnetic materials. For the most part, then, we shall be concerned with a linear response theory. In this section we shall define the wave-vector-dependent frequency-dependent linear susceptibility and investigate some of its properties.

Let us consider the magnetization $M(r, t)$ associated with a particular magnetic field $H(r, t)$. These quantities are related to their Fourier components by

$$M(r, t) = \frac{1}{2\pi V} \sum_k \int d\Omega\, M(k, \Omega) e^{i(k \cdot r - \Omega t)} \ ,\tag{1.50}$$

12   1. The Magnetic Susceptibility

$$H(r, t) = \frac{1}{2\pi V} \sum_q \int d\omega H(q, \omega) e^{i(q \cdot r - \omega t)}, \qquad (1.51)$$

where we have used the relations

$$\int dr\, e^{i(k-k') \cdot r} = V\Delta(k - k'), \qquad (1.52)$$

$$\int dt\, e^{-i(\Omega - \Omega')t} = 2\pi \delta(\Omega - \Omega'), \qquad (1.53)$$

$$\sum_k e^{ik \cdot (r-r')} = \frac{V}{(2\pi^3)} \int dk\, e^{ik \cdot (r-r')} = V\delta(r - r'). \qquad (1.54)$$

Here $\Delta(k - k')$ is the *Kronecker delta function* and $\delta(r - r')$ is the *Dirac delta function*.

We now define the generalized wave-vector-dependent frequency-dependent susceptibility by

$$M_\nu(k, \Omega) = \sum_q \int d\omega \sum_\mu \chi_{\nu\mu}(k, q; \Omega, \omega) H_\mu(q, \omega) \qquad (1.55)$$

where ν and $\mu = x, y$, and z. This may be written in the more convenient dyadic form

$$M(k, \Omega) = \sum_q \int d\omega\, \chi(k, q; \Omega, \omega) \cdot H(q, \omega).$$

In general $\chi(k, q; \Omega, \omega)$ will depend on the particular form of $H(r, t)$, or equivalently, $H(q, \omega)$; that is, the susceptibility is a functional of the field. The susceptibility is also a tensor. Furthermore, since the magnetization may be out of phase with the exciting field, the susceptibility is also complex. Substituting this expression into (1.50) gives

$$M(r, t) = \frac{1}{2\pi V} \sum_k \int d\Omega \sum_q \int d\omega\, \chi(k, q; \Omega, \omega) \cdot H(q, \omega) e^{i(k \cdot r - \Omega t)} \qquad (1.56)$$

or

$$M(r, t) = \iint dr'\, dt'$$
$$\times \left\{ \left[\frac{1}{2\pi V} \sum_k \int d\Omega \sum_q \int d\omega\, \chi(k, q; \Omega, \omega) e^{ik \cdot (r-r')} e^{-i\Omega(t-t')} \right] \right.$$
$$\left. e^{i(k-q) \cdot r'} e^{-i(\Omega - \omega)t'} \right\} \cdot H(r', t'), \qquad (1.57)$$

where the quantity in braces defines a general spatial-temporal susceptibility density $\chi(r, r'; t, t')$.

If the magnetic medium possesses translational invariance, then this susceptibility must be a function only of the relative coordinate $r - r'$. From the expression above we see that this implies that in the wave-vector-dependent susceptibility q is equal to k. Furthermore, if the medium is stationary, it can be shown that the temporal dependence is $t-t'$, which implies a monochromatic response to a monochromatic excitation with the same frequency, that is, $\Omega = \omega$. Therefore, when these conditions are satisfied, the susceptibility takes the form

$$\chi(k, q; \Omega, \omega) = \chi(q, \omega)\, \Delta(k - q)\delta(\Omega - \omega) \,.$$

Thus

$$M(r, t) = \iint dr'\, dt'\, \chi(r - r', t - t') \cdot H(r', t') \,, \tag{1.58}$$

where

$$\chi(r - r', t - t') = \frac{1}{2\pi V} \sum_q \int d\omega\, \chi(q, \omega)\, e^{iq\cdot(r-r')} e^{-i\omega(t-t')} \,, \tag{1.59}$$

and its Fourier transform is

$$\chi(q, \omega) = \int d(t - t') \int d(r - r')\chi(r - r', t - t') e^{-iq\cdot(r-r')} e^{i\omega(t-t')} \,. \tag{1.60}$$

The more general susceptibility is required whenever the presence of impurities destroys the translational invariance. The response of a typical paramagnet is such a case.

Since the susceptibility has such a general nature, it should not be surprising that there are various important theorems involving this quantity. We shall consider three of these now. The first theorem, known as the *Kramers–Kronig relations*, relates the real and imaginary parts of the susceptibility. The second is the *fluctuation-dissipation theorem*, which relates the susceptibility to thermal fluctuations in the magnetization. Finally, we shall present a derivation of the so-called *Onsager relation* that describes the symmetry of the susceptibility tensor.

1.3.1 The Kramers–Kronig Relations

As a consequence of some rather general properties of

$$\chi(q, \omega) = \chi'(q, \omega) + i\chi''(q, \omega) \,,$$

its real part $\chi'(q, \omega)$ and its imaginary part $\chi''(q, \omega)$ are connected on the real axis ω by integral relations known as the *Kramers–Kronig relations*, or just as *dispersion relations*. Let us consider a medium which is *linear* and *stationary* (and translationally invariant, although this is not a necessary condition). Then $\chi(q, \omega)$ is related to $\chi(r-r', t-t')$ by (1.60). If the system obeys the principle of *causality*, then $\chi(r - r', t - t') = 0$ for $t < t'$. Hence the time integral in (1.60) runs only from 0 to ∞; that is,

14   1. The Magnetic Susceptibility

$$\chi(\boldsymbol{q}, \omega) = \int_0^\infty dt\, \chi(\boldsymbol{q}, t) e^{i\omega t}\,. \tag{1.61}$$

Therefore the function $\chi(\boldsymbol{q}, \omega)$ is a complex function of ω which has no singularities at the ends of the real axis, *provided* that

$$\int_0^\infty \chi(\boldsymbol{q}, t)\, dt$$

is finite. This is equivalent to the assumption that the total response to a finite total excitation is finite. The finite values of $\chi(\boldsymbol{q}, \omega)$ at the ends of the real axis may be identified with the real part of the susceptibility $\chi'(\boldsymbol{q}, \infty)$. The fact that $\chi''(\boldsymbol{q}, \omega)$ vanishes as $\omega \to \infty$ may be obtained from the following physical argument. As we shall see in Chap. 5, the rate of energy absorption by a magnetic system is proportional to $\omega \chi''(\boldsymbol{q}, \omega)$. If this is to remain finite as $\omega \to \infty$, then $\chi''(\boldsymbol{q}, \omega)$ must go to 0 as $\omega \to \infty$. This result may also be derived from the finite-response assumption.

There is no reason for the real part of the susceptibility to vanish as $\omega \to \infty$. Therefore let us define $\lim_{\omega \to \infty} \chi'(\boldsymbol{q}, \omega) = \chi(\boldsymbol{q}, \infty)$. The quantity $\chi(\boldsymbol{q}, \omega) - \chi(\boldsymbol{q}, \infty)$ is then a complex function which vanishes at the ends of the real axis. The theory of complex variables tells us that the function $\chi(\boldsymbol{q}, z) - \chi(\boldsymbol{q}, \infty)$, where z is a complex variable, will be analytic in the upper half plane. The residue theorem then says

$$\oint_C \frac{\chi(\boldsymbol{q}, z) - \chi(\boldsymbol{q}, \infty)}{z - \omega}\, dz = 0 \tag{1.62}$$

where the contour C runs from $-\infty$ to $+\infty$ along the real axis and closes in the upper half plane. In terms of its *principal value*, this integral may be written as

$$\mathscr{P} \int_{-\infty}^{\infty} \frac{\chi(\boldsymbol{q}, \omega') - \chi(\boldsymbol{q}, \infty)}{\omega' - \omega}\, d\omega' - i\pi[\chi(\boldsymbol{q}, \omega) - \chi(\boldsymbol{q}, \infty)] = 0\,. \tag{1.63}$$

Equating the real and imaginary parts to 0 separately gives the result

$$\chi'(\boldsymbol{q}, \omega) - \chi(\boldsymbol{q}, \infty) = \frac{1}{\pi} \mathscr{P} \int \frac{\chi''(\boldsymbol{q}, \omega')}{\omega' - \omega}\, d\omega'\,, \tag{1.64}$$

$$\chi''(\boldsymbol{q}, \omega) = -\frac{1}{\pi} \mathscr{P} \int \frac{\chi'(\boldsymbol{q}, \omega') - \chi(\boldsymbol{q}, \infty)}{\omega' - \omega}\, d\omega'\,. \tag{1.65}$$

The usefulness of this result lies in the fact that χ'' is proportional to the absorption spectrum of the medium. Therefore (1.64) tells us, for example, that the static susceptibility may be obtained by integrating over the absorption spectrum. This is, in fact, an experimental technique used to obtain the static susceptibility of certain systems.

1.3.2 The Fluctuation-Dissipation Theorem

It is well known that a colloidal suspension of particles exhibits Brownian motion; that is, the particles move about irregularly because they are being bombarded by the molecules of the liquid. Now, suppose that these particles are charged, and we attempt to accelerate them with an external electric field. Because of the impacts with the molecules, the particles experience a resistive force which is proportional to their velocity. Thus the mechanism that produces the random fluctuations in the position of the particle is also responsible for its response to an external excitation. The relationship between the response of a system and its thermal fluctuation spectrum is called the *fluctuation-dissipation theorem*. This relationship is a very general one, and we shall consider only its specific application to a magnetic medium.

Let us consider a linearly polarized magnetic field of amplitude $H_1 \cos(\boldsymbol{q}\cdot\boldsymbol{r})$ oscillating at a frequency ω in the μ direction, $H_1 \cos(\boldsymbol{q}\cdot\boldsymbol{r})\cos\omega t$. Since we have a linear system, the principle of superposition applies. Therefore we may construct the response to an arbitrary field if we know the response to this particular field. The response in the ν direction to such an excitation is given by (1.55). Since

$$H_\mu(\boldsymbol{q}', \omega') = \frac{\pi H_1 V}{2}[\varDelta(\boldsymbol{q}' - \boldsymbol{q})\delta(\omega' + \omega) + \varDelta(\boldsymbol{q}' - \boldsymbol{q})\delta(\omega' - \omega)$$
$$+ \varDelta(\boldsymbol{q}' + \boldsymbol{q})\delta(\omega' + \omega) + \varDelta(\boldsymbol{q}' + \boldsymbol{q})\delta(\omega' - \omega)], \quad (1.66)$$

we obtain

$$M_\nu(\boldsymbol{k}, \Omega) = \frac{\pi H_1 V}{2}[\chi_{\nu\mu}(\boldsymbol{k}, \boldsymbol{q}; \Omega, -\omega) + \chi_{\nu\mu}(\boldsymbol{k}, \boldsymbol{q}; \Omega, \omega),$$
$$+ \chi_{\nu\mu}(\boldsymbol{k}, -\boldsymbol{q}; \Omega, -\omega) + \chi_{\nu\mu}(\boldsymbol{k}, -\boldsymbol{q}; \Omega, \omega)]. \quad (1.67)$$

Let us now compute $M_\nu(\boldsymbol{k}, \Omega)$, using the prescription given in Sect. 1.2. The magnetization is

$$M_\nu(\boldsymbol{r}, t) = \text{Tr}\{\rho \mathcal{M}_\nu(\boldsymbol{r})\}. \quad (1.68)$$

Although ρ is a function of time, we shall not display this dependence explicitly, for a reason that will be apparent later. Since the time-varying field now disrupts the thermodynamic equilibrium, we must solve for ρ. We write the total Hamiltonian as

$$\mathcal{H} = \mathcal{H}_0 + \mathcal{H}_1 \quad (1.69)$$

where $\mathcal{H}_1 = -\int d\mathbf{r}\, \mathcal{M}(\mathbf{r}) \cdot \mathbf{H}(\mathbf{r}, t)$. For the particular field we are considering this becomes

$$\mathcal{H}_1 = -H_1 \int d\mathbf{r}\, \mathcal{M}_\mu(\mathbf{r}) \cos(\mathbf{q} \cdot \mathbf{r}) \cos \omega t = -\frac{H_1}{2} [\mathcal{M}_\mu(\mathbf{q}) + \mathcal{M}_\mu(-\mathbf{q})] \cos \omega t. \tag{1.70}$$

The equation of motion for the density matrix is

$$\frac{\partial \rho}{\partial t} = \frac{i}{\hbar} [\rho, \mathcal{H}_0 + \mathcal{H}_1]. \tag{1.71}$$

It is now convenient to introduce

$$\rho(t) \equiv \exp\left(\frac{i\mathcal{H}_0 t}{\hbar}\right) \rho \exp\left(\frac{-i\mathcal{H}_0 t}{\hbar}\right). \tag{1.72}$$

Differentiating (1.72) and using (1.71) gives

$$\frac{d\rho(t)}{dt} = \frac{i}{\hbar} \left[\rho(t), \exp\left(\frac{i\mathcal{H}_0 t}{\hbar}\right) \mathcal{H}_1 \exp\left(\frac{-i\mathcal{H}_0 t}{\hbar}\right)\right]. \tag{1.73}$$

This has the solution

$$\rho(t) = \rho(-\infty) + \frac{i}{\hbar} \int_{-\infty}^{t} \left[\rho(t'), \exp\left(\frac{i\mathcal{H}_0 t'}{\hbar}\right) \mathcal{H}_1 \exp\left(\frac{-i\mathcal{H}_0 t'}{\hbar}\right)\right] dt'. \tag{1.74}$$

If the interaction is turned on adiabatically, then $\rho(-\infty) = \rho_0$, which is the equilibrium density matrix $\rho_0 = \exp(-\beta\mathcal{H}_0)/Z$ where $Z = \text{Tr}\{\exp(-\beta\mathcal{H}_0)\}$. Inverting (1.74), using (1.70), and replacing ρ within the commutator by ρ_0, we have

$$\rho \simeq \rho_0 - i\frac{H_1}{2\hbar} \int_0^\infty \left\{\rho_0, \exp\left(\frac{-i\mathcal{H}_0 t'}{\hbar}\right) [\mathcal{M}_\mu(\mathbf{q}) + \mathcal{M}_\mu(-\mathbf{q})] \exp\left(\frac{i\mathcal{H}_0 t'}{\hbar}\right)\right\}$$
$$\times \cos \omega(t - t') \, dt'. \tag{1.75}$$

The magnetization is obtained from (1.68). If the system is ordered in the absence of the applied field, then $\text{Tr}\{\rho_0 \mathcal{M}_\nu\} \equiv M_\nu(-\infty)$ is nonzero. The response of such a system is then defined by the difference $M_\nu(\mathbf{r}, t) - M_\nu(-\infty)$ resulting from the applied field. In the following we shall understand $M_\nu(\mathbf{r}, t)$ to be the response to the applied field. Then,

$$M_\nu(\mathbf{r}, t) = -i\frac{H_1}{2\hbar}$$
$$\times \text{Tr}\left\{\int_0^\infty \left\{\rho_0, \exp\left(\frac{-i\mathcal{H}_0 t'}{\hbar}\right) [\mathcal{M}_\mu(\mathbf{q}) + \mathcal{M}_\mu(-\mathbf{q})] \exp\left(\frac{i\mathcal{H}_0 t'}{\hbar}\right)\right\} \mathcal{M}_\nu(\mathbf{r})\right\}$$
$$\times \cos \omega(t - t') dt'. \tag{1.76}$$

Taking the Fourier transform of this equation gives

$$M_\nu(\mathbf{k}, \Omega) = -i\frac{\pi H_1}{2\hbar} \text{Tr}\left\{\int_0^\infty [\rho_0, \mathscr{M}_\mu(\mathbf{q}, -t')]\mathscr{M}_\nu(\mathbf{k})\right\} e^{i\omega t'}\, dt'\delta(\Omega + \omega)$$
$$+ \text{ (terms involving } -\mathbf{q} \text{ and } -\omega). \tag{1.77}$$

Here $\mathscr{M}_\mu(\mathbf{q}, t)$ is defined in a manner identical to (1.72). The delta function involving the frequency results from our having linearized the expression for ρ when we replaced ρ by ρ_0 within the commutator.

If we now commute the integral with the trace in (1.77) and make use of the cyclic invariance of the trace, we have

$$\text{Tr}\left\{\int_0^\infty [\rho_0, \mathscr{M}_\mu(\mathbf{q}, -t')]\mathscr{M}_\nu(\mathbf{k})\right\} e^{-i\omega t'}\, dt'$$
$$= \int \langle [\mathscr{M}_\mu(\mathbf{q}, -t'), \mathscr{M}_\nu(\mathbf{k})] \rangle e^{-i\omega t'}\, dt'. \tag{1.78}$$

By comparing the resulting expression for $M_\nu(\mathbf{k}, \Omega)$ with (1.67) we make the following identification:

$$\chi_{\nu\mu}(\mathbf{k}, \mathbf{q}; \Omega, \omega) = \frac{i}{\hbar V} \int_0^\infty \langle [\mathscr{M}_\nu(\mathbf{k}, t), \mathscr{M}_\mu(-\mathbf{q})] \rangle e^{i\omega t}\, dt\, \delta(\Omega - \omega). \tag{1.79}$$

Since the $\mathbf{q}$ component of the applied field couples to the $-\mathbf{q}$ component of the magnetization, let us consider $\chi_{\nu\mu}(\mathbf{q}, \mathbf{q}, \omega)$ which we write as $\chi_{\nu\mu}(\mathbf{q}, \omega)$. Therefore,

$$\boxed{\chi_{\nu\mu}(\mathbf{q}, \omega) = \frac{i}{\hbar V} \int_0^\infty \langle [\mathscr{M}_\nu(\mathbf{q}, t), \mathscr{M}_\mu(-\mathbf{q})] \rangle e^{i\omega t}\, dt} \tag{1.80}$$

The quantity

$$(i/\hbar)\langle [\mathscr{M}_\nu(\mathbf{q}, t), \mathscr{M}_\mu(-\mathbf{q})] \rangle,$$

or equivalently

$$\frac{i}{\hbar} \text{Tr}\{[\mathscr{M}_\mu(-\mathbf{q}), \rho_0]\mathscr{M}_\nu(\mathbf{q}, t)\},$$

is referred to in the literature as the *response function* of the system. The susceptibility may also be written as an integral over the full range of time by multiplying the integrand by the theta function $\theta(t)$ which equals 1 for $t > 0$ and 0 for $t < 0$. The product of the response function and this theta function is called the double-time-retarded Green's function (e.g., [1.4]),

$$\langle\langle \mathcal{M}_\nu(q,t), \mathcal{M}_\mu(-q)\rangle\rangle \equiv -i\langle[\mathcal{M}_\nu(q,t), \mathcal{M}_\mu(-q)]\rangle\theta(t).$$

These functions are very useful in calculating thermodynamic properties but are beyond the scope of this monograph.

Since the response function does not have a classical analog and is not a well-defined observable, it is more convenient to relate the susceptibility to the *correlation function* $\langle\{\mathcal{M}_\nu(q,t)\mathcal{M}_\mu(-q)\}\rangle$, where $\{\cdots\}$ is the symmetrized product, which is defined by

$$\{\mathcal{M}_\nu(q,t)\mathcal{M}_\mu(-q)\} \equiv \tfrac{1}{2}[\mathcal{M}_\nu(q,t)\mathcal{M}_\mu(-q) + \mathcal{M}_\mu(-q)\mathcal{M}_\nu(q,t)],$$

In order to relate the response function to the correlation function let us consider their Fourier transforms,

$$f_{\nu\mu}(q,\omega) = \frac{i}{\hbar}\int_{-\infty}^{\infty}\langle[\mathcal{M}_\nu(q,t), \mathcal{M}_\mu(-q)]\rangle e^{i\omega t}\,dt, \tag{1.81}$$

$$g_{\nu\mu}(q,\omega) = \int_{-\infty}^{\infty}\langle\{\mathcal{M}_\nu(q,t)\mathcal{M}_\mu(-q)\}\rangle e^{i\omega t}\,dt. \tag{1.82}$$

We can rewrite (1.81) by using the following relation:

$$\int_{-\infty}^{\infty} dt\, \langle \mathcal{M}_\mu(-q)\mathcal{M}_\nu(q,t)\rangle e^{i\omega t}$$

$$= \int_{-\infty}^{\infty} dt\, \mathrm{Tr}\left\{\exp(-\beta\mathcal{H}_0)\mathcal{M}_\mu(q)\exp\left(\frac{i\mathcal{H}_0 t}{\hbar}\right)\mathcal{M}_\nu(q)\exp\left(\frac{-i\mathcal{H}_0 t}{\hbar}\right)\right\}e^{i\omega t}$$

$$= \int_{-\infty}^{\infty} dt\, \mathrm{Tr}\left\{\exp(-\beta\mathcal{H}_0)\exp\left(\frac{i\mathcal{H}_0(t-i\hbar\beta)}{\hbar}\right)\mathcal{M}_\nu(q)\right.$$

$$\left.\times \exp\left[\frac{-i\mathcal{H}_0(t-i\hbar\beta)}{\hbar}\right]\mathcal{M}_\mu(-q)\right\}e^{i\omega t}$$

$$= e^{-\beta\hbar\omega}\int_{-\infty}^{\infty} dt\,\langle\mathcal{M}_\nu(q,t)\mathcal{M}_\mu(-q)\rangle e^{i\omega t}. \tag{1.83}$$

Therefore

$$f_{\nu\mu}(q,\omega) = \frac{i}{\hbar}(1 - e^{-\beta\hbar\omega})\int_{-\infty}^{\infty} dt\,\langle\mathcal{M}_\nu(q,t)\mathcal{M}_\mu(-q)\rangle e^{i\omega t}. \tag{1.84}$$

From the definition of $g_{\nu\mu}$ we see that its relation to $f_{\nu\mu}$ is

$$g_{\nu\mu}(q,\omega) = (\hbar/2i)\coth(\beta\hbar\omega/2)f_{\nu\mu}(q,\omega). \tag{1.85}$$

We can also relate $f_{\nu\mu}$ to the susceptibility by separating the time integral as follows:

$$f_{\nu\mu}(\boldsymbol{q},\omega) = \frac{i}{\hbar}\int_0^\infty dt\,\langle[\mathcal{M}_\nu(\boldsymbol{q},t),\mathcal{M}_\mu(-\boldsymbol{q})]\rangle e^{i\omega t}$$
$$+ \frac{i}{\hbar}\int_{-\infty}^0 dt\,\langle[\mathcal{M}_\nu(\boldsymbol{q}\,t),\mathcal{M}_\mu(-\boldsymbol{q})]\rangle e^{i\omega t}. \tag{1.86}$$

If we now make the transformation $t \to -t$ in the second integral and use the fact that $\chi_{\mu\nu}(-\boldsymbol{q},-\omega) = \chi_{\mu\nu}^*(\boldsymbol{q},\omega)$, which follows from (1.80), then

$$f_{\nu\mu}(\boldsymbol{q},\omega) = [\chi_{\nu\mu}(\boldsymbol{q},\omega) - \chi_{\mu\nu}^*(\boldsymbol{q},\omega)]V$$

and

$$g_{\nu\mu}(\boldsymbol{q},\omega) = (\hbar V/2i)\coth(\beta\hbar\omega/2)[\chi_{\nu\mu}(\boldsymbol{q},\omega) - \chi_{\mu\nu}^*(\boldsymbol{q},\omega)]$$

Therefore

$$\boxed{\int_{-\infty}^\infty dt\,\langle\{\mathcal{M}_\nu(\boldsymbol{q},t)\mathcal{M}_\mu(-\boldsymbol{q})\}\rangle_s e^{i\omega t} = \hbar V \coth(\beta\hbar\omega/2)\chi_{\nu\mu}''(\boldsymbol{q},\omega)_s} \tag{1.87}$$

where the subscript s indicates the symmetric part of the tensor.

This is the result we had set out to find. It tells us that the Fourier transform of the correlation function is proportional to the imaginary part of the susceptibility. The derivation has been presented here in some detail because we shall make frequent reference to this result throughout the text.

1.3.3 Onsager Relation

Generally, when we probe a magnetic system it is in the presence of a dc field $\boldsymbol{H}$. Therefore $\mathcal{H}_0$, and hence the response function, is a function of this field. In 1931 Onsager pointed out that microscopic reversibility requires the simultaneous reversal of both the magnetic field *and* time. The time reversal of an operator, $\bar{A}$, is defined by

$$\langle n|\bar{A}|m\rangle = \langle Tn|A|Tm\rangle^*, \quad \text{or} \quad \bar{A} = T^{-1}AT, \tag{1.88}$$

where T is the time reversal operator.

In particular, if the operator is explicitly time dependent and a function of the Hamiltonian $\mathcal{H}_0$, then

$$\overline{A(t,\mathcal{H}_0)} = \bar{A}(-t,\mathcal{H}_0).$$

Let us now use this to rewrite the response function,

20   1. The Magnetic Susceptibility

$$\langle[\mathcal{M}_\nu(\mathbf{q}, t), \mathcal{M}_\mu(-\mathbf{q})]\rangle = \sum_n \langle n|\rho_0(\mathcal{H}_0)[\mathcal{M}_\nu(\mathbf{q}, t), \mathcal{M}_\mu(-\mathbf{q})]|n\rangle ,$$

$$= \sum_n \langle Tn|\rho_0(\mathcal{H}_0)T\overline{[\mathcal{M}_\nu(\mathbf{q}, t), \mathcal{M}_\mu(-\mathbf{q})]}T^{-1}|Tn\rangle ,$$

$$= \sum_n \langle n|T^{-1}\rho_0(\mathcal{H}_0)T\overline{[\mathcal{M}_\nu(\mathbf{q}, t), \mathcal{M}_\mu(-\mathbf{q})]}|n\rangle^* ,$$

$$= \sum_n \langle n|\rho_0(\mathcal{H}_0)[\overline{\mathcal{M}}_\nu(\mathbf{q}, -t, \mathcal{H}_0), \overline{\mathcal{M}}_\mu(-\mathbf{q})]|n\rangle^* ,$$

$$= \sum_n \langle \rho_0(\mathcal{H}_0)[\overline{\mathcal{M}}_\nu(\mathbf{q}, -t, \mathcal{H}_0), \overline{\mathcal{M}}_\mu(-\mathbf{q})]n|n\rangle ,$$

$$= \sum_n \langle n|[\overline{\mathcal{M}}_\mu(-\mathbf{q})^\dagger, \overline{\mathcal{M}}_\nu(\mathbf{q}, -t, \mathcal{H}_0)^\dagger]\rho_0(\mathcal{H}_0)|n\rangle .$$

Since the magnetic-moment operator is Hermitian and changes sign under time reversal, $\overline{\mathcal{M}}_\mu(-\mathbf{q})^\dagger = -\mathcal{M}_\mu(\mathbf{q})$, for example, and

$$\langle[\mathcal{M}_\nu(\mathbf{q}, t), \mathcal{M}_\mu(-\mathbf{q})]\rangle = \sum_n \langle n|\rho_0(\mathcal{H}_0)[\mathcal{M}_\nu(\mathbf{q}, t, \mathcal{H}_0), \mathcal{M}_\eta(-\mathbf{q})]|n\rangle .$$

It therefore follows that

$$\boxed{\chi_{\nu\mu}(\mathbf{q}, \omega, H) = \chi_{\mu\nu}(-\mathbf{q}, \omega, -H)} \quad (1.89)$$

which is known as the Onsager relation. Note that this tells us immediately that the diagonal components of the susceptibility tensor must be even functions of the field.

1.4 Second Quantization

Magnetism, particularly in metals, is a many-body phenomenon. Consequently, we often find articles beginning with a second-quantized Hamiltonian. In this section we shall briefly develop the technique of second quantization and apply it to a free-electron system. We shall have occasion to use these results later, particularly in setting up the Anderson and Hubbard Hamiltonians.

Let us begin by considering a system of N interacting particles described by the Hamiltonian

$$\mathcal{H} = \sum_{i=1}^N T(\mathbf{r}_i, \dot{\mathbf{r}}_i) + \tfrac{1}{2} \sum_{\substack{i,j=1 \\ i\neq j}}^N V(\mathbf{r}_i, \mathbf{r}_j) \quad (1.90)$$

The many-body wave function $\psi(\mathbf{r}_1, \ldots, \mathbf{r}_N, t)$ satisfies the Schrödinger equation

$$i\hbar \frac{\partial}{\partial t} \psi(\mathbf{r}_1, \ldots \mathbf{r}_N, t) = \mathcal{H}\psi(\mathbf{r}_1, \ldots, \mathbf{r}_N, t) . \quad (1.91)$$

We now expand this wave function in terms of products of single-particle wave functions characterized by quantum numbers E_i,

$$\psi(\mathbf{r}_1, \ldots, \mathbf{r}_N, t) = \sum_{\{E_1, \ldots, E_N\}} c(E_1, \ldots, E_N, t)\varphi_{E_1}(\mathbf{r}_1)\varphi_{E_2}(\mathbf{r}_2) \ldots \varphi_{E_N}(\mathbf{r}_N) \quad (1.92)$$

where the sum is over all possible sets of quantum numbers. The statistical nature of the particles is contained in the coefficients $c(E_1, \ldots, E_N, t)$. For example, if the particles are bosons, the sign of the coefficient is invariant under particle interchange,

$$c(E_1, \ldots, E_k, \ldots, E_i, \ldots, E_N, t) = c(E_1, \ldots, E_i, \ldots, E_k, \ldots, E_N, t), \quad (1.93)$$

and any number of particles may occupy a given state. If the particles are fermions, then

$$c(E_1, \ldots, E_k, \ldots, E_i, \ldots, E_N, t) = -c(E_1, \ldots, E_i, \ldots, E_k, \ldots, E_N, t). \quad (1.94)$$

That is, there may not be more than one particle in a particular state.

Since we shall be concerned mainly with electrons, which are fermions, we shall be faced with the problem of keeping track of the minus sign introduced when two electrons are interchanged. It is to facilitate this book keeping that the concept of second quantization is introduced.

The coefficients in the expansion of ψ above are characterized by the set of N quantum numbers. We could just as well, however, have chosen coefficients characterized by the number of electrons in each of the possible states. That is, instead of the set of N numbers $\{E_1, \ldots, E_N\}$ we could have used the infinite set of numbers $\{n_1, \ldots, n_\infty\}$, where for fermions $n = 0$ or 1. We must be very careful in making this transcription. For example, suppose that the electron at $\mathbf{r}_i$ is in a state E_i and the electron at $\mathbf{r}_k$ is in a state E_k. In this case the occupation-number description would be the same as if the electron at $\mathbf{r}_i$ were in state E_k and the electron at $\mathbf{r}_k$ were in state E_i. However,

$$c(E_1, \ldots, E_k, \ldots, E_i, \ldots, E_N, t) = -c(E_1, \ldots, E_i, \ldots, E_k, \ldots, E_N, t). \quad (1.95)$$

If we wish to use the occupation-number scheme, we must account for this minus sign. This is done by arbitrarily assigning a certain order to the particular set of quantum numbers $\{E_1, \ldots, E_N\}$. Then the relative sign of any permutation of the electrons from this order is automatically given by writing the single-particle wave functions as a *Slater determinant:*

$$c(E_1, \ldots, E_N, t)\varphi_{E_1}(\mathbf{r}_1) \ldots \varphi_{E_N}(\mathbf{r}_N)$$
$$+ \text{(all permutations within the set } \{E_1, \ldots, E_N\})$$

$$= f(n_1, \ldots, n_\infty, t) \begin{vmatrix} \varphi_{E_1}(\mathbf{r}_1) & \cdots & \varphi_{E_1}(\mathbf{r}_N) \\ \cdots & \cdots & \cdots \\ \varphi_{E_N}(\mathbf{r}_1) & \cdots & \varphi_{E_N}(\mathbf{r}_N) \end{vmatrix} \quad (1.96)$$

where $f(n_1, \ldots, n_\infty, t)$ has the sign and magnitude of the first c. Summing over all sets $\{E_1, \ldots, E_N\}$ is equivalent to summing over all combinations of occupied states. Therefore

$$\psi(\mathbf{r}_1, \ldots, \mathbf{r}_N, t) = \sum_{\{n_1, \cdots, n_\infty\}} f(n_1, \ldots, n_\infty, t) \frac{1}{\sqrt{N!}} \begin{vmatrix} \varphi_{E_1}(\mathbf{r}_1) & \cdots & \varphi_{E_1}(\mathbf{r}_N) \\ \cdots & & \cdots \\ \varphi_{E_N}(\mathbf{r}_1) & \cdots & \varphi_{E_N}(\mathbf{r}_N) \end{vmatrix}. \quad (1.97)$$

The states used in constructing the determinant are, of course, those occupied.

By using this occupation-number description we have succeeded in moving the statistics from the expansion coefficients into the basis functions, which, in fact, form an orthonormal antisymmetric set.

Let us now define an abstract vector space, or *Hilbert space*, spanned by the vectors $|n_1, n_2, \ldots, n_\infty\rangle$. We introduce operators which satisfy the *anticommutation relations*

$$\{a_i, a_j^\dagger\} \equiv a_i a_j^\dagger + a_j^\dagger a_i = \delta_{ij}, \quad \{a_i, a_j\} = \{a_i^\dagger, a_j^\dagger\} = 0. \quad (1.98)$$

From these relations it can be shown that $a_i^\dagger$ creates an entry in position i (provided one does not already exist there) and a_i destroys an entry at i. Therefore we may represent a basis vector of our Hillbert space as

$$|n_1, \ldots, n_\infty\rangle = (a_1^\dagger)^{n_1}(a_2^\dagger)^{n_2} \cdots (a_\infty^\dagger)^{n_\infty}|0\rangle. \quad (1.99)$$

Now consider operating on this with a_k. If $n_k = 0$, then a_k can be commuted all the way over to the "vacuum," where it gives 0. Thus we let $n_k = 1$. Then a_k will commute until it comes to $a_k^\dagger$.

$$a_k|n_1, \ldots, n_k, \ldots, n_\infty\rangle = (-1)^{\Sigma_k}(a_1^\dagger)^{n_1} \cdots a_k a_k^\dagger \cdots (a_\infty^\dagger)^{n_\infty}|0\rangle. \quad (1.100)$$

Here $\Sigma_k = n_1 + n_2 + \cdots n_{k-1}$ accounts for all the sign changes that a_k left in its wake as it commuted over to $a_k^\dagger$. We now use $a_k a_k^\dagger = 1 - a_k^\dagger a_k$. In the second term a_k may again commute over to the vacuum to give 0. Thus we are left with

$$a_k|n_1, \ldots, n_k, \ldots, n_\infty\rangle = \begin{cases} 0 & n_k = 0 \\ (-1)^{\Sigma_k}|n_1, \ldots, n_k - 1, \ldots, n_\infty\rangle & n_k = 1 \end{cases}. \quad (1.101)$$

Similarly,

$$a_k^\dagger|n_1, \ldots, n_k, \ldots, n_\infty\rangle = \begin{cases} (-1)^{\Sigma_k}|n_1, \ldots, n_k + 1, \ldots, n_\infty\rangle & n_k = 0 \\ 0 & n_k = 1 \end{cases}. \quad (1.102)$$

Since it can be shown that $a_k^\dagger a_k$ has the eigenvalue n_k, we can simplify these results by writing

$$a_k|n_1, \ldots, n_k, \ldots, n_\infty\rangle = (-1)^{\Sigma_k}\sqrt{n_k}|n_1, \ldots, n_k - 1, \ldots, n_\infty\rangle,$$
$$a_k^\dagger|n_1, \ldots, n_k, \ldots, n_\infty\rangle = (-1)^{\Sigma_k}\sqrt{n_k + 1}|n_1, \ldots, n_k + 1, \ldots, n_\infty\rangle. \quad (1.103)$$

1.4 Second Quantization

Having developed the properties of the Hilbert space, we now use the expansion coefficients $f(n_1, \ldots, n_\infty, t)$ to define the abstract state vector

$$|\psi(t)\rangle = \sum_{\{n_1, \cdots, n_\infty\}} f(n_1, \ldots, n_\infty, t) |n_1, \ldots, n_\infty\rangle. \tag{1.104}$$

The reason for this becomes clear when we consider the equation of motion of this state vector. Taking the time derivative, we have

$$i\hbar \frac{\partial |\psi(t)\rangle}{\partial t} = i\hbar \sum_{\{n_1, \cdots, n_\infty\}} \frac{\partial f(n_1, \ldots, n_\infty, t)}{\partial t} |n_1, \ldots, n_\infty\rangle. \tag{1.105}$$

To evaluate this we go back to the real-space Schrödinger equation,

$$i\hbar \frac{\partial |\psi(\mathbf{r}_1, \ldots, \mathbf{r}_N, t)\rangle}{\partial t}$$

$$= i\hbar \sum_{\{n_1, \cdots, n_\infty\}} \frac{\partial f(n_1, \ldots, n_\infty, t)}{\partial t} \frac{1}{\sqrt{N!}} \begin{vmatrix} \varphi_{E_1}(\mathbf{r}_1) & \cdots & \varphi_{E_1}(\mathbf{r}_N) \\ \cdots & \cdots & \cdots \\ \varphi_{E_N}(\mathbf{r}_1) & \cdots & \varphi_{E_N}(\mathbf{r}_N) \end{vmatrix},$$

$$= \mathcal{H} \sum_{\{n_1, \cdots, n_\infty\}} f(n_1, \ldots, n_\infty, t) \frac{1}{\sqrt{N!}} \begin{vmatrix} \varphi_{E_1}(\mathbf{r}_1) & \cdots & \varphi_{E_1}(\mathbf{r}_N) \\ \cdots & \cdots & \cdots \\ \varphi_{E_N}(\mathbf{r}_1) & \cdots & \varphi_{E_N}(\mathbf{r}_N) \end{vmatrix}. \tag{1.106}$$

We now multiply through from the left by the conjugate Slater determinant appropriate for some particular set of occupation numbers $\{n_1, \ldots, n_\infty\}$. The left-hand side just gives

$$i\hbar \frac{\partial f(n_1, \ldots, n_\infty, t)}{\partial t}. \tag{1.107}$$

Consider the one-particle terms of the right-hand side of (1.106), which arise from the $T(\dot{\mathbf{r}}_i)$ term in (1.90). We write the Slater determinant as

$$\frac{1}{\sqrt{N!}} \sum_P (-1)^p P \varphi_{E_1}(\mathbf{r}_1) \varphi_{E_2}(\mathbf{r}_2) \cdots \varphi_{E_N}(\mathbf{r}_N), \tag{1.108}$$

where P is an operator which permutes the order of the electrons and p is the number of such permutations. Then the matrix element becomes

$$\frac{1}{N!} \sum_i \sum_{\{n'_1, \cdots, n'_\infty\}} \sum_{P, P'} (-1)^{p+p'} f(n'_1, \ldots, n'_\infty, t)$$
$$\times \int P \varphi^*_{E_1}(\mathbf{r}_1) \cdots T(\dot{\mathbf{r}}_i) P' \varphi_{E'_1}(\mathbf{r}_1) \cdots d\mathbf{r}_1 \cdots d\mathbf{r}_N. \tag{1.109}$$

Since $T(\dot{\mathbf{r}}_i)$ is a one-particle operator, the set of numbers $\{n'_1, \ldots, n'_\infty\}$ cannot differ from the particular set $\{n_1, \ldots, n_\infty\}$ by more than two numbers. In particular, let the set $\{n_1', \ldots, n_\infty'\}$ contain the state E_l and the set $\{n_1, \ldots, n_\infty\}$ contain the state E_k. The sums over i, P, and P' give $N!$, leaving us with

$$\sum_{k,l} (-1)^{\Sigma_k + \Sigma_l} \int dr\, \varphi_{E_k}^*(r) T(\dot{r}) \varphi_{E_l}(r) f(n_1, \ldots, n_k - 1, n_l + 1, \ldots, n_\infty, t). \quad (1.110)$$

Therefore

$$i\hbar \frac{\partial |\psi(t)\rangle}{\partial t} = \sum_{\{n_1, \ldots, n_\infty\}} \sum_{k,l} \langle k|T|l\rangle$$
$$\times f(n_1, \ldots, n_k = 0, \ldots, n_l = 1, \ldots, n_\infty, t)(-1)^{\Sigma_k + \Sigma_l}$$
$$|n_1, \ldots, n_k = 1, \ldots, n_l = 0, \ldots, n_\infty\rangle + \text{(interaction terms)}. \quad (1.111)$$

We now recall from above that

$$(-1)^{\Sigma_k + \Sigma_l} |n_1, \ldots, n_k, \ldots, n_l, \ldots, n_\infty\rangle$$
$$= a_k^\dagger a_l |n_1, \ldots, n_k - 1, \ldots, n_l + 1, \ldots, n_\infty\rangle. \quad (1.112)$$

Substituting this into the equation for $\partial |\psi(t)\rangle/\partial t$, we see that the sum over $\{n_1, \ldots, n_\infty\}$ just gives $|\psi(t)\rangle$. Carrying through the same arguments for the two-particle interaction terms, we find

$$i\hbar \frac{\partial |\psi(t)\rangle}{\partial t} = \mathcal{H}|\psi(t)\rangle \quad (1.113)$$

where

$$\mathcal{H} = \sum_{k,l} \langle k|T|l\rangle a_k^\dagger a_l + \tfrac{1}{2} \sum_{k,l,s,t} \langle kl|V|st\rangle a_k^\dagger a_l^\dagger a_t a_s. \quad (1.114)$$

Thus we have the important result that in this occupation-number space the state vector $|\psi(t)\rangle$, as defined above, also satisfies a Schrodinger-like equation, with the Hamiltonian expressed in this second-quantized form. It is now easy to show that the matrix elements of such second-quantized operators between occupation-number states are the same as the matrix elements of "first-quantized" operators between the usual states.

Since we shall often have occasion to express an operator in second-quantized form, let us develop a prescription for doing this. For this purpose it is convenient to define what is called the *field operator* in our Hilbert space,

$$\psi(r) = \sum_k \varphi_k(r) a_k. \quad (1.115)$$

Here again the $\varphi_k(r)$ are a complete set of single-particle states characterized by the quantum numbers k, and a_k is the fermion operator introduced above. To second quantize a one-particle operator such as $T(\dot{r}_i)$ we write $r_i \to r$, sandwich this operator between $\psi^\dagger(r)$ and $\psi(r)$, and integrate over all space. For a two-particle operator such as $V(\mathbf{r}_i, \mathbf{r}_j)$ we let $\mathbf{r}_i \to \mathbf{r}$ and $\mathbf{r}_j \to \mathbf{r}'$, sandwich it between $\psi^\dagger(\mathbf{r})\psi^\dagger(\mathbf{r}')$ and $\psi(\mathbf{r}')\psi(\mathbf{r})$, and integrate over $d\mathbf{r}$ and $d\mathbf{r}'$.

Example: The degenerate-electron gas. As an example of the use of this prescription let us second quantize the Hamiltonian for a gas of electrons moving in the field of a uniform positive charge distribution. The total Hamiltonian is

$$\mathcal{H} = \mathcal{H}_{\text{el-el}} + \mathcal{H}_{\text{el-n}} + \mathcal{H}_{n-n}. \tag{1.116}$$

The electron-electron Hamiltonian is

$$\mathcal{H}_{\text{el-el}} = \sum_i \frac{p_i^2}{2m} + \frac{e^2}{2} \sum_{i \neq j} \frac{\exp(-\mu|\mathbf{r}_i - \mathbf{r}_j|)}{|\mathbf{r}_i - \mathbf{r}_j|} \tag{1.117}$$

where a screening factor has been inserted for mathematical convenience. The interaction of the electrons with the positive background is

$$\mathcal{H}_{\text{el-n}} = -e^2 \sum_i \int \frac{\rho(\mathbf{r}')}{|\mathbf{r}_i - \mathbf{r}'|} \exp(-\mu|\mathbf{r}_i - \mathbf{r}'|) \, d\mathbf{r}' \tag{1.118}$$

where $\rho(\mathbf{r})$ is the positive charge density, which for a uniform distribution is

$$\rho(\mathbf{r}) = \frac{N}{V}. \tag{1.119}$$

Thus $\mathcal{H}_{\text{el-n}}$ becomes

$$\mathcal{H}_{\text{el-n}} = -e^2 \left(\frac{N}{V}\right) \sum_i \int \frac{\exp(-\mu|\mathbf{r}_i - \mathbf{r}'|)}{|\mathbf{r}_i - \mathbf{r}'|} \, d\mathbf{r}'. \tag{1.120}$$

If μ^{-1} is much smaller than L, where L is the sample dimension, we may replace the integral over $d\mathbf{r}'$ by one over $d(\mathbf{r}' - \mathbf{r}_i)$, which gives

$$\mathcal{H}_{\text{el-n}} = -e^2 \frac{N^2}{V} \frac{4\pi}{\mu^2}. \tag{1.121}$$

Finally, the self-energy of the background charge is

$$\mathcal{H}_{n-n} = -\tfrac{1}{2} e^2 \iint \frac{\rho(\mathbf{r})\rho(\mathbf{r}')}{|\mathbf{r} - \mathbf{r}'|} e^{-\mu|\mathbf{r}-\mathbf{r}'|} \, d\mathbf{r} \, d\mathbf{r}' = \frac{e^2}{2} \frac{N^2}{V} \frac{4\pi}{\mu^2}. \tag{1.122}$$

We must now decide what functions to use as a basis for our field operator. Since the eigenfunctions for a gas of *free* electrons are plane waves, we shall use these as our basis. These states are characterized by their wave vector $\mathbf{k}$ and spin quantum number σ. Thus

26   1. The Magnetic Susceptibility

$$\psi(r) = \sum_{k,\sigma} \frac{1}{\sqrt{V}} e^{i k \cdot r} \eta_\sigma a_{k\sigma} \tag{1.123}$$

where

$$\eta_\uparrow = \begin{pmatrix} 1 \\ 0 \end{pmatrix} \quad \text{and} \quad \eta_\downarrow = \begin{pmatrix} 0 \\ 1 \end{pmatrix}. \tag{1.124}$$

Notice that

$$\int \psi^\dagger(r)\psi(r) dr = \sum_{k\sigma} a^\dagger_{k\sigma} a_{k\sigma} = N. \tag{1.125}$$

Since the terms $\mathcal{H}_{el-n}$ and $\mathcal{H}_{n-n}$ do not involve any electron coordinates, they are carried over directly into our Hilbert space. The kinetic energy of the electron–electron Hamiltonian becomes

$$\sum_{k\sigma} \frac{\hbar^2 k^2}{2m} a^\dagger_{k\sigma} a_{k\sigma}. \tag{1.126}$$

For the Coulomb interaction we have

$$\frac{e^2}{2} \frac{1}{V^2} \sum_{k\sigma} \sum_{k'\sigma'} \sum_{k''\sigma''} \sum_{k'''\sigma'''} \int dr \int dr' \, e^{-i(k \cdot r + k' \cdot r')} \frac{e^{-\mu|r-r'|}}{|r-r'|} e^{i(k''' \cdot r' + k'' \cdot r)}$$
$$\times \eta^\dagger_\sigma(r) \eta^\dagger_{\sigma'}(r') \eta_{\sigma'''}(r') \eta_{\sigma''}(r) a^\dagger_{k\sigma} a^\dagger_{k'\sigma'} a_{k'''\sigma'''} a_{k''\sigma''}$$

$$= \frac{e^2}{2} \frac{1}{V^2} \sum_{k\sigma} \sum_{k'\sigma'} \sum_{k''\sigma''} \sum_{k'''\sigma'''} \delta_{\sigma,\sigma''} \delta_{\sigma',\sigma'''} \int dr \int dr' \, e^{-i(k+k'-k''-k''') \cdot r}$$
$$\times \frac{e^{i(k'-k''') \cdot (r-r')} e^{-\mu|r-r'|}}{|r-r'|} a^\dagger_{k\sigma} a^\dagger_{k'\sigma'} a_{k'''\sigma'''} a_{k''\sigma''}. \tag{1.127}$$

If we again treat r and $r - r'$ as independent variables, the integrations may be carried out separately to give

$$\frac{e^2}{2} \frac{1}{V} \sum_{k\sigma} \sum_{k'\sigma'} \sum_{k''\sigma''} \sum_{k'''\sigma'''} \delta_{\sigma,\sigma''} \delta_{\sigma',\sigma'''} \Delta(k+k'-k''-k''') \frac{4\pi}{(k'-k''')^2 + \mu^2}$$
$$\times a^\dagger_{k\sigma} a^\dagger_{k'\sigma'} a_{k'''\sigma'''} a_{k''\sigma''}. \tag{1.128}$$

If we define $k' - k''' \equiv q$ and collect all the terms, we obtain the total second-quantized Hamiltonian,

$$\mathcal{H} = \sum_{k\sigma} \frac{\hbar^2 k^2}{2m} a^\dagger_{k\sigma} a_{k\sigma} + \frac{e^2}{2V} \sum_k \sum_{k'} \sum_q \sum_{\sigma,\sigma'} \frac{4\pi}{q^2 + \mu^2}$$
$$\times a^\dagger_{k-q,\sigma} a^\dagger_{k'+q,\sigma'} a_{k',\sigma'} a_{k,\sigma} - \frac{e^2 N^2}{2V} \frac{4\pi}{\mu^2}. \tag{1.129}$$

It is convenient to introduce an equivalent electron radius r_0 by

$$\frac{4}{3}\pi r_0^3 = \frac{V}{N}. \tag{1.130}$$

This is made dimensionless by dividing it by the Bohr radius $a_0 = \hbar^2/me^2$. We also define $r_0/a_0 = r_s$ and $V/r_0^3 \equiv \Omega$. Since 1 Rydberg $= me^4/2\hbar^2 = e^2/2a_0$, the Hamiltonian expressed in Rydbergs is

$$\begin{aligned}\mathcal{H} = &\frac{1}{r_s^2}\sum_{k,\sigma}(r_0 k)^2 a_{k\sigma}^\dagger a_{k\sigma} + \frac{1}{r_s\Omega}\sum_{k,\sigma}\sum_{k',\sigma'}\sum_q \frac{4\pi}{r_0^2(q^2+\mu^2)}\\ &\times a_{k-q,\sigma}^\dagger a_{k'+q,\sigma'}^\dagger a_{k',\sigma'} a_{k,\sigma} - \frac{N^2}{r_s\Omega}\frac{4\pi}{(r_0\mu)^2}.\end{aligned} \tag{1.131}$$

Notice that the electron–electron interaction for $q = 0$ gives a term proportional to N^2, which cancels the contribution from the positive background, plus a term proportional to N, which vanishes by virtue of the condition $\mu^{-1} \ll L$. Thus we finally have

$$\mathcal{H} = \sum_{k,\sigma}\varepsilon_k a_{k,\sigma}^\dagger a_{k,\sigma} + \sum_{k,\sigma}\sum_{k',\sigma'}\sum_{q\neq 0} V(q) a_{k-q,\sigma}^\dagger a_{k'+q,\sigma'}^\dagger a_{k',\sigma'} a_{k,\sigma} \tag{1.132}$$

where $\varepsilon_k \equiv (r_0 k/r_s)^2$ and $V(q) \equiv 4\pi/[r_s\Omega(r_0 q)^2]$.

Aside from its advantages for handling particle statistics, the second-quantization formalism also lends itself readily to graphical interpretation. For example, the interaction term above corresponds to the destruction of two particles in states (k, σ) and (k', σ') and the creation of two particles in states $(k' + q, \sigma')$ and $(k - q, \sigma)$. This may be represented as

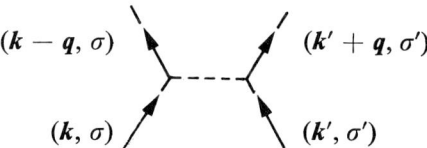

In general, we shall also have particle–hole interactions which have the form

$(k - q, \sigma)$ \qquad $(k' + q, \sigma')$

(k, σ) \qquad (k', σ')

We shall see that long-range magnetic order may be characterized as a coherent

electron–hole state just as superconductivity is characterized as a coherent electron–electron state.

Example: The Zeeman interaction. Finally, let us apply this second quantization prescription to the interaction of an electron spin with a magnetic field $H \cos(\mathbf{q} \cdot \mathbf{r})$. In the next chapter we shall show that this interaction has the form

$$\mathcal{H} = \mu_B \sigma_z H \cos(\mathbf{q} \cdot \mathbf{r}),$$

where μ_B is the Bohr magneton and σ_z the Pauli matrix

$$\sigma_z = \begin{pmatrix} 1 & 0 \\ 0 & -1 \end{pmatrix}.$$

In terms of the field operators (1.123) this becomes

$$\mathcal{H} = \tfrac{1}{2} \mu_B H \sum_k (a^\dagger_{k+q,\uparrow} a_{k\uparrow} - a^\dagger_{k+q,\downarrow} a_{k\downarrow} + a^\dagger_{k-q,\uparrow} a_{k\uparrow} + a^\dagger_{k-q,\downarrow} a_{k\downarrow}).$$

Since $\mathcal{H}_0 = 0$ in this simple example, $\sigma_z(t) = \sigma_z$ and the calculation of the longitudinal susceptibility using (1.80) simply involves terms of the form $\langle a^\dagger_{k\uparrow} a_{k+q\uparrow} a^\dagger_{k+q,\uparrow} a_{k\uparrow} \rangle$. This has the diagrammatic representation

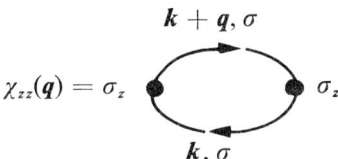

$$\chi_{zz}(\mathbf{q}) = \sigma_z \qquad \sigma_z$$

In more complex systems it is convenient to characterize various approximations in terms of their diagrammatic representations.

2. The Magnetic Hamiltonian

Most of the magnetic properties that we shall consider arise from electrons. In this chapter we shall develop the Hamiltonian which pertains to the magnetic behavior of a system of electrons. It has been found experimentally that the electron possesses an intrinsic magnetic moment, or *spin*. The existence of such a moment follows directly from relativistic considerations. Therefore it is essential that we look for a relativistic description of the motion of an electron. The *Dirac wave equation* offers just such a description. We shall limit our discussion of the Dirac equation to the origin of the spin and the form of the spin–orbit interaction (for a more thorough treatment see [2.1]).

2.1 The Dirac Equation

The objective in developing a relativistic theory of the electron is to ensure that space coordinates and time enter the theory symmetrically. If we start with the general wave equation

$$i\hbar \frac{\partial \psi(\mathbf{r}, t)}{\partial t} = \mathscr{H} \psi(\mathbf{r}, t) \tag{2.1}$$

where the first derivative with respect to time enters on the left, the Hamiltonian must contain a linear space derivative, that is, the Hamiltonian must be linear in the momentum. Thus we assume that the Hamiltonian has the form

$$\mathscr{H} = c\boldsymbol{\alpha} \cdot \mathbf{p} + \beta mc^2 \tag{2.2}$$

where $\boldsymbol{\alpha}$ and β are arbitrary coefficients. By imposing certain requirements on the solutions of (2.1), we obtain conditions on the quantities $\boldsymbol{\alpha}$ and β. These conditions may be satisfied by the representations

$$\beta = \begin{bmatrix} 1 & 0 \\ 0 & -1 \end{bmatrix} \quad \text{and} \quad \boldsymbol{\alpha} = \begin{bmatrix} 0 & \sigma \\ \sigma & 0 \end{bmatrix} \tag{2.3}$$

where the σ_i are the Pauli matrices

$$\sigma_x = \begin{bmatrix} 0 & 1 \\ 1 & 0 \end{bmatrix}, \quad \sigma_y = \begin{bmatrix} 0 & -i \\ i & 0 \end{bmatrix}, \quad \sigma_z = \begin{bmatrix} 1 & 0 \\ 0 & -1 \end{bmatrix}, \quad \mathbf{1} = \begin{bmatrix} 1 & 0 \\ 0 & 1 \end{bmatrix}. \tag{2.4}$$

2. The Magnetic Hamiltonian

Thus the wave function ψ must be a four-component object. Two of the components correspond to positive-energy solutions and the other two correspond to negative-energy solutions. Holes in the negative–energy spectrum correspond to positrons and require energies of the order of mc^2 for their production.

From a Lagrangian formulation we find that the effect of an external electromagnetic field described by the vector potential A and the scalar potential ϕ may be included by making the substitutions $p \to p - (e/c)A$ and adding $e\phi$ to the Hamiltonian. Thus the Dirac equation becomes

$$i\hbar \frac{\partial \psi}{\partial t} = \left[c\boldsymbol{\alpha} \cdot \left(\boldsymbol{p} - \frac{e}{c} \boldsymbol{A} \right) + \beta mc^2 + e\phi \right] \psi . \tag{2.5}$$

Since the energies encountered in magnetic phenomena are much smaller than mc^2, it is convenient to decouple the positive- and negative-energy solutions. This is accomplished by a canonical transformation due to Foldy and Wouthuysen [2.2]. The resulting Hamiltonian for the positive-energy solutions has the form

$$\mathcal{H} = \left[mc^2 + \frac{1}{2m}\left(\boldsymbol{p} - \frac{e}{c}\boldsymbol{A}\right)^2 - \frac{p^4}{8m^3c^2} \right] + e\phi - \frac{e\hbar}{2mc}\boldsymbol{\sigma}\cdot\boldsymbol{H}$$
$$- i\frac{e\hbar^2}{8m^2c^2}\boldsymbol{\sigma}\cdot\boldsymbol{\nabla}\times\boldsymbol{E} - \frac{e\hbar}{4m^2c^2}\boldsymbol{\sigma}\cdot\boldsymbol{E}\times\boldsymbol{p} - \frac{e\hbar^2}{8m^2c^2}\boldsymbol{\nabla}\cdot\boldsymbol{E} . \tag{2.6}$$

The interesting terms in this Hamiltonian are the last four. The term $-(e\hbar/2mc)\boldsymbol{\sigma}\cdot\boldsymbol{H}$ corresponds to the interaction of the intrinsic spin of the electron with the external field $\boldsymbol{H}$. The next two terms are spin–orbit terms. In a stationary vector potential and a spherically symmetric scalar potential $V(r)$ we have $\boldsymbol{\nabla}\cdot\boldsymbol{E} = 0$ and

$$\boldsymbol{\sigma}\cdot\boldsymbol{E}\times\boldsymbol{p} = -\frac{1}{r}\frac{\partial V}{\partial r}\boldsymbol{\sigma}\cdot\boldsymbol{r}\times\boldsymbol{p} = -\frac{\hbar}{r}\frac{\partial V}{\partial r}\boldsymbol{\sigma}\cdot\boldsymbol{l}$$

where $\hbar\boldsymbol{l} = \boldsymbol{r}\times\boldsymbol{p}$. These two terms give

$$\frac{e\hbar^2}{4m^2c^2}\frac{1}{r}\frac{\partial V}{\partial r}\boldsymbol{\sigma}\cdot\boldsymbol{l} . \tag{2.7}$$

This is what would be expected for an electron spin interacting with the field produced by its orbital motion, except that it is reduced by a factor of $\frac{1}{2}$ because of the Thomas precession. The last term in (2.6), the so-called *Darwin term*, represents a correction to the Coulomb interaction due to fluctuations (zitterbewegung) in the electron position arising from the presence of a negative–energy component in the wave function.

The term $p^4/8m^3c^2$ in (2.6) is very small, and along with the Darwin term, it may be neglected for our purposes. If we define the zero of energy as the rest-mass energy, the Hamiltonian which governs the magnetic behavior of an electron is

$$\mathcal{H} = \frac{1}{2m}\left(\boldsymbol{p} - \frac{e}{c}\boldsymbol{A}\right)^2 + e\phi - \frac{e\hbar}{2mc}\boldsymbol{\sigma}\cdot\boldsymbol{H} + \xi \boldsymbol{l}\cdot\boldsymbol{\sigma} \qquad (2.8)$$

where we have introduced the spin–orbit parameter

$$\xi = \frac{e\hbar^2}{4m^2c^2}\frac{1}{r}\frac{\partial V}{\partial r}.$$

2.2 Sources of Fields

In developing the general Hamiltonian for a single electron we found that the interaction of an electron with its environment is described by the scalar potential ϕ and the vector potential A. Both these potentials are functions of the position of the electron under consideration as well as of the coordinates and momenta of any other particles in the system, that is,

$$\phi(\boldsymbol{r}; \boldsymbol{r}_1, \boldsymbol{r}_2, \ldots, \boldsymbol{p}_1, \boldsymbol{p}_2, \ldots) \quad \text{and} \quad A(\boldsymbol{r}; \boldsymbol{r}_1, \boldsymbol{r}_2, \ldots, \boldsymbol{p}_1, \boldsymbol{p}_2, \ldots).$$

In this section we shall investigate the form these potentials take in a crystalline solid. Our objective is to catalog all the interactions that enter into the magnetic properties of solids so that we shall be free to draw on these results later.

2.2.1 Uniform External Field

The simplest potentials are those arising from uniform external fields. For an electric field E, uniform over all space, the interaction $e\phi$ becomes $-e\boldsymbol{r}\cdot\boldsymbol{E}$, where $e\boldsymbol{r}$ is the electric-dipole-moment operator.

For a uniform magnetic field H the magnetic vector potential is not uniquely defined. However, it is convenient to take $A = \frac{1}{2}\boldsymbol{H}\times\boldsymbol{r}$. In this gauge $\boldsymbol{\nabla}\cdot\boldsymbol{A} = 0$. Thus $(\boldsymbol{p} - e\boldsymbol{A}/c)^2/2m$ becomes

$$\frac{p^2}{2m} - \frac{e}{2mc}(\boldsymbol{r}\times\boldsymbol{p})\cdot\boldsymbol{H} + \frac{e^2}{8mc^2}(\boldsymbol{H}\times\boldsymbol{r})^2. \qquad (2.9)$$

The first term is the kinetic-energy term. The second term is a paramagnetic term, since $\boldsymbol{r}\times\boldsymbol{p} = \hbar\boldsymbol{l}$ is related to the electron's orbital moment $\boldsymbol{\mu}_l$ by

$$\boldsymbol{\mu}_l = \frac{e}{2mc}\boldsymbol{r}\times\boldsymbol{p} = -\frac{|e|\hbar}{2mc}\boldsymbol{l} = -\mu_\text{B}\boldsymbol{l}, \qquad (2.10)$$

where

2. The Magnetic Hamiltonian

$$\mu_B \equiv \frac{|e|\hbar}{2mc}$$

is the *Bohr magneton*. Thus the second term may be written as $\mu_B \boldsymbol{l} \cdot \boldsymbol{H}$. The third term is a diamagnetic term. When $\boldsymbol{H}$ is in the z direction, that is, $\boldsymbol{H} = H\hat{\boldsymbol{z}}$, then this term reduces to $(e^2 H^2/8mc^2)(x^2 + y^2)$. This gives, for the total Hamiltonian of an electron in a uniform magnetic field,

$$\mathcal{H} = \frac{p^2}{2m} + \mu_B(\boldsymbol{l} + \boldsymbol{\sigma}) \cdot \boldsymbol{H} + \frac{e^2 H^2}{8mc^2}(x^2 + y^2) + \zeta \boldsymbol{l} \cdot \boldsymbol{\sigma} \quad . \tag{2.11}$$

2.2.2 The Electric Quadrupole Field

Let us now look into the potentials the electron sees as it moves around or past a nucleus. If we assume that the electron remains outside the nuclear charge and current distributions, we may expand $|\boldsymbol{r} - \boldsymbol{r}_n|^{-1}$ in spherical harmonics, which results in a *multipole expansion*.

Let us first consider the charge distribution. If $\rho(\boldsymbol{r}_n)$ is the charge density at a point $\boldsymbol{r}_n$ inside the nucleus, the electrostatic potential becomes

$$\phi(\boldsymbol{r}) = \int d\boldsymbol{r}_n \frac{\rho(\boldsymbol{r}_n)}{|\boldsymbol{r} - \boldsymbol{r}_n|} = 4\pi \sum_{l=0}^{\infty} \sum_{m=-l}^{l} \frac{Y_l^m(\theta, \varphi)}{(2l+1)r^{l+1}} \int d\boldsymbol{r}_n \rho(\boldsymbol{r}_n) Y_l^{m*}(\theta_n, \varphi_n) r_n^l. \tag{2.12}$$

Writing out the first few terms explicitly, we have

$$\phi(\boldsymbol{r}) = 4\pi \frac{Y_0^0(\theta, \varphi)}{r} \int d\boldsymbol{r}_n \rho(\boldsymbol{r}_n) Y_0^{0*}(\theta_n, \varphi_n)$$

$$+ 4\pi \sum_{m=-1}^{1} \frac{Y_1^m(\theta, \psi)}{3r^2} \int d\boldsymbol{r}_n \rho(\boldsymbol{r}_n) r_n Y_1^{m*}(\theta_n, \varphi_n)$$

$$+ 4\pi \sum_{m=-2}^{2} \frac{Y_2^m(\theta, \varphi)}{5r^3} \int d\boldsymbol{r}_n \rho(\boldsymbol{r}_n) r_n^2 Y_2^{m*}(\theta_n, \varphi_n) + \cdots. \tag{2.13}$$

Since $Y_0^0(\theta, \varphi) = 1/\sqrt{4\pi}$, the first term becomes

$$\frac{1}{r} \int d\boldsymbol{r}_n \rho(\boldsymbol{r}_n) = \frac{Ze}{r}, \tag{2.14}$$

which is just the field arising from a point charge *at the origin*. We can make use of the spherical-harmonic addition theorem

$$\frac{4\pi}{2l+1} \sum_{m=-l}^{l} Y_l^{m*}(\theta_1, \varphi_1) Y_l^m(\theta_2, \varphi_2) = P_l(\cos \theta_{12}) \tag{2.15}$$

to write the second term as

$$\frac{\hat{\boldsymbol{r}}}{r^2}\cdot\int d\boldsymbol{r}_n\rho(\boldsymbol{r}_n)\boldsymbol{r}_n\,. \tag{2.16}$$

The integral is the electric-dipole-moment operator of the nucleus. If the nuclear states have definite parity, the diagonal matrix elements of this operator vanish by symmetry. The third term is the *quadrupole* term.

Since we shall eventually be interested in computing matrix elements of the quadrupole term, as well as various other similar operators, it is appropriate to digress for a moment to develop a technique for rewriting such operators in a form which greatly facilitates the evaluation of their matrix elements. This technique is based on the transformation properties of these operators.

Operator Equivalents. Suppose we consider a rotation of our coordinate system through some angle θ about an axis defined by $\hat{\boldsymbol{n}}$. Then the wave function of the original system, ψ, is related to the wave function in the rotated system, ψ', by a unitary transformation $R(\hat{\boldsymbol{n}}, \theta)$. It is convenient to write this unitary transformation in an exponential form with a Hermitian argument,

$$R(\hat{\boldsymbol{n}}, \theta) = \exp\left[\mathrm{i}S(\hat{\boldsymbol{n}}, \theta)\right].$$

For infinitesimal rotations ψ' will differ from ψ by an amount proportional to θ. If we write this proportionality as

$$\psi' - \psi = -\mathrm{i}\theta\hat{\boldsymbol{n}}\cdot\boldsymbol{J}\psi\,, \tag{2.17}$$

it follows that

$$R(n, \theta) = \exp\left(-\mathrm{i}\theta\hat{\boldsymbol{n}}\cdot\boldsymbol{J}\right). \tag{2.18}$$

This may be taken as a definition of the *angular momentum* $\hbar\boldsymbol{J}$ of the system. That is, the angular momentum characterizes the transformation properties of a system under rotations of the coordinates.

Now let $|JM\rangle$ be an eigenfunction of J^2 and J_z. Then

$$R|JM\rangle = \exp\left(-\mathrm{i}\theta\boldsymbol{n}\cdot\boldsymbol{J}\right)|JM\rangle\,. \tag{2.19}$$

Inserting the identity in the form

$$\sum_{M'}|JM'\rangle\langle JM'|$$

on the right gives

$$R|JM\rangle = \sum_{M'}\langle JM'|\exp\left(-\mathrm{i}\theta\hat{\boldsymbol{n}}\cdot\boldsymbol{J}\right)|JM\rangle|JM'\rangle = \sum_{M'}D^J_{MM'}(\alpha\beta\gamma)|JM'\rangle. \tag{2.20}$$

Thus the rotation operator transforms the function $|JM\rangle$ into a linear combination of the states $|JM'\rangle$ whose coefficients are the matrix elements of the rotation operator itself, $D^J_{MM'}(\alpha\beta\gamma)$, where α, β, and γ are the Euler angles that specify the rotation.

Under the rotation R an *operator O* is transformed into ROR^{-1}. If the operator O consists of $2J+1$ functions $T_{JM}(M = -J, -J+1, \ldots, J)$, and *if* it transforms according to

$$RT_{JM}R^{-1} = \sum_{M'} D^J_{MM'}(\alpha\beta\gamma)T_{JM'}, \tag{2.21}$$

then it is called an *irreducible tensor operator* of rank J. This may seem a rather restrictive definition. However, it turns out that many operators encountered in physical situations are, in fact, tensor operators. For example, a vector is a tensor of rank 1; moments of inertia and quadrupole moments are tensors of rank 2. An example of an operator which may not be a tensor is the density operator discussed earlier.

Tensors have their own algebra, including various theorems. One of the most useful of these for our purposes is the *Wigner–Eckart theorem*. This states that the matrix element of a tensor operator may be factored into a part which involves the projection quantum numbers but is independent of the tensor itself and a part not involving the projection quantum numbers, called the *reduced matrix element*. The first part is, in fact, just the *Clebsch–Gordan coefficient* encountered in the coupling of angular momenta. Thus we have

$$\langle J' \, M' | T_{J''M''} | JM \rangle = C(JJ''J'; MM''M') \langle J' \| T_{J''} \| J \rangle \tag{2.22}$$

where $C(JJ''J'; MM''M')$ is the Clebsch–Gordan coefficient and $\langle J' \| T_{J''} \| J \rangle$ is the reduced matrix element. Notice that if T' is also a tensor operator of the same rank as T, then the matrix elements of T are proportional to those of T'. This result has immense practical application to our magnetic Hamiltonian.

Let us return now to the quadrupole terms in $\phi(\mathbf{r})$. Since the quadrupole moment operators Q_2^M are proportional to the spherical harmonics, they are obviously tensor operators of rank 2. We can also form a tensor of rank 2 from the components of the total nuclear angular momentum $\mathbf{I}$. Thus if

$$T_2^{+1} = Q_2^{+1} = \frac{\sqrt{6}}{2} \sum_i z_i(x_i + iy_i), \tag{2.23}$$

where i runs over the protons, suggests that we form

$$(T')_2^{+1} = \frac{\sqrt{6}}{4}(I_z I^+ + I^+ I_z). \tag{2.24}$$

Notice that (2.24) is written in the symmetrized form. The reason for this is that the coordinates entering (2.23) commute with each other, whereas the angular momenta do not. Therefore, in order to preserve this symmetry, we must symmetrize the operator equivalent. By the Wigner–Eckart theorem, the matrix elements of these two operators must be proportional. Thus

$$\langle IM'|Q_2^{+1}|IM\rangle = \alpha\langle IM'|\frac{\sqrt{6}}{4}(I_zI^+ + I^+I_z)|IM\rangle, \tag{2.25}$$

$$\langle IM'|Q_2^{+1}|IM\rangle = \alpha\frac{\sqrt{6}}{4}(2M+1)\sqrt{(I-M)(I+M+1)}\delta_{M',M+1}. \tag{2.26}$$

It is customary to define the particular matrix element $\langle II|Q_2^0|II\rangle$ as eQ. The proportionality constant then becomes $eQ/I(2I-1)$.

As long as we remain within a manifold in which I is a good quantum number, we may also equate the operators themselves. Thus

$$Q_2^{+1} = \frac{eQ}{I(2I-1)}\frac{\sqrt{6}}{4}(I_zI^+ + I^+I_z) \tag{2.27}$$

with similar expressions for the other operators. The corresponding term in the quadrupole potential is therefore

$$-\sqrt{\frac{4\pi}{5}}\frac{Y_2^{-1}(\theta,\varphi)}{r^3}\frac{eQ}{I(2I-1)}\frac{\sqrt{6}}{4}(I_zI^+ + I^+I_z). \tag{2.28}$$

It is obvious that we could now write the *electron* part in terms of the total orbital angular momentum of the electron state [2.3, 4]. Thus,

$$e\phi(r) = -\frac{Ze^2}{r} + e^2Q\xi[3(\mathbf{l}\cdot\mathbf{I})^2 + 3/2(\mathbf{l}\cdot\mathbf{I}) - l(l+1)I(I+1)],$$

where ξ is a constant that is proportional to the reduced matrix element of the electronic angular momentum. Notice that an *s*-state electron is not affected by the quadrupole field of the nucleus. The quadrupole field is, in general, small compared with other fields acting on the electron. From the point of view of the nucleus, however, this interaction is very important. If the electron is in a nondegenerate state characterized by the orbital quantum numbers l, m_l, and the coordinates are chosen to lie along the principal axes of the tensor $l_\mu l_\nu$, then the nuclear Hamiltonian becomes

$$\boxed{\mathscr{H}_Q = \frac{e^2qQ}{4I(2I-1)}[3I_z^2 - I(I-1) + \tfrac{1}{2}\eta(I_+^2 + I_-^2)]} \tag{2.29}$$

when $q = \xi\langle l_z^2\rangle$ and $\eta = \dfrac{(\langle l_x^2\rangle - \langle l_y^2\rangle)}{\langle l_z^2\rangle}$.

2. The Magnetic Hamiltonian

The same expression also characterizes the interaction with a more general surrounding charge distribution. In this case q is proportional to the electric field gradient produced by this charge distribution.

2.2.3 The Magnetic Dipole (Hyperfine) Field

The vector potential arising from the nuclear currents may also be expanded to yield

$$A(r) = \int dr_n \frac{\rho(r_n)v(r_n)}{c|r - r_n|}$$

$$= \frac{1}{cr}\int dr_n \rho(r_n)v(r_n) + \frac{1}{2cr^3}\int dr_n \rho(r_n)\{(r \cdot r_n)v(r_n) + [r \cdot v(r_n)]r_n\}$$

$$- \frac{r}{r^3} \times \frac{1}{2c}\int dr_n \rho(r_n)[r_n \times v(r_n)] + \ldots . \tag{2.30}$$

If the current distribution is stationary with respect to the angular-momentum axis, then the first two terms vanish, leaving only the third. The integral in this term is the nuclear magnetic dipole moment μ_I, which is related to the nuclear angular momentum by

$$\mu_I = g_N \mu_N I = \gamma_N \hbar I,$$

where μ_N is the nuclear magneton, γ_N is the nuclear gyromagnetic ratio, and g_N is the nuclear g value.

Notice that the nuclear angular momentum I is in units of $\hbar$. Thus

$$A(r) = \mu_I \times \frac{r}{r^3}. \tag{2.31}$$

Substituting this into the expression

$$\frac{1}{2m}\left(p - \frac{e}{c}A\right)^2 - \frac{e\hbar}{2mc}\sigma \cdot \nabla \times A \tag{2.32}$$

and recognizing that $\nabla \cdot A(r) = 0$, we obtain

$$\frac{p^2}{2m} - \frac{e}{mc}\frac{(\mu_I \times r) \cdot p}{r^3} + \frac{e\hbar}{2mc}\sigma \cdot \left[\frac{\mu_I}{r^3} - 3\frac{(r \cdot \mu_I)r}{r^5}\right]. \tag{2.33}$$

Interchanging the dot and cross products in the second term gives the *orbital hyperfine interaction*

$$2\mu_B \frac{\mu_I \cdot I}{r^3}. \tag{2.34}$$

2.2 Sources of Fields

The last term in (2.33) containing the square brackets is the *dipolar hyperfine interaction*.

If the electron is in an s state, then the matrix elements of the orbital hyperfine interaction will clearly vanish. Similarly, the matrix elements of the dipolar hyperfine interaction also vanish for an s-state electron. However, there is an additional interaction for s-state electrons that is not included in expansion (2.30), since it is valid only for charge distributions which vanish at the nucleus. To obtain this we consider the matrix element of the hyperfine interaction for an electron orbital state $\psi(r)$:

$$-\frac{e\hbar}{2mc}\int_{\text{all space}} dr\, \psi^*(r)\boldsymbol{\sigma}\cdot\boldsymbol{\nabla}\times A(r)\psi(r)$$

$$= -\frac{e\hbar}{2mc}\int_{r<R} dr\, \psi^*(r)\boldsymbol{\sigma}\cdot\boldsymbol{\nabla}\times A(r)\psi(r)$$

$$-\frac{e\hbar}{2mc}\int_{r>R} dr\, \psi^*(r)\boldsymbol{\sigma}\cdot\boldsymbol{\nabla}\times A(r)\psi(r). \quad (2.35)$$

The radius R defines a sphere which encloses the nucleus. Outside this sphere $A(r)$ has the form $(\boldsymbol{\mu}_I \times r)/r^3$. The second term in (2.35) gives the dipolar hyperfine interaction derived previously. The first term is the additional interaction, which may be rewritten as

$$-\frac{e\hbar}{2mc}\int_{r<R} dr\, \boldsymbol{\nabla}\cdot(A\times\boldsymbol{\sigma})|\psi(r)|^2 = -\frac{e\hbar}{2mc}\boldsymbol{\sigma}\cdot\int dS \times A|\psi(R)|^2. \quad (2.36)$$

Because the sphere of integration has been chosen to lie outside the nucleus, $A(R)$ has the form $(\boldsymbol{\mu}_I \times R)/R^3$. Since $\psi(R)$ is essentially constant over this surface and is equal to $\psi(0)$, the interaction becomes

$$-\frac{e\hbar}{2mc}\boldsymbol{\sigma}\cdot\int \frac{R\times(\boldsymbol{\mu}_I\times R)}{R^2} d\Omega|\psi(0)|^2 = -\frac{8\pi}{3}\frac{e\hbar}{2mc}\boldsymbol{\sigma}\cdot\boldsymbol{\mu}_I|\psi(0)|^2,$$

$$= \frac{16\pi}{3}g_N\mu_B\mu_N\, \boldsymbol{I}\cdot\boldsymbol{\sigma}|\psi(0)|^2. \quad (2.37)$$

This is a *contact hyperfine interaction* often written as the operator

$$(8\pi/3)g_N\mu_B\mu_N\, \boldsymbol{I}\cdot\boldsymbol{\sigma}\delta(r).$$

Combining these results gives us the total hyperfine interaction,

$$\boxed{\mathcal{H}_{\text{hyper}} = 2g_N\mu_B\mu_N\frac{\boldsymbol{l}\cdot\boldsymbol{I}}{r^3} - g_N\mu_B\mu_N\boldsymbol{\sigma}\cdot\left[\frac{\boldsymbol{I}}{r^3} - 3\frac{(\boldsymbol{r}\cdot\boldsymbol{I})\boldsymbol{r}}{r^5}\right] \\ + \frac{8\pi}{3}g_N\mu_B\mu_N\, \boldsymbol{\sigma}\cdot\boldsymbol{I}\,\delta(r).} \quad (2.38)$$

The Hamiltonian (2.11) plus the interactions (2.29, 38) determine the behavior of a single electron in the presence of a nucleus.

2.2.4 Other Electrons on the Same Ion

Let us now consider the effect of other electrons. One of the most important sources of the electric field felt by an ionic electron is the Coulomb field arising from the other electrons on the same ion,

$$\phi(r) = \sum_i \frac{e}{|r - r_i|}. \qquad (2.39)$$

In ionic materials this interaction leads to the term levels (the determination of these many-electron states in terms of their Coulomb integrals is discussed in [2.5]). In itinerant-electron materials it is often assumed that the electrons experience a Coulomb repulsion only if they both happen to be in the same ionic cell. We shall consider the corresponding Hamiltonian for this situation later.

2.2.5 Crystalline Electric Fields

The Coulomb interactions between each electron and all the charges external to the ion are described by the electrostatic potential $V(r)$. In the case of iron-group ions the magnetic electrons (the $3d$ electrons) are outermost and hence are strongly affected by such a potential. In the case of rare-earth ions the magnetic $4f$ electrons are shielded by the $5s^2 5p^6$ shells and are less affected.

Since the charge distribution associated with neighboring ions may overlap that of the electron in question, the full treatment of this problem is very complex. The external charge distributions of these neighboring ions are called *ligands*, and their effects are computed by means of *ligand field theory* [2.6]. However, for our purposes, it will be sufficient to treat the neighboring ions as point charges; the problem may then be handled by *crystal field theory*. The advantage of using point charges is that $V(r)$ satisfies Laplace's equation and may be expanded in spherical harmonics as

$$V(r, \theta, \varphi) = \sum_{L'} \sum_{M'} A_{L'}^{M'} r^{L'} Y_{L'}^{M'}(\theta, \varphi). \qquad (2.40)$$

The number of terms that need be considered is greatly reduced, for the following reasons. Suppose we consider an iron-group ion in a crystal field. Then we shall eventually be interested in matrix elements of the form $\int \chi^* V \psi d\tau$, where χ and ψ are d-electron wave functions. Since the density $\chi^* \psi$, when expanded in spherical harmonics will not contain terms with $L' > 4$, the integrals with $L' > 4$ will vanish, by orthogonality of the spherical harmonics. Similarly, the integral vanishes for all terms in V which have L' odd. The term for $L' = 0$ is usually dropped, because it is an additive constant. If we are considering several $3d$ electrons within a term, then L'_{max} is given by the L value of this term (for example, $L' \leqslant 6$ for an F-state ion).

The potential energy of a charge q' at (r, θ, φ), in a potential due to charges q at a distance d from the origin and arranged in a *cubic coordination*, is

$$V_c(r, \theta, \varphi) = D'_4 \left\{ Y_4^0(\theta, \varphi) + \sqrt{\frac{5}{14}} [Y_4^4(\theta, \varphi) + Y_4^{-4}(\theta, \varphi)] \right\}$$
$$+ D'_6 \left\{ Y_6^0(\theta, \varphi) - \sqrt{\frac{7}{2}} [Y_6^4(\theta, \varphi) + Y_6^{-4}(\theta, \varphi)] \right\}, \quad (2.41)$$

where, for example, $D'_4 = +\frac{7}{3} \sqrt{\pi} \, qq'r^4/d^5$ for sixfold coordination. The coefficients in such expansions have been tabulated by *Hutchings* [2.7]. The potential may also be expressed in terms of cartesian coordinates. Thus the above potential may be written as

$$V_c(x, y, z) = C_4[(x^4 + y^4 + z^4) - \tfrac{3}{5}r^4] + C_6(x^6 + y^6 + z^6)$$
$$+ \tfrac{15}{4}(x^2y^4 + x^2z^4 + y^2x^4 + y^2z^4 + z^2x^4 + z^2y^4) - \tfrac{15}{14}r^6] \quad (2.42)$$

where $C_4 = +\frac{35}{4} qq'/d^5$ for sixfold coordination.

We are now faced with the problem of calculating the matrix elements of this potential. This is easily accomplished by the operator-equivalent method, which makes use of the fact that the matrix elements of operators involving x, y, and z within a given L or J manifold are proportional to those of L_x, L_y, and L_z or J_x, J_y, and J_z. As pointed our earlier, the fact that the angular-momentum operators do not commute necessitates some care in constructing the operator equivalents. Fortunately there are tables for these (probably the best source is [2.7]). For example, within a manifold where L is constant the sum of the potential energies of all the electrons contributing to L is

$$\sum (x^4 + y^4 + z^4 - \tfrac{3}{5}r^4)$$
$$\Rightarrow \frac{\beta \overline{r^4}}{20} [35L_z^4 - 30L(L+1)L_z^2 + 25L_z^2 - 6L(L+1) + 3L^2(L+1)^2]$$
$$+ \frac{\beta \overline{r^4}}{8} [(L^+)^4 + (L^-)^4] \equiv \frac{\beta \overline{r^4}}{20} O_4^0 + \frac{\beta \overline{r^4}}{4} O_4^4 = B_4^0 O_4^0 + B_4^4 O_4^4 \quad (2.43)$$

where $\overline{r^4}$ is the average value of the fourth power of the electron radius. The operators O_n^m appear frequently in the literature. The ground state β is a constant which depends on the term; for a ^2D or a ^5D term $\beta = \frac{2}{63}$.

Consider a single $3d$ electron. This has the term ^2D, which is fivefold orbitally degenerate, with states $^2D(L_z, S_z)$. The matrix elements of V_c are [2.7]

$$\frac{\overline{\beta r^4}}{20}\begin{array}{c} {}^2D(2) \quad {}^2D(1) \quad {}^2D(0) \quad {}^2D(-1) \quad {}^2D(-2) \\ \begin{bmatrix} 12 & 0 & 0 & 0 & 60 \\ 0 & -48 & 0 & 0 & 0 \\ 0 & 0 & 72 & 0 & 0 \\ 0 & 0 & 0 & -48 & 0 \\ 60 & 0 & 0 & 0 & 12 \end{bmatrix} \end{array}. \qquad (2.44)$$

The eigenvalues and eigenvectors are easily found to be

Energy   *Eigenfunctions*

$\frac{12}{5}\overline{\beta r^4}$
$$\begin{cases} {}^2D(1, S_z) \\ {}^2D(-1, S_z) \\ \dfrac{1}{\sqrt{2}}[{}^2D(2, S_z) - {}^2D(-2, S_z)] \end{cases}$$

$-\frac{18}{5}\overline{\beta r^4}$
$$\begin{cases} {}^2D(0, S_z) \\ \dfrac{1}{\sqrt{2}}[{}^2D(2, S_z) + {}^2D(-2, S_z)] \end{cases}$$

The quantity 6β is often written as Δ. Therefore we find that the ${}^2D$ term is split into two states separated by $C_4 \overline{r^4} \Delta$. Notice that C_4 can be positive or negative, depending on the coordination. This is illustrated in Fig. 2.1.

The nature of such splittings obviously depends on the symmetry of the crystal field. For this reason group theory is a powerful tool in determining the

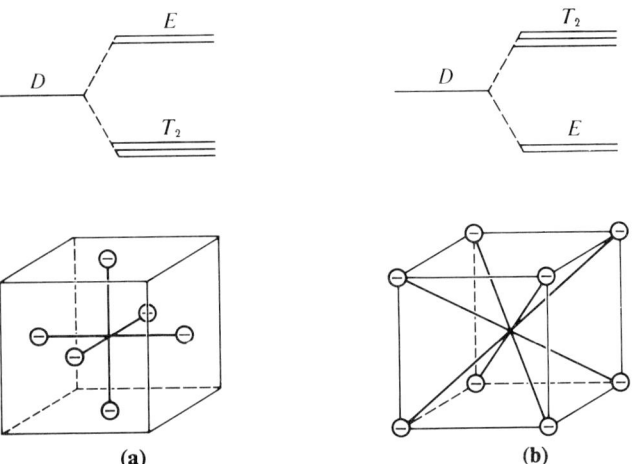

Fig. 2.1.a, b. Splitting of a D state in a cubic crystal field for (a) sixfold coordination and (b) eightfold coordination

degeneracies associated with various symmetries. Group theory as applied to crystal wave functions is discussed fully elsewhere [2.8], but it is worth our while to digress again briefly to introduce some of the group-theory terminology and notation which will enter our discussions from time to time.

Symmetry Representations. The symmetry of a system is generally specified by those operations which leave its physical appearance unchanged. For example, the symmetry operations which leave an equilateral triangle unchanged are listed in Table 2.1. A collection of symmetry operations that satisfies certain conditions is called a *group*. In order to take advantage of the powerful theorems associated with group theory we always work with those symmetry operations which do, in fact, constitute a group. These operations may be specifically represented by matrices which describe how a *coordinate point* transforms under the particular symmetry operation. Thus, if R represents a rotation and t represents a translation, the most general coordinate transformation is $x' = Rx + t$. Such a collection of operations is called a *space group*. The rotational part, obtained by setting $t = 0$, itself forms a group, called the *point group*. When we are dealing with noninteracting ions, the point group is sufficient to characterize the properties of the system. However, for interacting systems the full space group must be employed. Fortunately, since the point groups in a crystal must be compatible with translational symmetry, there are only 32 such groups [Ref. 2.8, p. 55]. Our equilateral triangle is characterized by the point group labeled D_3 in the so-called Schöflies notation or 32 in the international notation.

Now, let us consider a *function* whose form depends on the arrangement of the system. For example, suppose that three protons are located at the vertices of an equilateral triangle. The energy of an electron in such an environment depends on the positions of the protons, but this energy is unchanged under any permutation of the protons. Notice that there are 3!, or six, such permutations. These are just the results of the six symmetry operations which leave the triangle invariant. If a figure is defined by some arrangement of identical

Table 2.1. Symmetry operations associated with an equilateral triangle

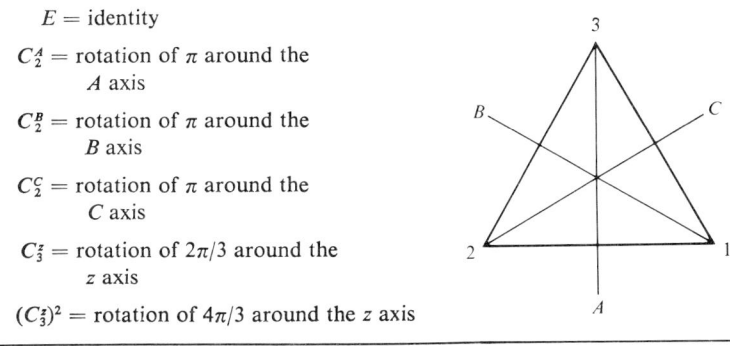

E = identity

C_2^A = rotation of π around the A axis

C_2^B = rotation of π around the B axis

C_2^C = rotation of π around the C axis

C_3^z = rotation of $2\pi/3$ around the z axis

$(C_3^z)^2$ = rotation of $4\pi/3$ around the z axis

particles, the operations which leave it invariant also leave the interaction energy between these particles and other particles invariant. Thus it is convenient to introduce a new group, isomorphic to the coordinate-transformation group, in which the group elements are operators which operate on *functions* rather than on *coordinates*. These operators are defined by

$$P_R f(x) \equiv f(R^{-1} x). \tag{2.45}$$

The particular function with which we shall be concerned is the energy in its operator form, the Hamiltonian $\mathscr{H}$. Those symmetry operations which leave the Hamiltonian invariant comprise the group of the Schrödinger equation. If P_R leaves $\mathscr{H}$ invariant, then it must commute with $\mathscr{H}$. Therefore

$$P_R \mathscr{H} \psi_n = \mathscr{H} P_R \psi_n = E_n P_R \psi_n. \tag{2.46}$$

Thus any function $P_R \psi_n$ obtained by operating on an eigenfunction ψ_n with a symmetry operator of the group of the Schrödinger equation will also be an eigenfunction having the same energy. Suppose that the state n is l_n-fold degenerate. Then the function $P_R \psi_n$ must be a linear combination of l_n orthonormal eigenfunctions, which we shall denote $\psi_\mu^{(n)}$ ($\mu = 1, \ldots, l_n$). Therefore

$$P_R \psi_\mu^{(n)} = \sum_{\nu=1}^{l} \Gamma^{(n)}(R)_{\mu\nu} \psi_\nu^{(n)}. \tag{2.47}$$

The transformation coefficients constitute a set of matrices which form an *irreducible representation* of the group of the Schrödinger equation. Furthermore, we see that the nth representation is associated with the nth eigenstate, and the *dimensionality of the representation is equal to the degeneracy* of this eigenstate. This representation is irreducible, since there is always an operator in the group that will transform each function into any other function. If this were not true, we could construct smaller sets of states which would, in general, have different eigenvalues, contradicting our original hypothesis.

Because of the relation (2.47), we speak of the $\psi^{(n)}$ as "transforming according to $\Gamma^{(n)}$". For this reason energy eigenstates are labeled by their irreducible representations. Also, since the representations are generated from the eigenfunctions, we say that the l_n degenerate eigenfunctions $\psi_\mu^{(n)}$ form a *basis* for an l_n-dimensional representation $\Gamma^{(n)}$ of the group of the Schrödinger equation.

The number and nature of the irreducible representations associated with the various symmetry groups have all been tabulated in what is known as a *character table*. The character $\chi_n(R)$ associated with the operation R belonging to the nth irreducible representation is merely the trace of the matrix of that representation, that is,

$$\chi_n(R) = \sum_\mu \Gamma^{(n)}(R)_{\mu\mu}. \tag{2.48}$$

One of the powerful features of group theory is that it enables us to determine the irreducible representations and all their characters without ever having to know specifically the basis functions. The character table for the equilateral-triangle symmetry group D_3 is given in Table 2.2. To see what this character table implies, let us suppose that we have a single electron, bound, say, to some ionic core giving rise to certain eigenstates. Since this system has complete rotational symmetry, these states are labeled by the familiar s, p, d, etc. Let us surround the system by three protons located at the vertices of an equilateral triangle. The symmetry of this system is D_3. The character table for D_3 tells us that the eigenfunctions of the electron are now labeled by the irreducible representations A_1, A_2, or E. The character associated with the identity operation, E, tells us the degeneracy of the various representations. For example, A_1 and A_2 nondegenerate states, and E is doubly degenerate.

Table 2.2. Character table for the point group D_3

Symmetry group D_3	Operations		
	E	$2C_3$	$3C'_2$
Irreducible representations		*Characters*	
A_1	1	1	1
A_2	1	1	−1
E	2	−1	0

Exactly how the original states decompose into the new types of states depends on the original states themselves and on the symmetry group characterizing the environment. Fortunately, such decompositions have been tabulated for a great number of situations. For example, in Fig. 2.1 we saw that a D state, when exposed to a cubic crystal field, splits into a doublet labeled by E and a triplet labeled by T_2. Group theory, however, does not tell us the ordering of the states or their relative separations. Such specific information can be obtained only by doing a calculation, as we did at the beginning of this section.

Quenching. At this point it is convenient to introduce a general property of angular momentum. This might be stated as a theorem:

The matrix element of the orbital angular momentum between non-degenerate *states has an arbitrary phase. In particular, it may be pure real or pure imaginary.*

To prove this let us consider the time-reversal operator T acting on a state ψ [Ref. 2.8, p. 141]. If we neglect the spin, then $T\psi = \psi^*$. Furthermore, if the Hamiltonian of the system is Hermitian, then ψ^* has the same eigenvalue as ψ, but if ψ is nondegenerate, then ψ and ψ^* must be linearly dependent. That is,

$\psi^* = c\psi$, operating on this relation with T gives $\psi = |c|^2\psi$ which requires that $|c|^2 = 1$ or $c = e^{i\varphi}$, where φ is a real quantity.

Now consider the matrix element $\langle n|L|m\rangle$. Inserting the identity operator $T^{-1}T$, we may write this as $\langle n|T^{-1}TLT^{-1}T|m\rangle$. Under time reversal the angular momentum changes sign, $TLT^{-1} = -L$. Also, since T satisfies $\langle n|T^{-1}|m\rangle = \langle Tn|m\rangle^*$, we obtain

$$\langle n|L|m\rangle = -\exp[i(\varphi_n - \varphi_m)]\langle n|L|m\rangle^*.$$

Since the phases are arbitrary we could, for example, choose $\varphi_n - \varphi_m = 0$, in which case this matrix element would be the negative of its complex conjugate, which would make it pure imaginary. If $\varphi_n - \varphi_m = \pi/2$, then it would be pure real.

This theorem has an important corollary:

The expectation value of L for any nondegenerate state is 0.

In the theorem, if $|m\rangle = |n\rangle$, then $\langle n|L|n\rangle$ must be pure imaginary. But this is a physical observable. Therefore it must be 0.

Thus if the crystal field has sufficiently low symmetry to remove all the orbital degeneracy, then, to lowest order, the orbital angular momentum is 0, and we say that the crystal field has completely *quenched* it. For this reason the static susceptibility of iron-group salts is found experimentally to arise predominantly from the spin.

2.2.6 Dipole–Dipole Interaction

The magnetic neighbors surrounding a given ion will contribute to the vector potential a term similar to that which we found for the electron-nucleus magnetic coupling. If the ions have moments μ_i, the dipole–dipole interaction has the form

$$\mathcal{H}_{\text{dip}} = \sum_{\substack{i,j \\ i \neq j}} \frac{1}{r_{ij}^3} [\boldsymbol{\mu}_i \cdot \boldsymbol{\mu}_j - 3(\boldsymbol{\mu}_i \cdot \hat{\boldsymbol{r}}_{ij})(\boldsymbol{\mu}_j \cdot \hat{\boldsymbol{r}}_{ij})]. \tag{2.49}$$

It is convenient to separate this into various terms, the meaning of which will become evident later:

$$\mathcal{H}_{\text{dip}} = g^2 \mu_B^2 \sum_{i>j} \left\{ -\frac{3\cos^2\theta_{ij} - 1}{r_{ij}^3} S_i^z S_j^z \right.$$

$$+ \frac{3\cos^2\theta_{ij} - 1}{4r_{ij}^3}(S_i^+ S_j^- + S_i^- S_j^+)$$

$$- \frac{3}{2} \frac{\sin\theta_{ij}\cos\theta_{ij}\exp(-i\varphi_{ij})}{r_{ij}^3}(S_i^z S_j^+ + S_i^+ S_j^z)$$

$$-\frac{3}{2}\frac{\sin\theta_{ij}\cos\theta_{ij}\exp(-i\varphi_{ij})}{r_{ij}^3}(S_i^z S_j^- + S_i^- S_j^z)$$

$$-\frac{3}{4}\frac{\sin^2\theta_{ij}}{r_{ij}^3}[\exp(-2i\varphi_{ij})S_i^+ S_j^+ + \exp(2i\varphi_{ij})S_i^- S_j^-]\Big\} \quad (2.50)$$

where θ_{ij} and φ_{ij} are the angles that $\mathbf{r}_{ij}$ makes with the fixed coordinate system.

2.2.7 Direct Exchange

The exchange energy is that contribution to the interaction energy of a system of electrons which arises from the use of antisymmetrized wave functions, as opposed to single products of one-electron wave functions. Under certain conditions the same effect may be achieved with single-product wave functions if an exchange-interaction term is added to the Hamiltonian. This effect was discovered simultaneously and independently by Dirac and Heisenberg in 1926. Since then a great deal of work has been done on this subject, particularly in developing the appropriate Hamiltonian [2.9]. In this section we shall discuss the origin of exchange and indicate the approximations under which it may be represented by an effective interaction Hamiltonian.

Let us begin by considering two electrons interacting with each other and a fixed positive point charge Ze. Let us assume that we know the eigenfunctions of the one-electron Hamiltonian $\mathcal{H}_0(\mathbf{r}, \boldsymbol{\sigma})$. For the time being let us assume that this does not include the spin–orbit interaction. Then $\mathcal{H}_0(\mathbf{r}, \boldsymbol{\sigma}) = \mathcal{H}_0(\mathbf{r})$, and we may write the eigenfunctions as products of an orbital function $\varphi_n(\mathbf{r})$ and a spinor $\eta_\mu(\boldsymbol{\sigma})$. We shall consider the modifications introduced by the spin–orbit interaction later.

The two-electron Hamiltonian is

$$\mathcal{H} = \mathcal{H}_0(\mathbf{r}_1) + \mathcal{H}_0(\mathbf{r}_2) + \frac{e^2}{|\mathbf{r}_1 - \mathbf{r}_2|}. \quad (2.51)$$

Let us assume that the electron–electron interaction is smaller than $\mathcal{H}_0$, so that it may be treated by perturbation theory. We must now determine what functions to use as the basis for computing the matrix elements of the electron–electron interaction. The fact that the Hamiltonian *without* the electron–electron interaction is separable suggests that we try product wave functions. Thus, if electron 1 is in an orbital state n with spin up and electron 2 is in an orbital state m, also with spin up, we might try $\varphi_n(\mathbf{r}_1)\,\alpha(\boldsymbol{\sigma}_1)\,\varphi_m(\mathbf{r}_2)\,\alpha(\boldsymbol{\sigma}_2)$, where α is the spin-up spinor. However, the Pauli exclusion principle requires that the wave functions be antisymmetric with respect to particle interchanges. This condition may be satisfied by writing the wave function as a normalized *Slater determinant*. If the single-electron wave functions are orthogonal, the appropriate determinantal wave function is

$$\frac{1}{\sqrt{2}} \begin{vmatrix} \varphi_n(r_1)\alpha(\sigma_1) & \varphi_n(r_2)\alpha(\sigma_2) \\ \varphi_m(r_1)\alpha(\sigma_1) & \varphi_m(r_2)\alpha(\sigma_2) \end{vmatrix}. \tag{2.52}$$

Since there are an infinite number of orbital states, we could construct an infinite number of such Slater determinants. The general wave function would be a linear combination of such determinants. However, if the electron–electron interaction is small, we may neglect this admixture of other orbital states. In particular, let us assume that electron 1 has a low-lying nondegenerate orbital state φ_a with energy E_a and electron 2 has a similar low-lying non-degenerate orbital state φ_b with energy E_b. If both spin functions are up, the determinantal wave function becomes

$$\psi_1 = \frac{1}{\sqrt{2}} \begin{vmatrix} \varphi_a(r_1)\alpha(\sigma_1) & \varphi_a(r_2)\alpha(\sigma_2) \\ \varphi_b(r_1)\alpha(\sigma_1) & \varphi_b(r_2)\alpha(\sigma_2) \end{vmatrix}. \tag{2.53}$$

If the spin function associated with orbital a is down, then

$$\psi_2 = \frac{1}{\sqrt{2}} \begin{vmatrix} \varphi_a(r_1)\beta(\sigma_1) & \varphi_a(r_2)\beta(\sigma_2) \\ \varphi_b(r_1)\alpha(\sigma_1) & \varphi_b(r_2)\alpha(\sigma_2) \end{vmatrix}. \tag{2.54}$$

There are two additional possible spin configurations which lead to wave functions ψ_3 and ψ_4. These four functions form a complete orthonormal set and therefore constitute an appropriate bais with which to evaluate the matrix elements of $\mathcal{H}$. The result is

$$H = \begin{bmatrix} E_a+E_b+K_{ab}-J_{ab} & 0 & 0 & 0 \\ 0 & E_a+E_b+K_{ab} & -J_{ab} & 0 \\ 0 & -J_{ab} & E_a+E_b+K_{ab} & 0 \\ 0 & 0 & 0 & E_a+E_b+K_{ab}-J_{ab} \end{bmatrix} \tag{2.55}$$

where

$$K_{ab} = \iint dr_1\, dr_2\, \frac{e^2}{r_{12}} |\varphi_a(r_1)|^2 |\varphi_b(r_2)|^2 \tag{2.56}$$

and

$$J_{ab} = \iint dr_1\, dr_2\, \varphi_a^*(r_1)\varphi_b^*(r_2) \frac{e^2}{r_{12}} \varphi_b(r_1)\varphi_a(r_2). \tag{2.57}$$

Diagonalizing this matrix gives a singlet with energy

$$E_s = E_a + E_b + K_{ab} + J_{ab} \tag{2.58}$$

and a triplet with energy

$$E_t = E_a + E_b + K_{ab} - J_{ab}. \tag{2.59}$$

Since J_{ab} is the self-energy of the charge distribution $e\varphi_a^*(r)\,\varphi_b(r)$, it is positive definite. Therefore the triplet always has a lower energy than the singlet. This is just the origin of *Hund's rule,* which says that the ground state of an atom has *maximum multiplicity.*

Dirac noticed that the eigenvalues (2.58, 59) could be obtained with a basis consisting only of products of spin functions if an *exchange interaction* were added to the Hamiltonian. To obtain the form of this effective interaction term we notice that just as any 2×2 matrix may be expressed as a linear combination of Pauli matrices plus the unit matrix, any 4×4 matrix may be written as a quadratic function of *direct* products of Pauli matrices [Ref. 2.8, p. 320]. For example, if

$$\sigma_{1x} = \begin{bmatrix} 0 & 1 \\ 1 & 0 \end{bmatrix} \quad \text{and} \quad \sigma_{2x} = \begin{bmatrix} 0 & 1 \\ 1 & 0 \end{bmatrix}, \tag{2.60}$$

then

$$\sigma_{1x} \otimes \sigma_{2x} = \begin{bmatrix} 0 & 0 & 0 & 1 \\ 0 & 0 & 1 & 0 \\ 0 & 1 & 0 & 0 \\ 1 & 0 & 0 & 0 \end{bmatrix}. \tag{2.61}$$

We are particularly interested in that quadratic form which gives three equal eigenvalues. Such a form is

$$\boldsymbol{\sigma}_1 \cdot \boldsymbol{\sigma}_2 = \begin{bmatrix} 1 & 0 & 0 & 0 \\ 0 & -1 & 2 & 0 \\ 0 & 2 & -1 & 0 \\ 0 & 0 & 0 & 1 \end{bmatrix}. \tag{2.62}$$

Therefore the Hamiltonian which will produce in a spinor basis the same eigenvalues as (2.51) evaluated in a fully antisymmetrized basis is

$$\mathscr{H} = \tfrac{1}{4}(E_s + E_t) - \tfrac{1}{4}(E_s - E_t)\boldsymbol{\sigma}_1 \cdot \boldsymbol{\sigma}_2 = \text{const} - \tfrac{1}{4}J\boldsymbol{\sigma}_1 \cdot \boldsymbol{\sigma}_2 \tag{2.63}$$

Thus the exchange interaction, which is a purely electrostatic effect, may be expressed as a spin–spin interaction. The exchange parameter J is $E_s - E_t$. If J is *positive,* we say that the interaction is *ferromagnetic.*

In obtaining the exchange interaction (2.63) we have made two important assumptions. The first was that we could restrict ourselves to a certain subset of nondegenerate orbital states. There is no real justification for this, as the Coulomb interaction does, in fact, couple different orbital states. We shall see this more clearly in our discussion of exchange in the N-electron system. The

2. The Magnetic Hamiltonian

second assumption was that the orbital functions were orthogonal. When we are dealing with wave functions that have a common origin, as in an atom, this is usually the case. However, as soon as we begin talking about electrons centered at different sites the problem becomes very complex.

The hydrogen molecule is perhaps the simplest example of such a two-center problem. This was first considered by *Heitler* and *London* in 1927 [2.10]. In the limit of infinite separation we expect to find one electron on each core. The Hamiltonian for one of these systems is $\mathcal{H}_\infty(r_1)$. The corresponding orbital eigenfunctions will be denoted by $\varphi(r - R_a) \equiv \varphi_a(r)$. The Hamiltonian for the finite system is

$$\mathcal{H} = \mathcal{H}_\infty(r_1) + \mathcal{H}_\infty(r_2) + \frac{(Ze)^2}{|R_a - R_b|} - \frac{Ze^2}{|r_1 - R_b|} - \frac{Ze^2}{|r_2 - R_a|}. \quad (2.64)$$

Again we take Slater determinants as our starting point. For both functions with up-spin parts,

$$\psi_1 = \frac{1}{\sqrt{2 - 2l^2}} \begin{vmatrix} \varphi_a(r_1)\alpha(\sigma_1) & \varphi_a(r_2)\alpha(\sigma_2) \\ \varphi_b(r_1)\alpha(\sigma_1) & \varphi_b(r_2)\alpha(\sigma_2) \end{vmatrix}, \quad (2.65)$$

where

$$l = \int dr\, \varphi_a^*(r)\varphi_b(r) \quad (2.66)$$

is the overlap integral.

If the function at site 1 has an up-spin part and that at site 2 has spin down, the Slater determinant is

$$\phi_1 = \begin{vmatrix} \varphi_a(r_1)\alpha(\sigma_1) & \varphi_a(r_2)\alpha(\sigma_2) \\ \varphi_b(r_1)\beta(\sigma_2) & \varphi_b(r_2)\beta(\sigma_2) \end{vmatrix}. \quad (2.67)$$

We obtain a similar function ϕ_2 for the spins reversed. However, owing to the nonzero overlap integral, ϕ_1 and ϕ_2 are *not* orthogonal, that is,

$$\langle \phi_1 | \phi_2 \rangle = -2l^2. \quad (2.68)$$

In order to have an orthonormal basis, we orthogonalize these functions to obtain

$$\psi_2 = \frac{1}{\sqrt{2 - 2l^2}}(\phi_1 + \phi_2). \quad (2.69)$$

$$\psi_3 = \frac{1}{\sqrt{2 + 2l^2}}(\phi_1 - \phi_2). \quad (2.70)$$

Finally, the fourth basis function is identical to (2.65), with β replacing α. In this basis the Hamiltonian matrix has the form

$$H = \begin{bmatrix} E_t & 0 & 0 & 0 \\ 0 & E_t & 0 & 0 \\ 0 & 0 & E_s & 0 \\ 0 & 0 & 0 & E_t \end{bmatrix} \tag{2.71}$$

where

$$E_t = (1 - l^2)^{-1}[(E_a + E_b + K_0)(1 - l^2) + K_{ab} - l^2 J_{ab}], \tag{2.72}$$

$$E_s = (1 + l^2)^{-1}[(E_a + E_b + K_0)(1 + l^2) + K_{ab} + l^2 J_{ab}], \tag{2.73}$$

$$K_{ab} = \iint dr_1 \, dr_2 \frac{e^2}{r_{12}} |\varphi_a(r_1)|^2 |\varphi_b(r_1)|^2 - \int dr_2 \frac{Ze^2}{|r_1 - R_b|} |\varphi_a(r_1)|^2$$
$$- \int dr_2 \frac{Ze^2}{|r_2 - R_a|} |\varphi_b(r_2)|^2, \tag{2.74}$$

and

$$J_{ab} = \iint dr_1 \, dr_2 \varphi_a^*(r_1) \varphi_b^*(r_2) \frac{e^2}{r_{12}} \varphi_b(r_1) \varphi_a(r_2)$$
$$- 2l \int dr_1 \frac{Ze^2}{|r_1 - R_b|} \varphi_a^*(r_1) \varphi_b(r_1). \tag{2.75}$$

We see from (2.71) that the basis functions are, in fact, eigenfunctions. Again we obtain a singlet and a triplet. The difference in their energies is

$$E_t - E_s = \frac{(l^2 K_{ab} - J_{ab})}{1 - l^4}. \tag{2.76}$$

Notice that $E_t - E_s$ may be positive or negative, depending on the relative sizes of K_{ab} and J_{ab}. Thus it is not obvious whether the ground state will be ferromagnetic or antiferromagnetic. Actual evaluation shows that for realistic separations the singlet lies lowest. However, at large distances the triplet becomes lower. This cannot be, for, as *Herring* pointed out, the lowest eigenvalue of a semibounded Sturm–Liouville differential operator, such as (2.64), must be free of nodes. This means it must always be a singlet. The problem arises from the oversimplified nature of the Heitler–London states. The exchange coupling measures the rate at which two identifiable electrons exchange places by tunneling through the barrier separating them. In the Heitler–London approximation this tunneling is uncorrelated. In reality, however, the two electrons will tend to avoid one another. This reduces the amplitude of the wave function for configurations with both electrons close to the internuclear line. Fortunately, however, (2.75) is a reasonably good approximation for separations generally encountered.

In considering more than two electrons we find that difficulties arise from the nonorthogonality of the wave functions we have been using thus far. When the Heitler–London method is applied to a very large system the nonorthogonality integrals enter the secular equation with high powers and lead to an apparent divergence. This "nonorthogonality catastrophy" is a purely mathematical difficulty [2.9], and *Herring* has reviewed various treatments that show that even for large systems the energies and eigenstates are given by the exchange interaction with exchange constants having the same values as for a two-site system.

One approach to the problem of exchange among many sites is to give up our well-defined but nonorthogonal functions and work with functions which are orthogonal. An example of such a set of orthogonal functions are *Wannier functions*. The Wannier function $\phi_{n\lambda}(r - R_\alpha)$ resembles the nth atomic orbital with spin λ near the αth lattice site, but it falls off throughout the crystal in such a way that it is orthogonal to similar functions centered at other sites. Since the exchange interaction is essentially a quantum-statistical effect the technique of second quantization discussed in Chap. 1 is very convenient for obtaining the exchange Hamiltonian.

Let us consider N electrons reasonably localized on N lattice sites. The Hamiltonian for such a system is

$$\mathcal{H} = \sum_i \frac{p_i^2}{2m} - \sum_{i,\alpha} \frac{Ze^2}{|r_i - R_\alpha|} + \tfrac{1}{2} \sum_{i,j} \frac{e^2}{|r_i - r_j|}. \tag{2.77}$$

In Chap. 1 we found that the prescription for second quantizing such a Hamiltonian entailed introducing a field operator which could be expanded in terms of a complete set of single-particle wave functions. In this case the appropriate set of functions are the Wannier functions. Thus the field operator (1.115) becomes

$$\psi(r) = \sum_{\alpha, n, \lambda} \phi_{n\lambda}(r - R_\alpha) a_{n\lambda}(R_\alpha) \tag{2.78}$$

where $a_{n\lambda}(R_\alpha)$ annihilates an electron in orbital state n and spin state λ at the lattice site α. The interaction part of the Hamiltonian (2.77) becomes

$$\frac{1}{2} \sum_{\substack{\alpha_1, \alpha_2, \alpha_3, \alpha_4, \\ n_1, n_2, n_3, n_4, \\ \lambda_1, \lambda_2}} \langle \alpha_1 n_1; \alpha_2 n_2 | V | \alpha_3 n_3; \alpha_4 n_4 \rangle a^\dagger_{n_1 \lambda_1}(R_{\alpha_1}) a^\dagger_{n_2 \lambda_2}(R_{\alpha_2})$$
$$\times a_{n_4 \lambda_2}(R_{\alpha_4}) a_{n_3 \lambda_1}(R_{\alpha_3}). \tag{2.79}$$

Since the Wannier states are localized to within a unit cell, the main contributions to (2.79) arise from those terms in which $\alpha_3 = \alpha_1$ and $\alpha_4 = \alpha_2$ or $\alpha_3 = \alpha_2$

and $\alpha_4 = \alpha_1$. The remaining terms involve various orbital excitations induced by the Coulomb interaction. These lead to *off-diagonal exchange*. Just as in the two-electron case, we shall restrict each electron to a definite orbital state. That is, we shall keep only those terms in which $n_3 = n_1$ and $n_4 = n_2$ or $n_3 = n_2$ and $n_4 = n_1$. If, for simplicity, we also neglect orbital-transfer terms, in which two electrons interchange orbital states, then (2.79) reduces to

$$\frac{1}{2} \sum_{\substack{\alpha,\alpha' \\ n,n' \\ \lambda,\lambda'}} [\langle \alpha, n; \alpha', n' | V | \alpha, n; \alpha', n' \rangle a_{n\lambda}^\dagger(\boldsymbol{R}_\alpha) a_{n'\lambda'}^\dagger(\boldsymbol{R}_{\alpha'}) a_{n'\lambda'}(\boldsymbol{R}_{\alpha'}) a_{n\lambda}(\boldsymbol{R}_\alpha)$$
$$+ \langle \alpha, n; \alpha', n' | V | \alpha', n'; \alpha, n \rangle a_{n\lambda}^\dagger(\boldsymbol{R}_\alpha) a_{n'\lambda'}^\dagger(\boldsymbol{R}_{\alpha'}) a_{n,\lambda'}(\boldsymbol{R}_\alpha) a_{n'\lambda}(\boldsymbol{R}_{\alpha'})] \ . \quad (2.80)$$

The first term is called the *direct term* and the second is the *exchange term*. Using the fermion anticommutation relation

$$\{a_{n\lambda}^\dagger(\boldsymbol{R}_\alpha), a_{n'\lambda'}(\boldsymbol{R}_{\alpha'})\} = \delta_{\alpha\alpha'}\delta_{nn'}\delta_{\lambda\lambda'} \ , \quad (2.81)$$

we may write the exchange term as

$$-\frac{1}{2} \sum_{\substack{\alpha,\alpha' \\ n,n' \\ \lambda,\lambda'}} J_{nn'}(\boldsymbol{R}_\alpha, \boldsymbol{R}_{\alpha'}) a_{n\lambda}^\dagger(\boldsymbol{R}_\alpha) a_{n\lambda'}(\boldsymbol{R}_\alpha) a_{n'\lambda'}^\dagger(\boldsymbol{R}_{\alpha'}) a_{n'\lambda}(\boldsymbol{R}_{\alpha'}) \ . \quad (2.82)$$

When the spin sum is expanded, we obtain four terms. These may be written in a particularly revealing way by noting the following. First of all, if we allow only one electron to occupy each orbital, then

$$N_{n\uparrow}(\boldsymbol{R}_\alpha) + N_{n\downarrow}(\boldsymbol{R}_\alpha) = 1 \quad (2.83)$$

where $N_{n\lambda}(\boldsymbol{R}_\alpha) = a_{n\lambda}^\dagger(\boldsymbol{R}_\alpha) a_{n\lambda}(\boldsymbol{R}_\alpha)$ is the number operator associated with the nth orbital with spin λ at the site α. We also have

$$N_{n\uparrow}(\boldsymbol{R}_\alpha) - N_{n\downarrow}(\boldsymbol{R}_\alpha) = \sigma_z(\boldsymbol{R}_\alpha) \ . \quad (2.84)$$

Combining these two relations gives

$$N_{n\uparrow}(\boldsymbol{R}_\alpha) N_{n'\uparrow}(\boldsymbol{R}_{\alpha'}) + N_{n\downarrow}(\boldsymbol{R}_\alpha) N_{n'\downarrow}(\boldsymbol{R}_{\alpha'}) = \tfrac{1}{2}\sigma_z(\boldsymbol{R}_\alpha)\sigma_z(\boldsymbol{R}_{\alpha'}) + \tfrac{1}{2} \quad (2.85)$$

We also note that

$$a_{n\uparrow}^\dagger(\boldsymbol{R}_\alpha) a_{n\downarrow}(\boldsymbol{R}_\alpha) = \tfrac{1}{2}\sigma^+(\boldsymbol{R}_\alpha) \quad (2.86)$$

and

$$a_{n\downarrow}^\dagger(\boldsymbol{R}_\alpha) a_{n\uparrow}(\boldsymbol{R}_\alpha) = \tfrac{1}{2}\sigma^-(\boldsymbol{R}_\alpha) \ . \quad (2.87)$$

With these results the exchange interaction (2.82) becomes

$$-\sum_{\substack{\alpha\alpha'\\nn'}} J_{nn'}(R_\alpha, R_{\alpha'}) \left[\tfrac{1}{4} + \tfrac{1}{4}\sigma(R_\alpha)\cdot\sigma(R_{\alpha'})\right]. \tag{2.88}$$

This is the many-electron generalization of our earlier result for two electrons. Since the wave functions used in $J_{nn'}$ are orthogonal, this exchange is always ferromagnetic. In the section on superexchange we shall see that when we allow for the fact that an electron can hop onto a neighboring site giving it a double occupancy, the result is an exchange of the form (2.88) but with an antiferromagnetic coupling.

In most practical situations we do not have just N electrons each localized on one of N lattice sites, but rather Nh electrons, where h is the number of unpaired electrons on each ion. If these h electrons all have the same exchange integrals with all the other electrons, then the interaction may be expressed in terms of the total ionic spin,

$$\boxed{\mathcal{H}_{ex} = -\sum_{\substack{\alpha,\alpha'\\n,n'}} J_{nn'}(R_\alpha, R_{\alpha'})\left[\tfrac{1}{4} + S(R_\alpha)\cdot S(R_{\alpha'})\right].} \tag{2.89}$$

As *Van Vleck* has pointed out, this form is also valid if the unfilled shells of each ion are half full and the atom is in its state of maximum multiplicty [2.11]. Such a situation is always true for S-state ions but may also arise as a result of crystal fields. We shall consider an example of this in our discussion on effective exchange. Equation (2.89), usually referred to as the *Heisenberg exchange interaction*, often forms the starting point for discussions of ferromagnetism or antiferromagnetism in insulators. The fact that it is valid only under certain conditions does not seem to deter its application. In fact, it works surprisingly well, as we shall see in later chapters.

2.2.8 Superexchange

The exchange constant (2.89) involves essentially an eigenfunction of the whole crystal. Needless to say, such a function is difficult to obtain. Let us therefore consider an approach which has proven useful in discussing exchange in insulators. The transition-metal fluorides MnF_2, FeF_2, and CoF_2 are all observed to be antiferromagnets at low temperatures, with crystal structure and spin configuration indicated in Fig. 2.2. This exchange is difficult to understand in terms of direct exchange between the cations because of the intervening fluorine anions. A similar situation arises in the case of magnetic oxides. In 1934 Kramers proposed the explanation that the cation wave functions were being strongly admixed with the fluorine wave functions, enabling the cations to couple indirectly with each other. Kramers applied perturbation theory to obtain the effective exchange resulting from this mechanism. Let us consider two Mn^{2+} ions and an intervening F^- ion. Because of the overlap of their wave functions,

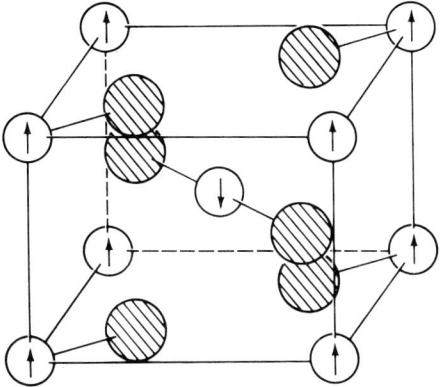
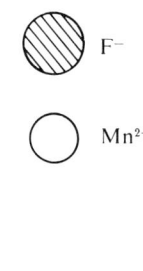

Fig. 2.2. Spin configuration of the transition-metal fluorides

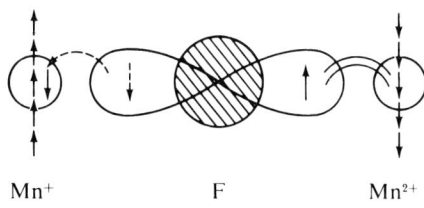

Mn⁺   F   Mn²⁺

Fig. 2.3. Schematic representation of one of the intermediate states in superexchange

one of the p electrons from the F^- hops over to one of the Mn^{2+} ions. The remaining unpaired p electron on the F then enters into a direct exchange with the other Mn^{2+} ion. This excited state is illustrated in Fig. 2.3 for the case in which the exchange between the unpaired p electron and the Mn^{2+} is antiferromagnetic. By using such excited states in a perturbation calculation of the total energy of the system we obtain an effective exchange between the Mn^{2+} ions. The sign of this exchange depends on the nature of the orbitals involved. However, a number of general features that have evolved through the work of *Goodenough* [2.12] and *Kanamori* [2.13] enable us to qualitatively predict the nature of the superexchange. Two such features are that the electron transfer can take place only if the cation and anion orbitals are nonorthogonal, and that if the cation–anion orbitals are orthogonal, the direct exchange referred to above is positive (ferromagnetic); otherwise it is negative (antiferromagnetic).

As an example of the application of these rules, let us consider the antiferromagnet $CaMnO_3$. In this material the manganese occurs in the tetravalent state Mn^{4+}, which means that we have three d electrons. The crystal field at the Mn^{4+} sites is cubic. The d electrons are strongly affected by this crystal field. In fact, the approximation is often made that the crystal field is stronger than the intraionic Coulomb interaction, so that the latter may be neglected. Then each electron may be considered separately. The effect of a cubic field on the fivefold orbitally degenerate d state of a single electron is to split this state into a threefold degener-

54   2. The Magnetic Hamiltonian

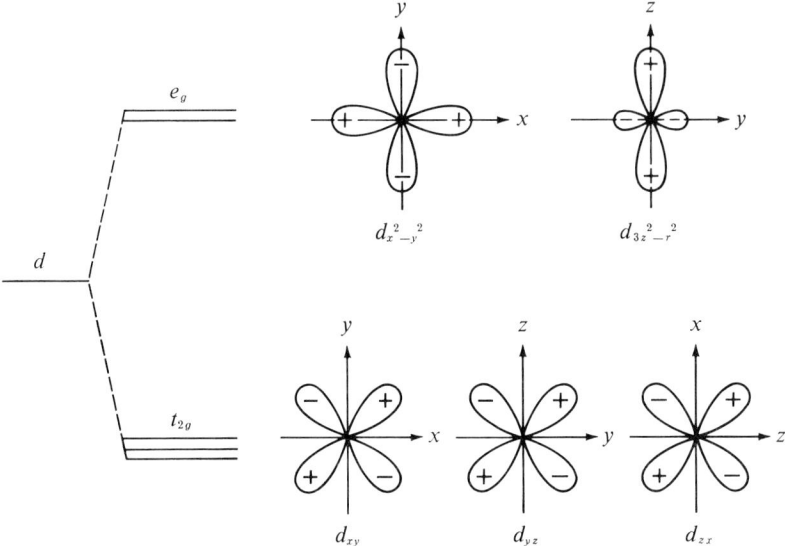

Fig. 2.4. Representation of the eigenvalues and eigenfunctions of a d electron in a cubic crystal field

ate state labeled t_{2g} and a twofold degenerate state labeled e_g. This splitting with the associated wave functions is shown in Fig. 2.4. There is, in fact, some intraionic Coulomb interaction, which leads to a *Hund's rule* coupling of the spins. Therefore, since the t_{2g} state lies lowest for the particular coordination in CaMnO$_3$, the three electrons will each go into one of the t_{2g} orbitals with their spins up.

The superexchange in this case involves the p electrons of the O^{2-}. The p orbitals are illustrated in Fig. 2.5. Examination of the wave functions in Figs. 2.4, shows that the p_σ orbital is orthogonal to all the cation orbitals except $d_{x^2-y^2}$. Therefore, if a p_σ electron hops over to a Mn^{4+}, it must go into this e_g

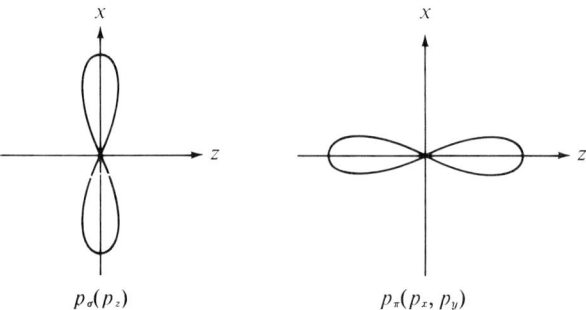

Fig. 2.5. Representation of the p orbitals

orbital. Since Hund's rule requires that the total spin be a maximum, it is the up spin from the p_σ orbital that transfers if the spins of the Mn^{4+} to which it is going are up. The remaining p_σ down spin, since it is orthogonal to e_g orbitals, couples ferromagnetically to the other Mn^{4+}. As a result, we find that the p_σ orbital has produced a net antiferromagnetic coupling between the cations themselves. It turns out that the contributions from the p_π orbitals are much smaller.

Anderson [2.14] has reformulated Kramer's theory in an attempt to avoid the high order perturbation expansion, and his results suggest that antiferromagnetism may be more common than Kramer's theory implies. Anderson worked in a basis of ligand wave functions which are a covalent admixture of cation and anion functions. In Anderson's theory magnetism is the result of the interplay between two effects—the hopping of electrons between ligand complexes, characterized by a hopping matrix element $t_{aa'}$, and an average Coulomb interaction U between electrons on the same complex. In the limit where the hopping may be treated as a perturbation, Anderson found that the super-exchange interaction has the same spin dependence as (2.89) but with a coefficient $-2t_{aa'}^2/U$.

In recent years chemists have recast superexchange in terms of molecular orbitals. *Hay* et al. [2.15], for example, have shown that the exchange interaction between two spin $\frac{1}{2}$ ions can be expressed as

$$J = J_{ab} - \frac{(\epsilon_1 - \epsilon_2)^2}{K_{aa} - K_{bb}}.$$

The exchange J_{ab} involves orthogonalized molecular orbitals and is therefore inherently ferromagnetic. The energies ϵ_1 and ϵ_2 are the bonding and antibonding energies associated with the molecular orbitals on the two metal ions. Since $K_{aa} > K_{bb}$ the nature of the exchange depends upon the magnitude of $\epsilon_1 - \epsilon_2$. This quantity turns out to be a sensitive function of the metal–ligand–metal angle. *Willett* [2.16] and his co-workers have exploited this fact to synthesize a variety of pseudo-one-dimensional magnetic systems based on copper dimeric species.

In semiconductors the anion energies form bands. As a result the superexchange can extend to distant neighbors as illustrated in Fig. 2.6. These results were deduced from the measured spin–wave dispersion relations[2.17] just as atomic force constants are deduced from phonon dispersion relations.

2.3 The Spin Hamiltonian

The Hamiltonian developed in the preceding section is completely general, and a knowledge of its eigenvalues would accurately describe the magnetic properties of any material. Unfortunately, because of the large number of particles in-

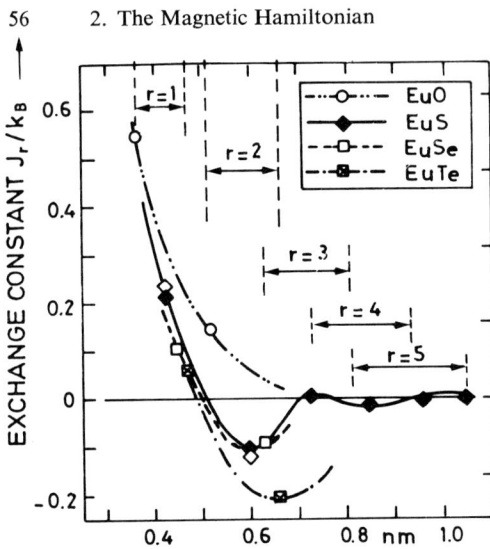

Fig. 2.6. Exchange interactions between Eu–Eu pairs in Eu monochalcogenides versus the pair separation R_r. The horizontal arrows indicate the range of R_r values from the oxides to the tellurides. Open symbols indicate results deduced assuming only second-nearest-neighbor coupling

volved, such a knowledge is beyond us at this time. Therefore we try to project out of the Hamiltonian those terms which adequately describe the situation and yet are amenable to calculation. Experimentalists in particular often propose "phenomenological" Hamiltonians to explain certain observations, leaving to theoreticians the job of establishing the legitimacy of such forms. In the remainder of this chapter we shall indicate the origin of such phenomenological Hamiltonians.

2.3.1 Transition-Metal Ions

The first row of transition-metal ions and their electronic configurations are listed in Table 2.3. The important feature about transition-metal ions is that the magnetic, or unpaired, electrons lie in the outermost shell of the ion. Therefore they are easily influenced by any external field produced by neigh-

Table 2.3. Configurations of the iron-group ions

	Ti^{3+}, V^{4+}	$3d^1$	^2D
	V^{3+}	$3d^2$	^3F
	Cr^{3+}, V^{2+}	$3d^3$	^4F
	Mn^{3+}, Cr^{2+}	$3d^4$	^5D
	Fe^{3+}, Mn^{2+}	$3d^5$	^6S
	Fe^{2+}	$3d^6$	^5D
	Co^{2+}	$3d^7$	^4F
	Ni^{2+}	$3d^8$	^3F
	Cu^{2+}	$3d^9$	^2D

boring ligands. That is, the crystal field is likely to be one of the largest terms in the Hamiltonian. Thus we might expect that the contributions to the Hamiltonian, in order of descending strength, are

$$\mathcal{H} = \mathcal{H}_{\text{intraatomic Coulomb}} + \mathcal{H}_{\text{crystal field}} + \mathcal{H}_{\text{spin-orbit}} + \mathcal{H}_{\text{Zeeman}} \,. \tag{2.90}$$

Of course, depending on the situation, we may have to consider additional terms, such as the hyperfine interaction. However, let us consider the eigenstates of the Hamiltonian (2.90). First of all, the intraatomic Coulomb interaction leads to spectroscopic energy levels, the lowest of which is determined by Hund's rules. This ground state is indicated in the last column of Table 2.3. For most magnetic properties it is sufficient to consider only the lowest term. This is because term energies are of the order of tens of thousand of wave numbers, whereas magnetic energies are at most tens of wave numbers.

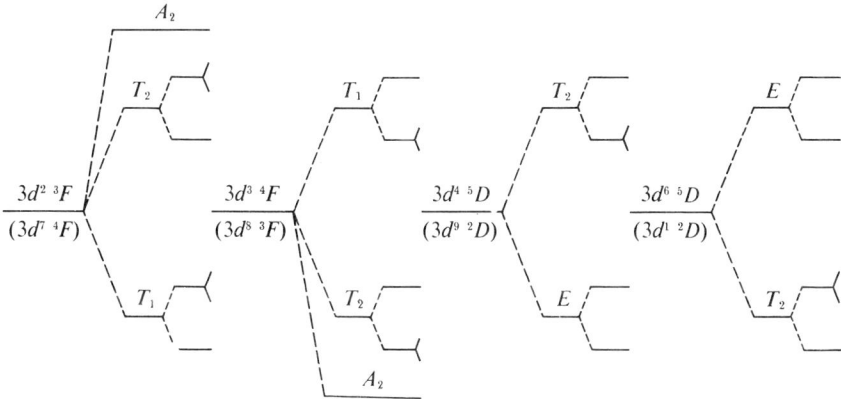

Fig. 2.7. Crystal-field splittings of the iron-group ion ground states. The first set of splittings are the result of a cubic crystal field. Subsequent splittings are due to an additional tetragonal distortion

The behavior of a given term in the crystal field may be calculated by the technique developed in the last section. Cubic symmetry is the predominant symmetry encountered in most crystals, and Fig. 2.7 indicates how group theory predicts splitting of the various terms of Table 2.3 in such a case. The eigenfunctions of $\mathcal{H}_{\text{intra}} + \mathcal{H}_{\text{cryst}}$ will be denoted as $|\Gamma, \gamma; S, M_S\rangle$, where Γ is the irreducible representation of the point-group symmetry.

Consider, for example, the $3d^3$ configuration. From Hund's rules, the ground state is

$$|L, M_L; S, M_S\rangle = |3, M_L; \tfrac{3}{2}, M_S\rangle \tag{2.91}$$

which is $(2L + 1)(2S + 1) = 28$-fold degenerate. In the presence of a cubic crystal field this state splits as shown in Fig. 2.7. Thus the ground state would be denoted as $|A_2, \gamma; \tfrac{3}{2}, M_S\rangle$.

The g Tensor. Now consider the spin–orbit and Zeeman terms. Since we are considering matrix elements only within a given LS term, the matrix elements of

$$\sum_i \zeta(r_i) l_i \cdot s_i$$

are proportional to those of $\boldsymbol{L} \cdot \boldsymbol{S}$, by the Wigner–Eckhart theorem. Thus the *spin–orbit Hamiltonian* is

$$\mathcal{H}_{\text{sp-orb}} = \lambda \boldsymbol{L} \cdot \boldsymbol{S}, \tag{2.92}$$

and similarly, the *Zeeman Hamiltonian* is

$$\mathcal{H}_z = \mu_B (\boldsymbol{L} + 2\boldsymbol{S}) \cdot \boldsymbol{H}. \tag{2.93}$$

We now transform $\mathcal{H}_{\text{sp-orb}} + \mathcal{H}_z$ into the so-called *spin Hamiltonian* by a method proposed by Pryce, in which we project out the orbital dependence. Since neither $\mathcal{H}_{\text{intra}}$ nor $\mathcal{H}_{\text{cryst}}$ has mixed orbital and spin states, our eigenfunctions are products of the form $|\Gamma, \gamma\rangle |S, M_S\rangle$. Let us evaluate the expectation value of $\mathcal{H}_{\text{sp-orb}} + \mathcal{H}_z$ for an *orbitally nondegenerate* ground state $|\Gamma, \gamma\rangle$. To second order in perturbation theory,

$$\begin{aligned}\mathcal{H}_{\text{eff}} &= \langle \Gamma, \gamma | \mathcal{H}_{\text{sp-orb}} + \mathcal{H}_z | \Gamma, \gamma \rangle \\ &= 2\mu_B \boldsymbol{H} \cdot \boldsymbol{S} - \sum_{\Gamma', \gamma'} \frac{|\langle \Gamma', \gamma' | \mu_B \boldsymbol{H} \cdot \boldsymbol{L} + \lambda \boldsymbol{L} \cdot \boldsymbol{S} | \Gamma, \gamma \rangle|^2}{E_{\Gamma', \gamma'} - E_{\Gamma, \gamma}}.\end{aligned} \tag{2.94}$$

Expanding the square gives

$$\begin{aligned}\mathcal{H}_{\text{eff}} = {}&2\mu_B \boldsymbol{H} \cdot \boldsymbol{S} - 2\mu_B \lambda \sum_{\mu,\nu} \Lambda_{\mu\nu} S_\mu H_\nu - \lambda^2 \sum_{\mu,\nu} \Lambda_{\mu\nu} S_\mu S_\nu \\ &- \mu_B^2 \sum_{\mu,\nu} \Lambda_{\mu\nu} H_\mu H_\nu\end{aligned} \tag{2.95}$$

where

$$\Lambda_{\mu\nu} = \sum_{\Gamma' \gamma'} \frac{\langle \Gamma, \gamma | L_\mu | \Gamma', \gamma' \rangle \langle \Gamma', \gamma' | L_\nu | \Gamma, \gamma \rangle}{E_{\Gamma',\gamma'} - E_{\Gamma,\gamma}}. \tag{2.96}$$

This may be written as

$$\mathcal{H}_{\text{eff}} = \sum_{\mu,\nu} (\mu_B g_{\mu\nu} H_\mu S_\nu - \lambda^2 \Lambda_{\mu\nu} S_\mu S_\nu - \mu_B^2 \Lambda_{\mu\nu} H_\mu H_\nu) \tag{2.97}$$

where $g_{\mu\nu}$ is the g tensor

$$g_{\mu\nu} = 2(\delta_{\mu\nu} - \lambda\Lambda_{\mu\nu}). \tag{2.98}$$

The fact that $g_{\mu\mu}$ differs from 2 tells us that owing to the spin–orbit interaction, the magnetization is now no longer spin only. That is, a small amount of orbital angular momentum has been admixed back into the ground state.

The g value of the free electron is not precisely 2. There are quantum electrodynamic corrections which lead to the value

$$g = 2.002319.$$

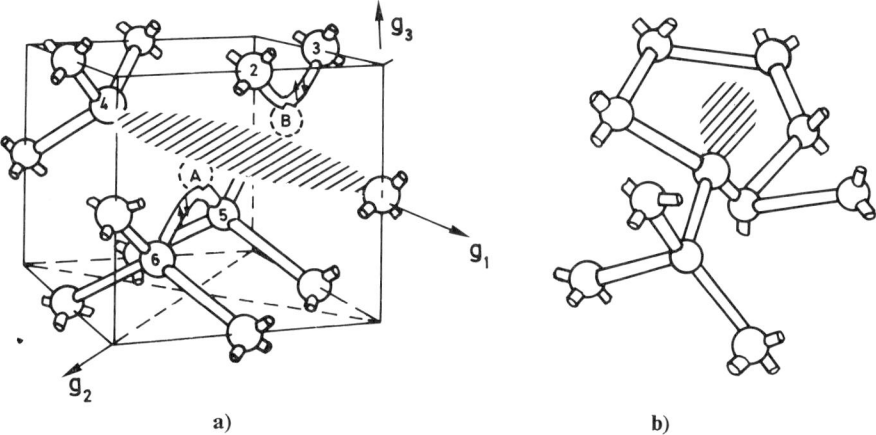

Fig. 2.8. a.b Defects in (**a**) crystalline and (**b**) amorphous silicon having nearly free electron g values

As an example of how this value is changed in a solid let us consider the semiconductor silicon. In the ideal crystalline state each silicon atom has four valence electrons which form bonds with four neighboring silicon atoms. If this ideal crystal is irradiated, with high-energy electrons, for example, defects are introduced. One of these is the divacancy illustrated in Fig. 2.8a. The removal of the two atoms A and B leaves six broken bonds, or unpaired electrons. Those electrons associated with atoms 2 and 3 and those associated with 5 and 6 reconstruct bonds as indicated in the figure. The remaining two electrons from atoms 1 and 4 then form an "extended" bond across the vacancy. If one of these latter two electrons is removed by some means, leaving the defect with a net positive charge, the remaining electron is found to have the g value [2.18]

$$g_1 = 2.0004,$$
$$g_2 = 2.0020,$$
$$g_3 = 2.0041.$$

If the silicon is prepared as an *amorphous* film, by the decomposition of silane gas, for example, then the atomic disorder sometimes leaves a silicon atom with only three other silicon neighbors as illustrated schematically in Fig. 2.8b. The fourth unpaired electron is referred to as a "dangling bond." This electron is found [2.19] to have the average g value

$$g = 2.0055.$$

Although these g values differ by less than 1% these differences are easily measured, as we shall discuss in Chap. 5, and provide an extremely important characterization of the electronic center.

Anisotropy. The second term in (2.97) represents the fine-structure or single-ion *anisotropy*. Notice that $\Lambda_{\mu\nu}$ reflects the symmetry of the crystal. The spin Hamiltonian must also display this symmetry; for example, in a cubic crystal $\Lambda_{xx} = \Lambda_{yy} = \Lambda_{zz}$. Thus the anisotropy term reduces to a constant. For axial symmetry $\Lambda_{xx} = \Lambda_{yy} = \Lambda_\perp$ and $\Lambda_{zz} = \Lambda_\parallel$. Thus, if we neglect the last term, the effective axial Hamiltonian is

$$\mathscr{H}_{\text{eff}} = g_\parallel \mu_B H_z S_z + g_\perp \mu_B (H_x S_x + H_y S_y) + D[S_z^2 - \tfrac{1}{3}S(S+1)]$$
$$+ \tfrac{1}{3}S(S+1)(2\Lambda_\perp + \Lambda_\parallel) \tag{2.99}$$

where $D = \lambda^2(\Lambda_\parallel - \Lambda_\perp)$. A Hamiltonian of this form based on the symmetry of the crystal is usually taken as the starting point in describing paramagnetic systems involving transition-metal ions. Thus such ions in crystals are characterized by their g, D, etc., parameters.

An important experimental fact is that these crystal-field parameters do not change appreciably for concentrated versions of the same salt. Hence, we shall find these paramagnetic states useful when we discuss ferromagnetism. For example, we can see from the discussion above that the anisotropy constants depend on the energy levels, which in turn depend on the positions of the neighboring ions. Thus, if these ions move because of the presence, say, of phonons, then we have a coupling between this motion and the spins.

The last term in the Hamiltonian (2.97) will survive the two field derivatives leading to the static susceptibility, resulting in the so-called *Van Vleck susceptibility*, which is temperature independent.

As an example of the application of the spin Hamiltonian, let us consider a spin $\tfrac{3}{2}$ in an axially symmetric system with an external field applied along the c axis. If we assume that the crystal field is of sufficiently low symmetry to remove any orbital degeneracy of the ground state, then (2.99) applies. Dropping the constant part, we have

$$\mathscr{H}_{\text{eff}} = g_\parallel \mu_B H S_z + D[S_z^2 - \tfrac{1}{3}S(S+1)]. \tag{2.100}$$

The matrix of $\mathscr{H}_{\text{eff}}$ in the basis $|\tfrac{3}{2}, M_S\rangle$ is

2.3 The Spin Hamiltonian

$$\mathcal{H}_{\text{eff}} = \begin{array}{c} \langle -\tfrac{3}{2}| \\ \langle -\tfrac{1}{2}| \\ \langle \tfrac{1}{2}| \\ \langle \tfrac{3}{2}| \end{array} \begin{bmatrix} |-\tfrac{3}{2}\rangle & |-\tfrac{1}{2}\rangle & |\tfrac{1}{2}\rangle & |\tfrac{3}{2}\rangle \\ D - \tfrac{3}{2} g_\parallel \mu_B H & 0 & 0 & 0 \\ 0 & -D - \tfrac{1}{2} g_\parallel \mu_B H & 0 & 0 \\ 0 & 0 & -D + \tfrac{1}{2} g_\parallel \mu_B H & 0 \\ 0 & 0 & 0 & D + \tfrac{3}{2} g_\parallel \mu_B H \end{bmatrix}.$$

(2.101)

The eigenvalues are shown in Fig. 2.9. We see that there is a zero-field splitting of 2D.

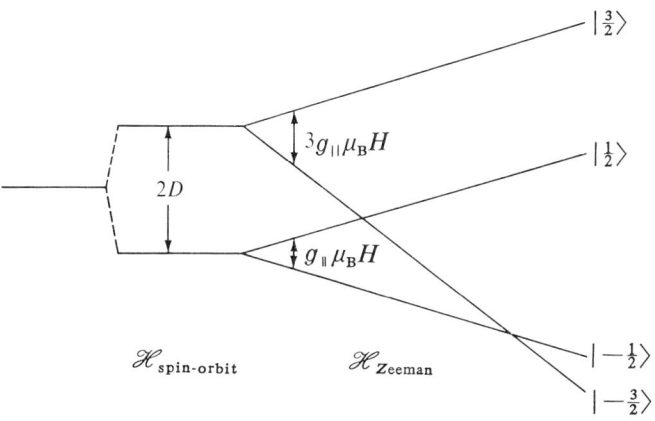

Fig. 2.9. Representation of the effect of the spin–orbit and Zeeman interactions on an orbital singlet with spin $\tfrac{3}{2}$

Effective Exchange. In our discussion of exchange in the last section we neglected spin–orbit effects as well as off-diagonal exchange effects. We found above that the spin–orbit interaction leads to important contributions to the spin Hamiltonian. It can also modify the form of the exchange interaction. To illustrate these effects, let us consider the exchange between an Mn^{2+} impurity and a neighboring Co^{2+} ion in $CoCl_2 \cdot 2H_2O$ [2.20]. Co^{2+} has seven d electrons which correspond to the three *holes* in the d shell. These holes behave just as electrons would. Therefore we find that the intraatomic Coulomb interaction leads to a ^4F ground state with a spin of $\tfrac{3}{2}$. The crystal field in $CoCl_2 \cdot 2H_2O$ is predominantly cubic, with a small tetragonal distortion. The cubic part splits the ^4F into three states, the lowest being a $^4T_1$, which has a threefold orbital degeneracy. As mentioned above, crystal-field effects on the transition-metal ions are rather large. Therefore, in so far as we are concerned only with magnetic properties, we may restrict our considerations to this $^4T_1$ ground state.

We now investigate how the tetragonal crystal-field component and the spin-orbit interaction affect this level. Any threefold orbitally degenerate state behaves as though it had an *effective* quantum number $L = 1$. We saw earlier that if we restrict ourselves to a manifold of states in which, say, L is a good quantum number, then we may express the crystal-field Hamiltonian in powers of L_x, L_y, and L_z. The same argument applies even if L is only a good effective quantum number. Thus the effect of a tetragonal crystal field upon our $^4T_1$ ground state may be described by the Hamiltonian

$$\mathcal{H}_{\text{tetra}} = -\delta(L_z^2 - \tfrac{3}{2}) \tag{2.102}$$

where δ is a phenomenological crystal-field parameter and $L = 1$. The spin–orbit interaction may also be expressed in terms of the effective orbital angular momentum,

$$\mathcal{H}_{\text{sp-orb}} = \lambda' \mathbf{L} \cdot \mathbf{S}_{\text{Co}}, \tag{2.103}$$

where λ' is an effective spin–orbit parameter. The operators L^+ and L^- are the raising and lowering operators for the components of the $^4T_1$, just as the corresponding real operators would connect the components of a real P state. This technique of writing Hamiltonians in terms of effective-angular-momentum operators is an extremely useful tool for understanding the qualitative features of magnetic spectra.

The basis functions for our $^4T_1$ state have the form $|M_L, M_S\rangle$, where $M_L = -1, 0,$ or $+1$, and $M_S = \pm\tfrac{1}{2}$ or $\pm\tfrac{3}{2}$. The effect of $\mathcal{H}_{\text{tetra}} + \mathcal{H}_{\text{sp-orb}}$ is to split the $^4T_1$ into six doublets. Again, for magnetic considerations, we consider the lowest of these. The eigenfunctions have the form

$$\psi^\pm = a|\mp 1, \pm\tfrac{3}{2}\rangle + b|0, \pm\tfrac{1}{2}\rangle + c|\pm 1, \mp\tfrac{1}{2}\rangle, \tag{2.104}$$

where a, b, and c are certain mixing coefficients which may, in general, be complex. We now inquire how the exchange with the neighboring Mn^{2+} affects this doublet. Since it is a doublet, the exchange should be expressible as an effective spin of $\tfrac{1}{2}$ interacting with the Mn^{2+} spin. The purpose of this example is to show that when the exchange is in fact expressed in terms of an effective spin of $\tfrac{1}{2}$, it does not have the simple isotropic form we have so far been using.

The first thing we must do is establish an expression for the exchange which we know to be correct. Recall that the $^4T_1$ state has a threefold orbital degeneracy, and that the three d holes can be treated just as electrons. Therefore we have a situation in which there is only one state of maximum multiplicity, which means that the total spin is a good quantum number. Also, since the ground state of Mn^{2+} is an S state, the total spin of this ion is a good quantum number. Therefore the exchange interaction may be written as

$$\mathcal{H}_{\text{ex}} = -\sum_{M_L M'_L} J(M_L, M'_L) \mathbf{S}_{\text{Mn}} \cdot \mathbf{S}_{\text{Co}}. \tag{2.105}$$

Notice that we are explicitly including the possibility of off-diagonal exchange. Thus, in terms of actual electron interchange, the exchange integral $J(M_L, M_L')$ would characterize the interchange in which an electron from a Co^{2+} in the state M_L jumps over to the Mn^{2+}, while an electron from the Mn^{2+} jumps back to the Co^{2+} and in so doing changes it to the M_L' state. These off-diagonal exchange integrals may be complex. As long as the orbital states so connected lie within the $^4T_1$ manifold, the scalar spin product may be used even to describe this off-diagonal exchange.

We now wish to know what form this exchange takes when expressed in terms of the effective spin $\frac{1}{2}$ of the Co^{2+} ground state. The most general interaction has the form

$$\mathscr{H}_{\text{eff}} = - \boldsymbol{S}_{\text{eff}} \cdot \mathscr{J} \cdot \boldsymbol{S}_{\text{Mn}} \tag{2.106}$$

where $\mathscr{J}$ is an exchange dyadic. If we evaluate the matrix elements of (2.105) in the basis $|\psi^{\pm}, M_S\rangle$, where M_S is the Mn^{2+} spin quantum number ($S = \frac{5}{2}$), and compare them with the matrix elements of (2.106) evaluated in the basis $|\pm\frac{1}{2}, M_S\rangle$, we find that *all* the elements of $\mathscr{J}_{\mu\nu}$ are nonzero. For example,

$$\mathscr{J}_{xy} = -2c^2 \, \text{Im} \{J(1, -1)\} \,. \tag{2.107}$$

The general matrix $\mathscr{J}_{\mu\nu}$ may be separated into a symmetric and an antisymmetric part. The *antisymmetric exchange* may be written as

$$\boldsymbol{D} \cdot \boldsymbol{S}_{\text{eff}} \times \boldsymbol{S}_{\text{Mn}} \tag{2.108}$$

where the vector coupling coefficient may be related to $\mathscr{J}_{\mu\nu}$. If the elements of the symmetric exchange are different, we speak of this as *anisotropic exchange*. There are two limiting forms that are popularly used: the Ising model

$$\mathscr{H}_{\text{Ising}} = -2J \sum_{i,\delta} S_i^z S_{i+\delta}^z \,,$$

and the XY model

$$\mathscr{H}_{XY} = -2J \sum_{i,\delta} (S_i^x S_{i+\delta}^x + S_i^y S_{i+\delta}^y) \,.$$

Here δ refers to a nearest neighbor. Thus we see that when the exchange interaction is expressed in terms of the effective spin of the ground state, it may have a very general form owing to the presence of orbital effects.

2.3.2 Rare-Earth Ions

The electronic configurations of the rare-earth ions are listed in Table 2.4. For these ions we see that the unpaired electrons lie inside the $5s^2p^6$ shells. Con-

Table 2.4. Configurations of the rare-earth ions

Ion	Configuration	Term
Ce^{3+}	$4f^1 5s^2 p^6$	$^2F_{5/2}$
Pr^{3+}	$4f^2 5s^2 p^6$	$^3H_4$
Nd^{3+}	$4f^3 5s^2 p^6$	$^4I_{9/2}$
Pm^{3+}	$4f^4 5s^2 p^6$	$^6I_4$
Sm^{3+}	$4f^2 5s^2 p^6$	$^6H_{5/2}$
Eu^{3+}	$4f^6 5s^2 p^6$	$^7F_0$
Gd^{3+}	$4f^7 5s^2 p^6$	$^8S_{7/2}$
Tb^{3+}	$4f^8 5s^2 p^6$	$^7F_6$
Dy^{3+}	$4f^9 5s^2 p^6$	$^6H_{15/2}$
Ho^{3+}	$4f^{10} 5s^2 p^6$	$^5I_8$
Er^{3+}	$4f^{11} 5s^2 p^6$	$^4I_{15/2}$
Tm^{3+}	$4f^{12} 5s^5 p^6$	$^3H_6$
Yb^{3+}	$4f^{13} 5s^2 p^6$	$^2F_{7/2}$

sequently they are not very strongly affected by crystal fields, and we might expect the Hamiltonian to consist of the following terms, in order of descending strength:

$$\mathcal{H} = \mathcal{H}_{\text{intraatomic Coulomb}} + \mathcal{H}_{\text{spin-orbit}} + \mathcal{H}_{\text{crystal field}} + \mathcal{H}_{\text{Zeeman}}. \quad (2.109)$$

The intraatomic Coulomb interaction produces states characterized by L, M_L, S, and M_S. When the spin–orbit interaction is added, only the total angular momentum $\boldsymbol{J} = \boldsymbol{L} + \boldsymbol{S}$ is conserved. Therefore the states have the form $|J, M_J; L, S\rangle$.

Let us consider, for the moment, the effect of the Zeeman term in the absence of any crystal field,

$$\mathcal{H}_Z = \mu_B(\boldsymbol{L} + 2\boldsymbol{S}) \cdot \boldsymbol{H}. \quad (2.110)$$

Since the states are characterized by the eigenvalues of $\boldsymbol{J}$, we rewrite this as

$$\mathcal{H}_{\text{Zeeman}} = g_J \mu_B \boldsymbol{J} \cdot \boldsymbol{H}, \quad (2.111)$$

where g_J is the *Landé g value*, defined by $g_J \boldsymbol{J} = \boldsymbol{L} + 2\boldsymbol{S}$. This has the value

$$g_J = 1 + \frac{J(J+1) + S(S+1) - L(L+1)}{2J(J+1)}. \quad (2.112)$$

Thus the state $|J, M_J\rangle$ is split into $2J + 1$ equally spaced states with the separation $g_J \mu_B H$. For Ce^{3+}, for example, $g_J = \frac{6}{7}$.

In the presence of a crystal field the splitting of the state $|J, M_J\rangle$ is easily determined by expressing the crystal field in operator equivalents of $\boldsymbol{J}$. Thus for a crystal field of D_2 symmetry we have

2.3 The Spin Hamiltonian

$$\mathcal{H}_{\text{cryst}} = B_2^0 O_2^0 + B_2^2 O_2^2 + B_4^0 O_4^0 + B_4^2 O_4^2 + B_4^4 O_4^4$$
$$+ B_6^0 O_6^0 + B_6^2 O_6^2 + B_6^4 O_6^4 + B_6^6 O_6^6 . \tag{2.113}$$

The matrix of $\mathcal{H}_{\text{cryst}}$ in the basis $|J, M_J; L, S\rangle$ is easily constructed from tables [2.7]. The result for Ce^{3+} is

$$\begin{array}{c} \phantom{\langle\tfrac{5}{2}|} \quad |\tfrac{5}{2}\rangle \quad |\tfrac{1}{2}\rangle \quad |-\tfrac{3}{2}\rangle \quad |-\tfrac{5}{2}\rangle \quad |-\tfrac{1}{2}\rangle \quad |\tfrac{3}{2}\rangle \\ \begin{array}{c} \langle\tfrac{5}{2}| \\ \langle\tfrac{1}{2}| \\ \langle-\tfrac{3}{2}| \\ \langle-\tfrac{5}{2}| \\ \langle-\tfrac{1}{2}| \\ \langle\tfrac{3}{2}| \end{array} \left[\begin{array}{cc} B & 0 \\ \\ 0 & B \end{array} \right] \end{array} \tag{2.114}$$

where

$$B \equiv \begin{bmatrix} 10B_2^0 + 60B_4^0 & \sqrt{10}B_2^2 + 9\sqrt{10}B_4^2 & 12\sqrt{5}\,B_4^4 \\ \sqrt{10}B_2^2 + 9\sqrt{10}B_4^2 & -8B_2^0 + 120B_4^0 & 3\sqrt{2}\,B_2^2 - 15\sqrt{2}\,B_4^2 \\ 12\sqrt{5}\,B_4^4 & 3\sqrt{2}\,B_2^2 - 15\sqrt{2}\,B_4^2 & -2B_2^0 - 180B_4^0 \end{bmatrix} .$$

This obviously gives three doublets which have the form

$$\psi_n^+ = a_n |\tfrac{5}{2}\rangle + b_n |\tfrac{1}{2}\rangle + c_n |-\tfrac{3}{2}\rangle , \tag{2.115a}$$

$$\psi_n^- = a_n |-\tfrac{5}{2}\rangle + b_n |-\tfrac{1}{2}\rangle + c_n |\tfrac{3}{2}\rangle . \tag{2.115b}$$

In terms of the states $|m_l, m_s\rangle$, these may be written as

$$\psi_n^+ = \varphi_1 \alpha + \varphi_2 \beta , \tag{2.116a}$$

$$\psi_n^- = \varphi_2^* \alpha - \varphi_1^* \beta . \tag{2.116b}$$

where $\varphi_1^* = \varphi_1$ and $\varphi_2^* = -\varphi_2$. Under the time-reversal operator $T = iK\sigma_y$, where K is the conjugation operator, these states are related by

$$T\psi_n^+ = \psi_n^- , \tag{2.117a}$$

$$T\psi_n^- = -\psi_n^+ . \tag{2.117b}$$

Such a pair of states is said to form a *Kramers doublet*. The states (2.104) also formed a Kramers doublet.

Let us now consider the behavior of the lowest Kramers doublet in an external magnetic field. The Hamiltonian is

$$\mathcal{H}_Z = \mu_B (l + 2s) \cdot H . \tag{2.118}$$

66   2. The Magnetic Hamiltonian

Since we are within a manifold where J is a good quantum number, this may be written as

$$\mathcal{H}_Z = g_J \mu_B J_z H . \tag{2.119}$$

The Zeeman matrix for the lowest doublet is

$$\begin{bmatrix} g_J\mu_B H(\tfrac{5}{2}a_1^2 + \tfrac{1}{2}b_1^2 - \tfrac{3}{2}c_1^2) & 0 \\ 0 & -g_J\mu_B H(\tfrac{5}{2}a_1^2 + \tfrac{1}{2}b_1^2 - \tfrac{3}{2}c_1^2) \end{bmatrix} . \tag{2.120}$$

Therefore the doublet splits linearly in the field. An effective g value is often introduced by defining the splitting of two levels in a field as $g_{\text{eff}}\mu_B H$. In this case the effective g value would be

$$g_{\text{eff}} = g_J(5a_1^2 + b_1^2 - 3c_1^2) . \tag{2.121}$$

Notice that the g value depends directly on the coefficients in the wave function (2.115). For this reason measurements of the g value provide a sensitive test of the ground-state wave function.

2.3.3 Semiconductors

It is interesting to carry out the calculation for the g value of an electron moving in the periodic potential, $V(\mathbf{r})$, of a crystal lattice. The Hamiltonian is

$$\mathcal{H} = \mathcal{H}_{\text{kinetic}} + \mathcal{H}_{\text{crystal}} + \mathcal{H}_{\text{spin-orbit}} + \mathcal{H}_{\text{Zeeman}} . \tag{2.122}$$

The eigenfunctions of an electron moving in a periodic potential have the Bloch form $u_{n\mathbf{k}}(\mathbf{r}) \exp(i\mathbf{k}\cdot\mathbf{r})$ which give rise to energy bands characterized by band indices n and wave vector $\mathbf{k}$. In the absence of the Zeeman term the function $u_{n\mathbf{k}}(\mathbf{r})$ satisfies

$$\left[\frac{1}{2m}(\mathbf{p} + \hbar\mathbf{k})^2 + V(\mathbf{r}) + \frac{\hbar}{4m^2c^2}\boldsymbol{\sigma}\times\nabla V(\mathbf{r})\cdot(\mathbf{p} + \hbar\mathbf{k})\right]u_{n\mathbf{k}}(\mathbf{r}) = \epsilon_{n\mathbf{k}} u_{n\mathbf{k}}(\mathbf{r}) . \tag{2.123}$$

In the case of the "three–five" semiconductors, which are composed of elements from the third and fifth columns of the periodic table (e.q., GaAs, InSb) the valence band is p like which the conduction band is s like. The band structure in the vicinity of $\mathbf{k} = 0$ is illustrated in Fig. 2.10. The labels Γ_6, etc., indicate the irreducible representations according to which the wave functions at $\mathbf{k} = 0$ transform. The first feature to note is that the spin–orbit interaction has split the threefold degeneracy of the p-like valence band. Secondly, the curvatures of these bands are different. This curvature is a measure of the effective mass of the electron. We may obtain an estimate of this mass by treating the term $(\hbar/m)$

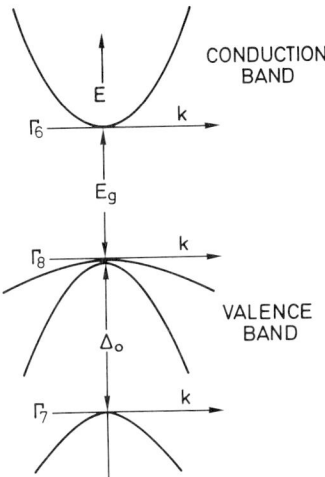

Fig. 2.10. Energy bands near $k=0$ for zinc-blende crystals

$k \cdot p$ in (2.123) as a perturbation. Calculating the second-order correction to the energy of the conduction band leads to the effective mass tensor [2.21],

$$\left(\frac{m}{m^*}\right)_{\mu\nu} = \delta_{\mu\nu} + \frac{2}{m}\sum_{\Gamma}\frac{\langle\Gamma_6|p_\mu|\Gamma\rangle\langle\Gamma|p_\nu|\Gamma_6\rangle}{\epsilon_{\Gamma_6} - \epsilon_\Gamma}. \tag{2.124}$$

The valence band wave functions are linear combinations of the usual x, y, and z components of the p state. Since the III–V compounds have the cubic zinc-blende structure, if we restrict our consideration only to the bands shown in Fig. 2.10 then there is only one parameter,

$$\frac{2}{m}|\langle\Gamma,\mu|p_\mu|s\rangle|^2 \equiv P^2,$$

and the effective mass becomes

$$\frac{m}{m^*} = 1 + \frac{P^2}{3}\frac{3E_g + 2\Delta}{E_g(E_g + \Delta)}. \tag{2.125}$$

Let us now consider the orbital moment associated with an electron at the bottom of the conduction band,

$$\langle\Gamma_6|L_z|\Gamma_6\rangle = \sum_{\Gamma}[\langle\Gamma_6|\,x|\Gamma\rangle\langle\Gamma|p_y|\Gamma_6\rangle - \langle\Gamma_6|y|\Gamma\rangle\langle\Gamma|p_x|\Gamma_6\rangle]. \tag{2.126}$$

This may be rewritten by noting that

$$[x, \mathscr{H}] = \frac{i}{m}p_x + \frac{i\hbar}{4m^2c^2}(\boldsymbol{\sigma}\times\nabla V)_x. \tag{2.127}$$

Neglecting the spin–orbit contribution, which can be shown to be small, the matrix elements of (2.127) satisfy

$$\langle \Gamma | x | \Gamma_6 \rangle = \frac{1}{m} \frac{\langle \Gamma | p_x | \Gamma_6 \rangle}{\epsilon_{\Gamma_6} - \epsilon_\Gamma}. \tag{2.128}$$

This enables us to convert the orbital angular-momentum matrix element into a form similar to the effective mass. In particular, eliminating the linear-momentum matrix element between these two expressions leads to the relation [2.22]

$$g^* = \frac{\mu^*}{\mu_B} = 1 + \langle \Gamma_6 | L_z | \Gamma_6 \rangle = 2\left[1 - \frac{\Delta}{3E_g + 2\Delta}\left(\frac{m}{m^*} - 1\right)\right]. \tag{2.129}$$

In general, one must include additional bands, in which case the relationship between the effective g value and the effective mass is not as simple. Nevertheless, when the effective mass is very small the g value can become quite large. In InSb, for example, $m^*/m = -0.014$ and $g^* = -51.4$. Herring [2.23] has shown that the angular momentum (2.126) may be spatially decomposed into intraatomic and interatomic contributions. The large g values were shown to be associated with the latter, corresponding to interatomic circulating currents.

In this chapter we have seen the origins of various terms in the Hamiltonian that may influence the magnetic response of a system. In the remaining chapters we shall investigate how these terms manifest themselves when the system is excited with a space-and/or time-varying magnetic field.

3. The Static Susceptibility of Noninteracting Systems

In this chapter we shall investigate the static susceptibility of systems described by Hamiltonians which may be written as the sum of individual terms. Our approach here, and in later discussions, will be to perturb the system with a magnetic field and then compute the response to this field. In this chapter we shall discuss the effect of a time-independent field H.

In the literature of magnetism a distinction is often made between insulators and conductors. This distinction is basically an operational one. That is, insulators are characterized by charge distributions that are reasonably well localized to unit cells; such systems may be described by localized effective spins. Metals are characterized by itinerant electrons which require an entirely different description. Of course, there are materials, such as metals containing rare-earth ions, in which both types of charge distributions coexist. Furthermore, there are many magnetic alloys in which the actual nature of the moment is not yet understood. We shall come back to this situation in Chap. 7. For the present, however, we shall follow tradition and discuss the response to local moments and itinerant systems separately.

3.1 Localized Moments

Let us consider a system of N noninteracting identical ions or molecules. In the absence of an external field each ion or molecule is characterized by a Hamiltonian $\mathcal{H}_i^0$. We are interested in the response of this system to a small static magnetic field. Therefore we add a Zeeman term to the Hamiltonian to give a total Hamiltonian,

$$\mathcal{H} = \sum_i \mathcal{H}_i = \sum_i (\mathcal{H}_i^0 + \mathcal{H}_i^Z). \tag{3.1}$$

The magnetization is given by (1.49) as

$$M(r) = \text{Tr}\{\rho \mathcal{M}(r)\}, \tag{3.2}$$

where

$$\mathcal{M}(r) = \tfrac{1}{2} \sum_{i=1}^{N} \sum_{\alpha=1}^{h} [\mu_{i\alpha}\delta(r - r_{i\alpha}) + \delta(r - r_{i\alpha})\mu_{i\alpha}]. \tag{3.3}$$

Here $\mu_{i\alpha}$ is the magnetic-moment operator associated with the αth electron on the ith ion. Since the ionic electrons interact with one another, the appropriate basis in which to evaluate (3.2) consists of the ionic eigenfunctions. Therefore we make use of the Wigner–Eckhart theorem and replace (3.3) by

$$\mathcal{M}(r) = \sum_{i=1}^{N} \boldsymbol{\mu}_i \delta(\boldsymbol{r} - \boldsymbol{R}_i). \tag{3.4}$$

We have assumed that the ionic moments have no spatial extent. As long as the wavelength of the applied field is larger than an atomic dimension, (3.4) is valid. However, when we talk about neutron scattering in Chap. 8 this assumption will have to be modified.

Since we are applying a static field, the density matrix is independent of time. Therefore $[\mathcal{H}, \rho] = 0$. The equilibrium solution of this equation may be written as

$$\rho = \exp\left[\beta(F - \mathcal{H})\right] \tag{3.5}$$

where F is the free energy

$$F = -k_{\mathrm{B}} T \ln Z \tag{3.6}$$

and Z is the partition function

$$Z = \mathrm{Tr}\left\{\exp\left(-\beta \mathcal{H}\right)\right\}. \tag{3.7}$$

The solution (3.5) for the density matrix, known as the *canonical distribution*, is the quantum-mechanical generalization of the classical Gibbs distribution. Using. (3.6,7), we may write this as

$$\rho = \frac{\exp\left(-\beta \mathcal{H}\right)}{\mathrm{Tr}\left\{\exp\left(-\beta \mathcal{H}\right)\right\}} \tag{3.8}$$

which, because of the form of the Hamiltonian (3.1), may be factored as

$$\rho = \frac{\prod_i \exp\left(-\beta \mathcal{H}_i\right)}{\mathrm{Tr}\left\{\prod_i \exp\left(-\beta \mathcal{H}_i\right)\right\}} = \prod_i \rho_i. \tag{3.9}$$

Therefore

$$\boldsymbol{M}(r) = \sum_i \mathrm{Tr}\left\{\rho_i \boldsymbol{\mu}_i \delta(\boldsymbol{r} - \boldsymbol{R}_i)\right\}, \tag{3.10}$$

which defines the ionic moment

$$\boldsymbol{M}_i = \mathrm{Tr}\left\{\rho_i \boldsymbol{\mu}_i\right\}. \tag{3.11}$$

In writing (3.10) we have assumed that we may interchange the operations of taking the trace and summing over the ions. This neglects the possibility that an electron on one ion may hop over to a neighboring ion. If the electrons were distinct, as labeled by their positions in the lattice, such hopping would be forbidden. Therefore our assumption of interchangeability is, in this sense, equivalent to the assumption that the ionic electrons are *distinguishable*. The importance of this assumption will become evident in Sect. 3.2.

So far we have not said anything about the spatial variation of the applied field. Let us assume that it has the wavelike form

$$H(r) = H \cos(q_0 \cdot r). \tag{3.12}$$

We shall always assume that the wave vector q_0 is one of the discrete wave vectors lying within the first Brillouin zone associated with the lattice under consideration. Let us keep in mind that we may have moments at only a few of the available lattice sites. If the applied field does not vary appreciably over a unit cell, the ionic eigenvalues will correspond to those of a uniform field with the value $H_i = H \cos(q_0 \cdot R_i)$. The corresponding moment is

$$\mu_i = -\frac{\partial \mathcal{H}_i}{\partial H_i}. \tag{3.13}$$

If the eigenvalues associated with the ith ion are E_n, then the ionic moment in the direction of the applied field is

$$M_i \equiv \mathrm{Tr}\{\rho_i \mu_i\} = -\frac{\sum_n \exp(-\beta E_n)(\partial E_n/\partial H_i)}{\sum_n \exp(-\beta E_n)}. \tag{3.14}$$

Since the eigenvalues E_n are functions of the field H_i, let us expand in powers of the field:

$$E_n = E_n^{(0)} + E_n^{(1)} H_i + E_n^{(2)} H_i^2 + \dots. \tag{3.15}$$

Since the magnetic field is only a probe, we may make it as small as we like. In particular, let us assume that the field-dependent corrections to the energy in (3.15) are smaller than the thermal energy. Then

$$\exp(-\beta E_n) = \exp(-\beta E_n^{(0)})(1 - \beta E_n^{(1)} H_i - \dots), \tag{3.16}$$

and (3.14) becomes

$$M_i = -\frac{\sum_n \exp(-\beta E_n^{(0)})(1 - \beta E_n^{(1)} H_i - \dots)(E_n^{(1)} + 2E_n^{(2)} H_i + \dots)}{\sum_n \exp(-\beta E_n^{(0)})(1 - \beta E_n^{(1)} H_i - \dots)}. \tag{3.17}$$

Assuming that the ion has no moment in the absence of the field and retaining only those terms linear in the field, we have

$$M_i = \frac{\sum_n (\beta E_n^{(1)2} - 2E_n^{(2)}) \exp(-\beta E_n^{(0)})}{\sum_n \exp(-\beta E_n^{(0)})} H_i \equiv \chi_i H_i \qquad (3.18)$$

which defines the quantity χ_i. Substituting this expression back into (3.10) and taking the spatial Fourier transform, we obtain

$$M_z(\mathbf{k}) = \sum_i M_i \exp(-i\mathbf{k}\cdot\mathbf{R}_i).$$

which may be written as

$$M_z(\mathbf{k}) = \sum_q \left[\frac{1}{V}\sum_i \chi_i \exp[-i(\mathbf{k}-\mathbf{q})\cdot\mathbf{R}_i]\right] H(\mathbf{q})$$

where

$$H(\mathbf{q}) = \frac{HV}{2}[\Delta(\mathbf{q}+\mathbf{q}_0) + \Delta(\mathbf{q}-\mathbf{q}_0)].$$

Therefore

$$\chi(\mathbf{k},\mathbf{q}) = \frac{1}{V}\sum_i \chi_i \exp[-i(\mathbf{k}-\mathbf{q})\cdot\mathbf{R}_i] \qquad (3.19)$$

If there is a moment at each lattice site, then the sum in this expression gives $N\Delta(\mathbf{k}-\mathbf{q})$, and

$$\chi(\mathbf{q}) = \frac{N}{V}\frac{\sum_n (\beta E_n^{(1)2} - 2E_n^{(2)}) \exp(-\beta E_n^{(0)})}{\sum_n \exp(-\beta E_n^{(0)})}.$$

The fact that this is independent of $\mathbf{q}$ is a consequence of our assumption that the magnetization is everywhere proportional to the local field. Thus it is not surprising that

$$\chi(\mathbf{r}-\mathbf{r}') = \frac{1}{V}\sum_q \exp[i\mathbf{q}\cdot(\mathbf{r}-\mathbf{r}')]\chi(\mathbf{q}) = \frac{N}{V}\chi\delta(\mathbf{r}-\mathbf{r}'). \qquad (3.20)$$

Let us now consider some specific applications of (3.19).

3.1.1 Diamagnetism

In Sect. (2.2) we found that the Zeeman Hamiltonian for an ionic or molecular system of electrons was

$$\mathcal{H}_i^z = \sum_\alpha \left[\mu_B \boldsymbol{l}_\alpha \cdot \boldsymbol{H}_i + \frac{e^2 H_i^2}{8mc^2}(x_\alpha^2 + y_\alpha^2) + \mu_B \boldsymbol{\sigma}_\alpha \cdot \boldsymbol{H}_i \right]. \tag{3.21}$$

The effect of this interaction on the eigenvalues may be obtained from perturbation theory, with the eigenfunctions of $\mathcal{H}_i^0$ as the unperturbed states. A diamagnetic system is one in which all the matrix elements of the orbital and spin angular momenta are 0. Therefore the only correction to the energy arises from the second term in (3.21). If the eigenfunctions of $\mathcal{H}_i^0$ are written as $|n\rangle$, then

$$E_n^{(2)} = \frac{e^2}{8mc^2} \langle n | \sum_\alpha (x_\alpha^2 + y_\alpha^2) | n \rangle . \tag{3.22}$$

Notice that the susceptibility involves the statistical average of

$$\sum_\alpha (x_\alpha^2 + y_\alpha^2) .$$

This is usually written as $\frac{3}{2}Z$ times the mean square radius $\langle r^2 \rangle$. Therefore the diamagnetic contribution to the quantity χ in the generalized susceptibility (3.19) is

$$\boxed{\chi_{\text{diamag}} = -\frac{Ze^2}{6mc^2} \langle r^2 \rangle .} \tag{3.23}$$

This negative response is just a manifestation of Lenz's law, which states that in the presence of a changing flux a system of charges will develop currents in such a direction as to oppose the flux change. These currents are the source of the diamagnetic moment.

In dielectric solids the molar susceptibility is of the order of -10^{-5} cm³/mole. This corresponds to an induced moment per atom of $10^{-8} H_i$ Bohr magneton, which is very small.

Notice that the diamagnetic susceptibility appears to depend on the choice of the origin for computing $\langle r^2 \rangle$. This ambiguity is, unfortunately, real. *Van Vleck* has shown that the change in diamagnetic susceptibility resulting from a change in origin is just compensated for by a corresponding change in the *Van Vleck paramagnetic susceptibility*, to be discussed in the next section. [3.1] For a single atom or ion the origin is usually taken at the nucleus. In this case the Van Vleck contribution is small, and the diamagnetic susceptibility is well described by (3.23). However for polyatomic molecules both contributions to the susceptibility must be considered together.

3.1.2 Paramagnetism of Transition-Metal Ions

In the preceding discussion we assumed that all the matrix elements, both diagonal and off-diagonal, of the orbital and spin angular momenta were 0. The interesting aspects of magnetism arise, however, when these matrix elements are nonzero. Let us now examine the response of such ions to a static field. Since crystal fields have important effects on the orbital angular momentum, it is convenient to discuss transition-metal ions and rare-earth ions separately.

In Sect. 2.3 we found that a transition-metal ion could be described by a spin Hamiltonian. Let us consider the particular case where the crystal field has removed all the orbital degeneracy. The spin Hamiltonian which describes the lowest orbital state then has the form

$$\mathcal{H}_{\text{eff}} = g_\| \mu_B H S_z - \mu_B^2 \Lambda_\| H^2 . \tag{3.24}$$

If we take the eigenfunctions of S_z as our unperturbed states, then our orbital singlet splits into the $2S + 1$ states characterized by the spin-projection quantum number M_S,

$$E_{M_S} = E_{M_S}^0 + g_\| \mu_B M_S H - \mu_B^2 \Lambda_\| H^2 . \tag{3.25}$$

Making the identifications with (3.15) and using (3.14), we find that the ($k = 0$, $q = 0$) susceptibility may be written as the sum of the *Langevin susceptibility* and the *Van Vleck susceptibility*,

$$\chi(0, 0) = \chi_{\text{Langevin}} + \chi_{VV} , \tag{3.26}$$

where

$$\chi_{\text{Langevin}} = \frac{N}{V} \frac{g_\|^2 \mu_B^2}{k_B T} \frac{\sum_{M=-S}^{S} M_S^2}{2S + 1} = \frac{N}{V} \frac{g_\|^2 \mu_B^2 S(S + 1)}{3 k_B T} \equiv \frac{C}{T} , \tag{3.27}$$

which defines the Curie constant C, and

$$\chi_{VV} = \frac{N}{V} 2 \mu_B^2 \Lambda_\| . \tag{3.28}$$

The Van Vleck susceptibility arises from that part of the oribtal moment which has been admixed back into the ground state by the orbital Zeeman effect. The fact that this contribution is temperature independent is a result of the assumption that the energy difference between the ground state and the higher-lying orbtal state entering $\Lambda_\|$ is larger than $k_B T$. If this condition does not hold, then the energy shift of this upper state must be considered, with the result that the Van Vleck susceptibility will acquire a temperature dependence.

The inverse temperature dependence of the Langevin susceptibility was first observed experimentally by Pierre Curie in 1895 and is referred to as *Curie's law*. The Curie constant C is used experimentally to determine an effective moment in units of Bohr magnetons by the relation

3.1 Localized Moments

Table 3.1. Effective moments associated with transition-metal ions

	Ti^{3+}	V^{3+}	Cr^{3+}	Mn^{3+}	Mn^{2+}, Fe^{3+}	Fe^{2+}	Co^{2+}	Ni^{2+}	Cu^{2+}
$2\sqrt{S(S+1)}$	1.73	2.83	3.87	4.0	5.92	4.90	3.87	2.83	1.73
$\mu_{\text{eff}}(\exp)$	1.8	2.8	3.8	4.9	5.9	5.4	4.8	3.2	1.9
$g_J\sqrt{J(J+1)}$	1.55	1.63	0.77	0	5.92	6.70	6.63	5.59	3.55
λ[cm^{-1}]	154	105	91	88	—	-103	-178	-325	-829

$$\mu_{\text{eff}} = \sqrt{\frac{3k_B C}{(N/V)\mu_B^2}}. \tag{3.29}$$

If the experimental values of μ_{eff} are compared with the spin-only theoretical values in Table 3.1, we see that the quenching is nearly complete for the first half of the periodic table. Orbital contributions become more important in the last half of the row because the spin–orbit coupling constant is becoming larger [3.2].

Because of the availability of high-static magnetic fields and low temperatures it is relatively easy to reach the nonlinear region of saturation. To describe this situation we must use the more general expression for the ionic moment given by (3.14). The Van Vleck susceptibility is unchanged. However, the Langevin susceptibility becomes

$$\begin{aligned}\chi_{\text{Langevin}} &= -\frac{Ng_\parallel \mu_B S}{HV} \frac{\left[\sum_{M_S=-S}^{S} \exp(-\beta g_\parallel \mu_B H M_S)\right] M_S/S}{\sum_{M_S=-S}^{S} \exp(-\beta g_\parallel \mu_B H M_S)} \\ &= \frac{Ng_\parallel \mu_B S}{HV} B_S\left(\frac{g_\parallel \mu_B SH}{k_B T}\right),\end{aligned} \tag{3.30}$$

where $B_S(g_\parallel \mu_B SH/k_B T)$ is called the *Brillouin function*. Notice that for $S = \frac{1}{2}$, $B_{1/2}(x) = \tanh(x)$

It is customary for experimentalists to plot the inverse susceptibility against temperature. In Fig. 3.1 we have plotted $1/\chi$ for a spin of $\frac{3}{2}$ in a field of 30,000 Oe. Since the Curie constant is of the order of unity, the Van Vleck contribution becomes important only for $T > \chi_{VV}^{-1}$. Normally, the Van Vleck susceptibility is of the order of 10^{-5}. However, the larger value of 10^{-3} has been chosen in Fig. 3.1 to illustrate its effect.

The only approximation in obtaining (3.17) was that the Zeeman energy be smaller than the thermal energy. Therefore this expression also applies if the thermal energy exceeds the crystal-field energy. In this case we must be careful to use the proper energies E_n. We shall consider an example of how to include excited-state effects in the next section.

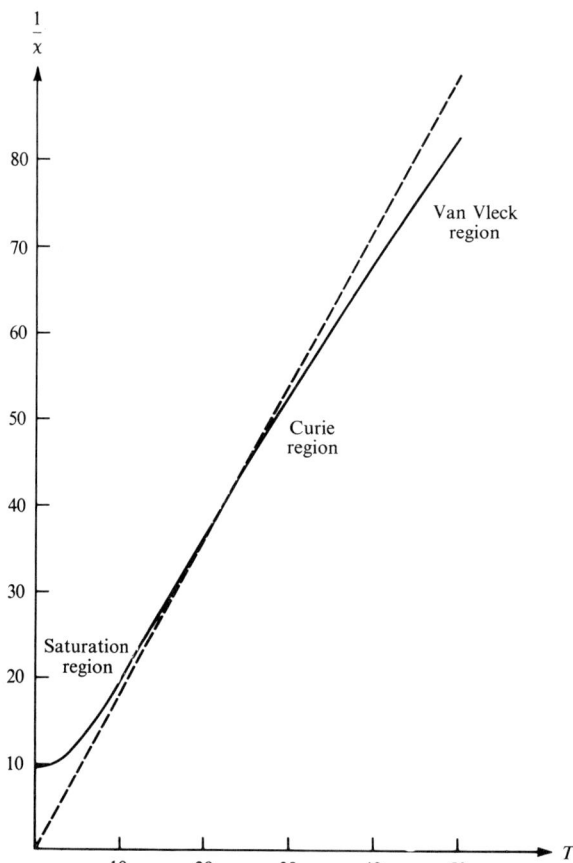

Fig. 3.1. Inverse susceptibility as a function of temperature for a spin $\frac{3}{2}$ in a field of 30,000 Oe

3.1.3 Paramagnetism of Rare-Earth Ions

The effect of crystal fields on rare-earth ions is generally quite small in comparison with the spin–orbit interaction. Let us therefore consider our rare-earth ion as described by eigenstates characterized by the quantum numbers J, M_J, L, and S. The perturbing Zeeman Hamiltonian has the form

$$\mathcal{H}_Z = \mu_B(\mathbf{L} + 2\mathbf{S}) \cdot \mathbf{H}, \tag{3.31}$$

If we restrict ourselves to the lowest J manifold and define the z direction as the direction of the applied field, then this Hamiltonian becomes

$$\mathcal{H}_Z = g_J \mu_B H J_z. \tag{3.32}$$

The effect of this interaction is to split the J state into its $2J + 1$ Zeeman components, with energies

$$E_{J,M_J} = g_J\mu_B M_J H .\tag{3.33}$$

From (3.15, 19) we find

$$\chi(0,0) = \frac{N}{V}\frac{g_J^2\mu_B^2 J(J+1)}{3k_B T},\tag{3.34}$$

and from (3.29) we obtain the effective moment,

$$\mu_{\text{eff}} = g_J\sqrt{J(J+1)}.\tag{3.35}$$

Most of the rare-earth ions exhibit moments in good agreement with (3.35). However, a notable exception is Eu^{3+}. This has a $4f^6$ configuration, which corresponds to a $^7F_0$ ground state. Therefore $J=0$, and μ_{eff} should also be 0. Experimentally, however, a moment of 3.4 Bohr magnetons has been found at room temperature. To account for this difference we note that the $^7F_1$ excited state lies $\Delta/k_B \simeq 350$ K above the ground state. Therefore we must include the influence of this state on our susceptibility. There are two effects to be considered, the addition of the terms in (3.19) associated with the excited state and the shift in the *ground-state energy* produced by the Zeeman coupling to the excited state

From perturbation theory we find that the first-order corrections to the energies are

$$E_{J,M_J}^{(1)} H = g_J\mu_B H\langle J,M_J|J_z|J,M_J\rangle ,\tag{3.36}$$

and the second-order corrections are

$$E_{J,M_J}^{(2)} H^2 = \mu_B^2 H^2 \sum_{J',M'_{J'}} \frac{|\langle J,M_J|L_z + 2S_z|J',M'_{J'}\rangle|^2}{E_{J,M_J}^{(0)} - E_{J',M'_{J'}}^{(0)}} .\tag{3.37}$$

From the properties of angular momentum [3.3] we find that the matrix element in (3.37) is nonvanishing only for $J' = J\pm 1$ and $M'_{J'} = M_J$. Therefore only the $|0,0\rangle$ and $|1,0\rangle$ states are coupled in second order. This matrix element is $\langle 0,0|L_z + 2S_z|1,0\rangle = 2$. The resulting eigenvalues are

$$E_{0,0} = -\frac{4\mu_B^2 H^2}{\Delta},$$

$$E_{1,-1} = \Delta - g_1\mu_B H ,$$

$$E_{1,0} = \Delta + \frac{4\mu_B^2 H^2}{\Delta},$$

$$E_{1,1} = \Delta + g_1\mu_B H ,\tag{3.38}$$

where g_1 is the Landé g value (2.112) for $J=1$. These eigenvalues are indicated in Fig. 3.2. Substituting these results into (3.19) leads to an effective moment at

78   3. The Static Susceptibility of Noninteracting Systems

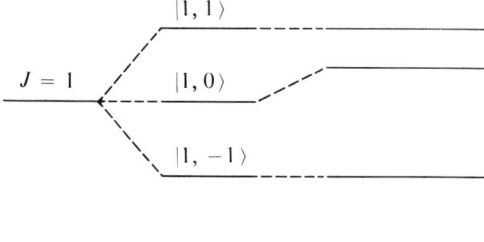
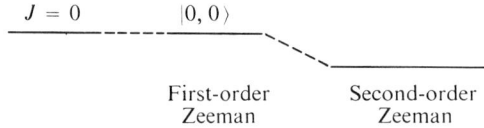

Fig. 3.2. Effect of the Zeeman interaction on the low-lying states of Eu^{3+}

room temperature of 3.2 Bohr magnetons, in much better agreement with the experimental value.

In conclusion, we have found that the susceptibility of an insulator containing noninteracting ionic moments is

$$\chi(T) = \chi_{\text{diamag}} + \chi(T)_{\text{Langevin}} + \chi_{VV}. \tag{3.39}$$

The temperature dependence arises primarily from the Langevin or Curie part.

3.2 Metals

We now turn to the response of a nonmagnetic metal to an applied field. The static response to a uniform field was first investigated by Pauli, who considered the spin paramagnetism, and by Landau, who considered the orbital diamagnetism.

3.2.1 Landau Diamagnetism

Let us consider the orbital moment induced in a free-electron metal by a static applied field. For the time being we shall neglect the spin associated with the electrons in a metal. As mentioned in Sect. 1.2, the classical approach to this problem leads to the conclusion that the metal shows no response to the applied field. *Landau* was the first to show that a quantum-mechanical approach indicates that a free-electron system does in fact respond to the applied field, and that this response is diamagnetic [3.4]. What he did was to solve the Schrödinger equation for a single electron in a magnetic field; with the resulting eigenvalues he was able to compute the free energy, from which he obtained the magnetization. The corresponding susceptibility was found to be $-\frac{1}{3}$ that obtained by Pauli for the spin contribution. This relationship is valid for all temperatures in degenerate as well as nondegenerate free-electron systems.

Peierls extended Landau's calculation to the case of electrons that are nearly bound by a periodic lattice potential [3.5], and *Wilson* considered nearly free electrons [3.6]. It was found that the effect of the lattice was, essentially, to modify Landau's result by a factor m/m^*, where m^* is the electron effective mass at the Fermi surface.

It is interesting to consider why Miss van Leeuwen's theorem does not hold in the quantum-mechanical case. The reason is that those electrons whose orbits are interrupted by the boundary have, in general, high quantum numbers. Therefore, their energy is also high which makes them thermo-dynamically less important than the "bulk" electrons. Consequently, the diamagnetic moments now dominate the paramagnetic moments arising from the "boundary" electrons.

The problem of boundary effects arises when one calculates the susceptibility by first solving the Schrödinger equation for the energy levels explicitly and then using them to calculate the partition function. *Sondheimer* and *Wilson* have pointed out that the whole question of boundary effects may be avoided by calculating the partition function without explicit knowledge of the energy levels ref [3.7].

We shall now calculate the Landau susceptibility by another method [3.8] that is more in keeping with our linear response philosophy. Let us assume that we have a free-electron system characterized by a spherical Fermi surface. To this system we apply a static external magnetic field $\boldsymbol{H}_0(\boldsymbol{r})$, which arises from an external current density $\boldsymbol{J}_0(\boldsymbol{r})$ according to the Maxwell equation

$$\boldsymbol{\nabla} \times \boldsymbol{H}_0(\boldsymbol{r}) = \frac{4\pi}{c} \langle \boldsymbol{J}_0(\boldsymbol{r}) \rangle . \tag{3.40}$$

This field induces a current density $\boldsymbol{J}(\boldsymbol{r})$ in the electron system, which in turn acts as the source for an additional magnetic field. The resulting total local field, denoted by $\boldsymbol{H}(\boldsymbol{r})$, is also related to its current sources by the Maxwell equation

$$\boldsymbol{\nabla} \times \boldsymbol{H}(\boldsymbol{r}) = \frac{4\pi}{c} \langle \boldsymbol{J}_0(\boldsymbol{r}) + \boldsymbol{J}(\boldsymbol{r}) \rangle . \tag{3.41}$$

The local field defined in this manner has zero divergence. Therefore it may be expressed in terms of a vector potential as

$$\boldsymbol{H}(\boldsymbol{r}) = \boldsymbol{\nabla} \times \boldsymbol{A}(\boldsymbol{r}) . \tag{3.42}$$

In the presence of this potential the electron velocity becomes

$$v_i = \frac{1}{m}\left[\boldsymbol{p}_i - \frac{e}{c} \boldsymbol{A}(\boldsymbol{r}_i) \right]. \tag{3.43}$$

The induced-current-density operator is related to this velocity by

$$\boldsymbol{J}(\boldsymbol{r}) = \frac{e}{2} \sum_i [v_i \delta(\boldsymbol{r} - \boldsymbol{r}_i) + \delta(\boldsymbol{r} - \boldsymbol{r}_i) v_i] . \tag{3.44}$$

80   3. The Static Susceptibility of Noninteracting Systems

According to (1.8), the average value of this current density determines the magnetization,

$$\langle J(r) \rangle = c \nabla \times M(r) . \tag{3.45}$$

Using the Fourier transforms of (3.41, 42, 45) and the definition of the susceptibility, $M(q) = \chi(q) \cdot H_0(q)$, we obtain

$$\chi(q) = -\frac{1}{4\pi} \frac{\langle J(q) \rangle}{\langle J(q) \rangle - (cq^2/4\pi)A(q)} . \tag{3.46}$$

Therefore, if we can compute the average current density to lowest order in the vector potential, we shall have an expression for the susceptibility.

Introducing the Fourier momentum operator p_q, defined by

$$p_q \equiv \tfrac{1}{2} \sum_i [p_i \exp(-i q \cdot r_i) + \exp(-i q \cdot r_i) p_i] , \tag{3.47}$$

we find from (3.44) that

$$\langle J(q) \rangle = \frac{e}{m} \langle p_q \rangle - \frac{Ne^2}{mcV} A(q) . \tag{3.48}$$

To obtain the diagonal components of χ we must find the average value of that component of p_q parallel to $A(q)$ to the lowest order in the vector potential. If n is a unit vector parallel to $A(q)$, then we must compute $\langle p_q \cdot \hat{n} \rangle$. This we shall do by perturbation theory.

In the presence of the field the Hamiltonian is

$$\mathcal{H} = \sum_i \frac{1}{2m} \left[p_i - \frac{e}{c} A(r_i) \right]^2 . \tag{3.49}$$

Since the kinetic-energy part gives rise to our unperturbed Fermi sphere, the perturbation, to lowest order, is

$$\mathcal{H}_1 = -\frac{e}{mc} \sum_i A(r_i) \cdot p_i . \tag{3.50}$$

Expanding $A(r_i)$ in a Fourier series and making use of (3.47), we obtain

$$\mathcal{H}_1 = -\frac{e}{mcV} \sum_k A(k) \cdot p_k . \tag{3.51}$$

Writing the eigenstate of the perturbed system as $|n\rangle$ and that of the unperturbed system as $|n)$, we have

$$|0\rangle = |0) + \sum_{n \neq 0} \frac{(n|\mathcal{H}_1|0)}{E_0 - E_n} |n) . \tag{3.52}$$

Therefore

$$\langle p_q \cdot \hat{n} \rangle \hat{n} = \frac{2e}{mcV} \sum_{n \neq 0} \frac{|\langle n | p_q \cdot \hat{n} | 0 \rangle|^2}{E_n - E_0} A(q) . \tag{3.53}$$

Since the states $|n\rangle$ are many-electron states and the operator p_q is a many-electron operator, it is convenient to second quantize p_q:

$$p_q \cdot \hat{n} = \sum_k \hbar k \cdot \hat{n} c^\dagger_{k+q} c_k . \tag{3.54}$$

Therefore the state $|n\rangle$ corresponds to the one in which we have taken an electron in state k inside the Fermi surface and moved it to the state $k + q$ outside the Fermi surface. We may thus write $\langle p_q \cdot \hat{n} \rangle$ as

$$\langle p_q \cdot \hat{n} \rangle \hat{n} = \frac{2e}{cV} \sum_{\substack{k < k_F \\ |k+q| > k_F}} \frac{(k \cdot n)^2}{k \cdot q + q^2/2} A(q) . \tag{3.55}$$

Converting the sum to an integral leads to the susceptibility

$$\chi_\mathrm{L}(q) = \chi_\mathrm{L} \frac{3 k_F^2}{2 q^2} \left[1 + \frac{q^2}{4 k_F^2} - \frac{k_F}{q} \left(1 - \frac{q^2}{4 k_F^2} \right)^2 \ln \left| \frac{2 k_F + q}{2 k_F - q} \right| \right], \tag{3.56}$$

where

$$\chi_\mathrm{L} = -\frac{(N/V) e^2}{4 m c^2 k_F^2} = -\frac{e^2 k_F}{12 \pi^2 m c^2}. \tag{3.57}$$

The second expression was arrived by using $N/V = k_F^3/3\pi^2$, which assumes that each point in k space has a twofold spin degeneracy. This result for $q = 0$ was first obtained by Landau, and as we shall see in the next section, it is $-\frac{1}{3}$ the Pauli susceptibility. The q dependence of $\chi_\mathrm{L}(q)$ is shown in Fig. 3.3.

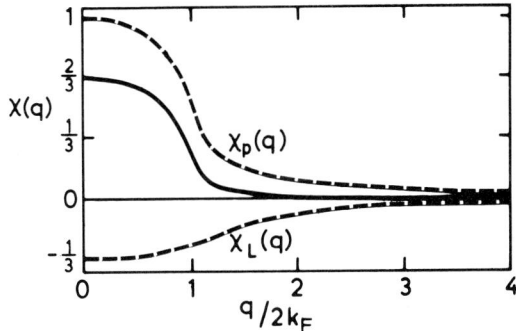

Fig. 3.3. Wave-vector dependence of the Landau and Pauli susceptibilities. The sum is shown by the solid curve

As we mentioned above, *Peierls* considered the effect of a periodic potential upon the orbital susceptibility. His result, which neglects interband effects, takes the form [3.9].

$$\chi_{\rm LP} = -\frac{e^2}{48\pi^3 \hbar^2 c^2} \int \left[\frac{\partial^2 \epsilon}{\partial k_x^2} \frac{\partial^2 \epsilon}{\partial k_y^2} - \left(\frac{\partial^2 \epsilon}{\partial k_x \partial k_y} \right)^2 \right] \frac{dS}{|\boldsymbol{V}_k \epsilon|}. \quad (3.58)$$

If $\epsilon(\boldsymbol{k}) = \hbar^2 k^2 / 2m^*$ we obtain the factor m/m^* mentioned above.

The general form of the orbital susceptibility of noninteracting Bloch electrons was established in the early 60's. However, the formulas were quite complicated. In 1970 *Fukayama* [3.10] reformulated these results in a way which allows calculations to be carried out for realistic energy bands. The susceptibility is found to be a sensitive function of the position of the Fermi level relative to critical points in the Brillouin zone. This is illustrated by Fig. 3.4 which shows the susceptibility, in units of the Landau susceptibility, for a simple model which is periodic in the x direction and free-electron-like in the other directions. The large diamagnetism of bismuth is due to such an interband effect and has the same physical interpretation of large interatomic circulating currents that give rise to large g values (see p. 68).

In dealing with electrons in bands other than s bands the possibility arises of a *paramagnetic* orbital contribution. Such a contribution was first pointed out by *Kubo* and *Obata* [3.11] and has the same origin as the Van Vleck susceptibility in the case of localized electrons. That is, it is a second-order effect involving the orbital part of the Zeeman interaction. In a metal this takes the form

$$\chi_{\rm orb} = \frac{2\mu_B^2}{(2\pi)^3} \int d\boldsymbol{k} \sum_{m' \neq m} \frac{f(\epsilon_{mk}) - f(\epsilon_{m'k})}{\epsilon_{m'k} - \epsilon_{mk}} |\langle m'\boldsymbol{k} | L | m\boldsymbol{k} \rangle|^2, \quad (3.59)$$

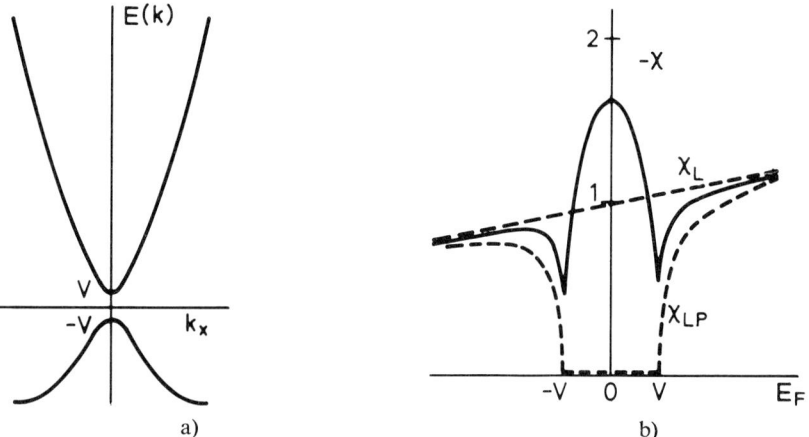

Fig. 3.4a, b. Orbital susceptibility (**b**) as a function of the Fermi level for the band structure shown in (**a**)

where ϵ_{mk} is the energy of an electron in the band m with wave vector k. The factor 2 in the numerator arises from the two spin states.

Finally, it is worth mentioning that in a superconductor the presence of a gap in the energy spectrum leads to the result that $\langle p_q \cdot \hat{n} \rangle = 0$. Therefore

$$\chi(q)_{\text{super}} = -\frac{1}{4\pi} \frac{\omega_p^2/c^2}{q^2 + \omega_p^2/c^2} \tag{3.60}$$

where $\omega_p^2 = 4\pi n e^2/m$ is the plasma frequency. The corresponding screening length is the London penetration length $(mc^2/4\pi n e^2)^{1/2}$. As $q \to 0$ this reduces to the susceptibility of a perfect diamagnet, $-1/4\pi$.

3.2.2 The de Haas–van Alphen Effect

Just as in the ionic case, when the temperature becomes very low or the field very large, such that $\mu_B H > k_B T$, nonlinear effects begin to appear. In the case of metallic diamagnetism this nonlinearity manifests itself as a periodic variation of the induced moment as a function of $1/H$. Such behavior was suspected by *Landau* [3.4], worked out in detail by *Peierls* [3.5], and observed by *de Haas* and *van Alphen* [3.12]. This *de Haas–van Alphen effect*, as it is now known, has become important in determining the nature of Fermi surfaces.

The perturbation approach used to compute the Landau susceptibility is not sufficient to describe the de Haas–van Alphen effect. Therefore, we shall compute the susceptibility from the free energy which is, in turn, obtained from an explicit knowledge of the energy levels. It should be mentioned that *Sondheimer* and *Wilson* [3.7] have shown that for a degenerate electron system the free energy is given by the inverse Laplace transform of the classical partition function $Z(\beta)$ where $\beta = 1/k_B T$ is regarded as a complex variable. The interesting feature of this approach is that $Z(\beta)$ contains a branch point at the origin which is responsible for the Landau diamagnetism and a row of poles on the imaginary axis that lead to the de Haas–van Alphen oscillations.

The Schrödinger equation for an electron in a uniform magnetic field is

$$\frac{1}{2m}\left(p - \frac{e}{c}A\right)^2 \psi = E\psi. \tag{3.61}$$

Landau noticed that it is particularly convenient to choose the gauge $A = Hx\hat{y}$. Equation (3.61) then becomes

$$\frac{\partial^2 \psi}{\partial x^2} + \left(\frac{\partial}{\partial y} - i\frac{eH}{\hbar c}x\right)^2 \psi + \frac{\partial^2 \psi}{\partial z^2} + \frac{2mE}{\hbar^2}\psi = 0. \tag{3.62}$$

If we assume a solution of the form

$$\psi = \exp[i(k_y y + k_z z)]\,\varphi(x),$$

then $\varphi(x)$ must satisfy

$$\frac{\partial^2 \varphi}{\partial x^2} + \left[\frac{2mE}{\hbar^2} - k_z^2 - \left(k_y - \frac{eH}{\hbar c}x\right)^2\right]\varphi = 0. \tag{3.63}$$

But this is just the Schrödinger equation for a one-dimensional harmonic oscillator centered at $x_0 = \hbar c k_y/eH$. Therefore we may immediately write the total eigenfunction as

$$\psi_{n,k_y,k_z} = \frac{N_n}{\sqrt{L_y L_z}} \exp[i(k_y y + k_z z)] \exp[-(\alpha^2/2)(x-x_0)^2] H_n[\alpha(x-x_0)], \tag{3.64}$$

where N_n is a normalization factor, H_n is a Hermite polynomial, and $\alpha = (m\omega_c/\hbar)^{1/2}$, ω_c being the cyclotron frequency $|e|H/mc$. This has the eigenvalue

$$E_{n,k_y,k_z} = (n + \tfrac{1}{2})\hbar\omega_c + \frac{\hbar^2 k_z^2}{2m}. \tag{3.65}$$

We could now proceed to compute the moment by means of the partition function (3.7). However, in computing the trace we must be careful to preserve the total number of electrons. This was not a problem with insulators, since each of the ions or electrons was distinguishable. In a metal, however, we have no ionic cores on which to "trap" electrons and thereby identify them. Thus we are left with a system of N truly identical particles. In such a situation it is convenient to sum over *all* possible states, irrespective of the number of particles involved, but to include in the distribution function a factor that ensures that the total number of particles is conserved. In mathematics such a factor is called a *Lagrange multiplier*. In our particular application it is referred to as the *chemical potential*. Thus we shall take as the equilibrium density matrix

$$\rho = \exp[\beta(\Omega - \mathscr{H} + \lambda\mathscr{N})], \tag{3.66}$$

where Ω is the thermodynamic potential

$$\Omega = -k_B T \ln Q \tag{3.67}$$

and Q is the grand partition function

$$Q = \text{Tr}\{\exp[-\beta(\mathscr{H} - \lambda\mathscr{N})]\}. \tag{3.68}$$

Here $\mathscr{N}$ is the total number operator and λ is the chemical potential, which is to be determined by the condition

$$\langle \mathscr{N} \rangle = N. \tag{3.69}$$

The trace in (3.68) is taken without any restriction on the total number of particles. Equation (3.66) is referred to as the *grand canonical distribution*.

Since the total number of particles is not dependent on the magnetic field, it is apparent that we may generalize (1.22) to

$$M = \frac{k_B T}{V} \frac{\partial}{\partial H} \ln \Omega \,. \tag{3.70}$$

In computing the trace indicated in (3.68) we shall take as our basis states those of the form (1.99). Since there is no restriction on the total number of particles, there will be a state with no particles, $|0_1, 0_2, ..., 0_i, ..., 0_\infty\rangle$; infinitely many with one particle, $|1_1, 0_2, ..., 0_i ...\rangle$; etc. Here i stands for the set of quantum numbers $\{n, k_y, k_z\}$. Thus Q takes the form

$$Q = 1 + \sum_i \exp[-\beta(E_i - \lambda)] + \sum_{i \neq j} \exp[-\beta(E_j - \lambda)]$$
$$\times \exp[-\beta(E_j - \lambda)] + ... \tag{3.71}$$

This may be rewritten as

$$Q = \prod_i (1 + \exp[-\beta(E_i - \lambda)]) \,. \tag{3.72}$$

Therefore the thermodynamic potential becomes

$$\Omega = -k_B T \sum_i \ln(1 + \exp[-\beta(E_i - \lambda)]) \,. \tag{3.73}$$

In the absence of a magnetic field the states would be characterized by their wave vector $\mathbf{k}$, and the sum in (3.73) would be converted to an integral over $\mathbf{k}$ space. In the presence of the field, however, we find that the states are characterized by n, k_y, and k_z. What has happened is that the original distribution of points in the $k_x k_y$ plane has condensed onto "tubes" parallel to the k_z axis labeled by the quantum number n. Since this total number of states must remain the same, we find that the number of states along a length dk_z of such a tube is $(eHV/2\pi^2 \hbar c) dk_z$. Therefore the sum over i in (3.73) may be replaced by a sum over n plus an integral along k_z [3.13],

$$\Omega = k_B T \left(\frac{eHV}{2\pi^2 \hbar c}\right) \int_{-\infty}^{\infty} dk_z \sum_{n=0}^{\infty} \ln(1 + \exp\{-\beta[E(n, k_z) - \lambda]\}) \,. \tag{3.74}$$

The resulting the thermodynamic potential leads to two contributions to the susceptibilities. The first is just the Landau result we have already obtained. The second has the oscillatory form

$$\Omega^{(2)} \simeq k_B T \sum_{s=1}^{\infty} (-1)^s \left(\frac{eH}{2\pi s c \hbar}\right)^{3/2} \frac{|A_0''|^{-1/2}}{\sinh(2\pi^2 s k_B T/\hbar \omega_c)} \cos\left(\frac{sc\hbar}{eH} A_0 \pm \frac{\pi}{4}\right). \tag{3.75}$$

In this expression A_0 is the extremal cross-sectional area of the Fermi surface at some height $k_z = k_0$. The expansion of the cross-sectional area about this point has the form

$$A = A_0 \pm \tfrac{1}{2}(k_z - k_0)^2 A_0'' + \cdots \tag{3.76}$$

which defines A_0''. Equation (3.75) implies that the magnetic susceptibility has a contribution which oscillates as a function of $1/H$. Furthermore, the period of this oscillation directly measures the extremal area of the Fermi surface perpendicular to the direction of the magnetic field.

In Fig. 3.5a we show the oscillatory component of the magnetization in iron. Since both the "up" and "down" spins have their own Fermi surfaces, and since both have a complex topology, there are numerous areas contributing to the de Haas–van Alphen effect. These are identified by frequency analyzing the magnetization spectrum as shown in Fig. 3.5b. The cross-sectional area in atomic units is related to the frequency in megagauss by $A = 2.673 \cdot 10^{-3}$ F. Combining such data with band calculations enables one to construct the Fermi surface.

3.2.3 Quantized Hall Conductance

The existence of Landau levels has dramatic consequences in two dimensions. The equation of motion for an electron in electric and magnetic fields is

$$m^* \frac{dv}{dt} = e\left(\boldsymbol{\varepsilon} + \frac{1}{c} \boldsymbol{v} \times \boldsymbol{H}\right) - m^* \frac{v}{\tau}.$$

In steady state ($dv/dt = 0$) and in high magnetic fields ($\omega_c \tau \gg 1$), this equation reduces to

$$0 = e\varepsilon_x + \frac{eH}{c} v_y.$$

The current density in the y direction is Nev_y where N is the density of electons. Consequently, the Hall conductivity is

$$\sigma_{xy} = \frac{j_y}{\varepsilon_x} = -\frac{Nec}{H}.$$

If we assume that all the electrons are describable by the conducting Landau states and that the Fermi level is somehow pinned between the n and $n+1$ Landau levels then the density N is simply the product of the number of filled Landau levels, n, times the degeneracy of each level, eH/hc, per unit area. Thus the Hall conductivity becomes

$$-\sigma_{xy} = n \frac{e^2}{h},$$

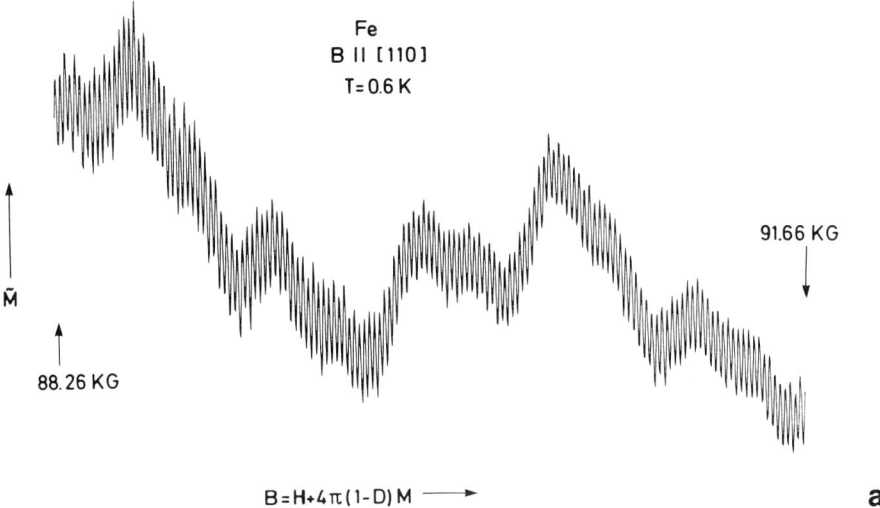
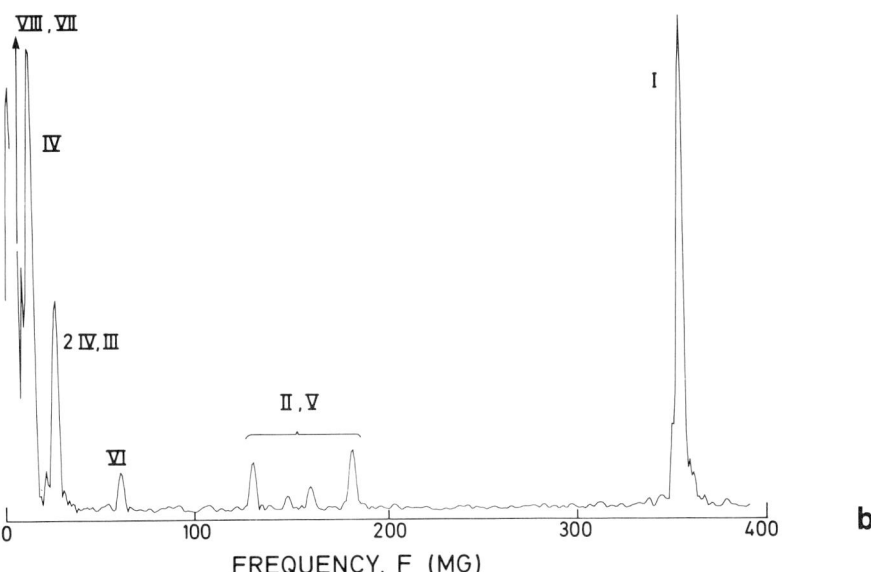

Fig. 3.5a,b. Oscillatory component of the diamagnetic moment in iron (**a**) with its frequency spectrum (**b**) (courtesy of G. G. Lonzarich)

that is, it is quantized in multiples of e^2/h. Furthermore, as long as the Fermi level lies between the Landau levels, the density of states is zero and $\sigma_{xx} = 0$. The fact that $\sigma_{xx} = 0$ while $\sigma_{xy} \neq 0$ means that the diagonal resistivity

$$\rho_{xx} = \frac{\sigma_{xx}}{\sigma_{xx}^2 + \sigma_{yy}^2}$$

also vanishes and the two-dimensional system is in a zero-resistance state! This two-dimensional behavior has, in fact, been observed by v. *Klitzing* et al. [3.14] in the inversion layer of a silicon MOSFET (metal-oxide-semiconductor field-effect transistor). As one changes the gate voltage on such a device, the Fermi level is swept through the Landau levels. Figure 3.6 shows the longitudinal voltage V associated with a constant current as well as the Hall voltage V_{Hall} as functions of the gate voltage. The plateaus in the Hall voltage correspond to a Hall resistance $R_{\text{H}} = h/e^2n$ to an accuracy of better than one part in 10^5. *Tsui* and *Gossard* [3.15] have also observed this effect in gallium arsenide heterojunctions as a function of an applied magnetic field.

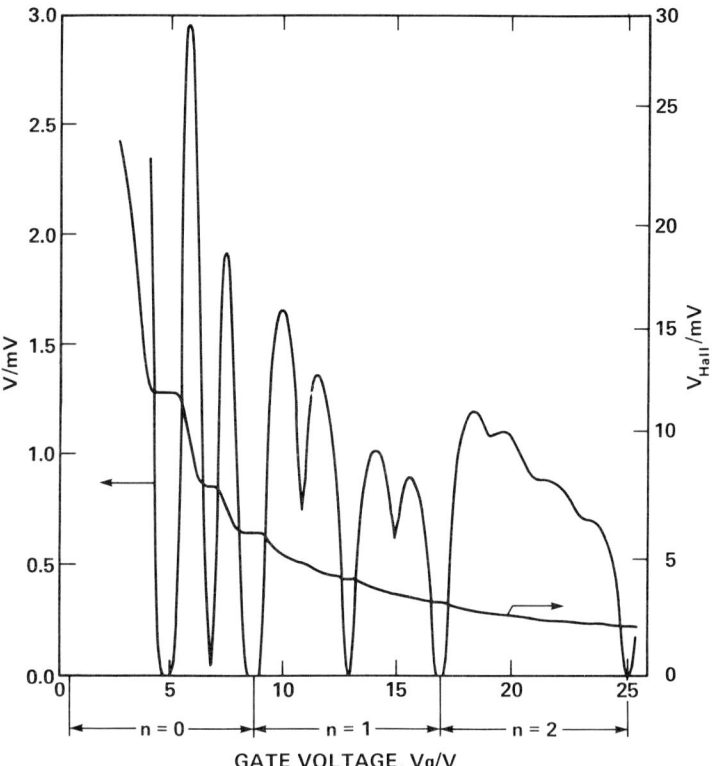

Fig. 3.6. Longitudinal and Hall voltages of a silicon MOSFET with a source drain current of 1 μA as a function of gate voltage in a magnetic field of 18 T. The index n refers to the Landau level occupied. Structure within a given level is due to spin and band structure degeneracies

What is puzzling about these observations is the extraordinary accuracy with which the plateaus in σ_{xy} give e^2/h. The fact that the plateaus exist suggests that there are localized states between the Landau levels through which the Fermi level can be moved, as illustrated in Fig. 3.7. Since such states do not carry cur-

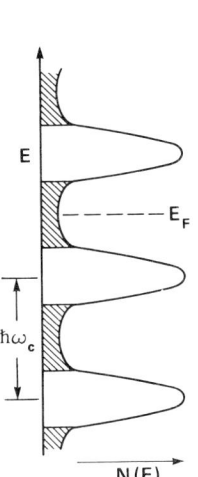
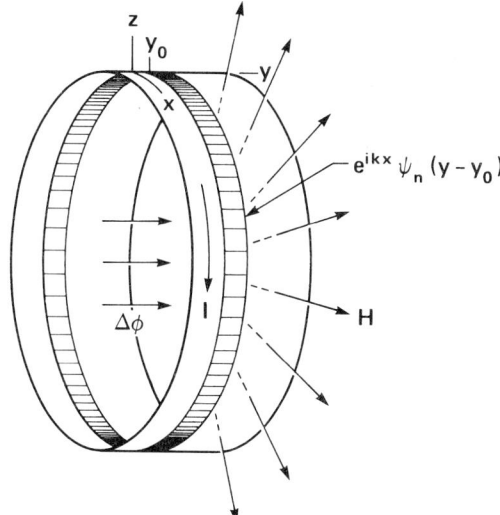

Fig. 3.7. Density of states showing broadened Landau levels and localized states (shaded)

Fig. 3.8. Electron wavefunction associated with a metal loop in a radial magnetic field

rent, we might have expected σ_{xy} to be reduced from its ideal value. *Prange* [3.16] as well as *Tsui* and *Allen* [3.17], however, have shown that even if localized states exist, the remaining nonlocalized states carry an extra Hall current which exactly compensates for that not carried by the localized states. Then there are also interaction effects and edge effects to worry about. *Laughlin* [3.18] has suggested that the quantization is more fundamental than that suggested above, being the consequence of gauge invariance. Consider a conducting strip connected in a ring, as shown in Fig. 3.8. Let us assume there is a magnetic field normal to the strip. The electron wave functions are then plane waves in the x direction, with a harmonic oscillator envelope in the y direction centered at y_0. If we now change the flux $\Delta\varphi$ which threads the ring, this corresponds to a change in the vector potential in the direction around the ring; i.e., $A \to A + \Delta A = A + (\Delta\varphi/L)\hat{x}$. This gauge change shifts the center of the wave function by an amount $\Delta A/H_0$. In particular, if the flux change is a quantum of flux, hc/e, then the shift moves each state precisely to an adjacent center. The net effect is to move one state per Landau level from one edge to the other. Assuming there is a voltage V across the strip, this corresponds to a total increase in energy of neV, where n is the number of filled Landau levels. This change in energy ΔU is related to the current flowing around the strip by the general relation

$$I = \frac{c\Delta U}{\Delta\varphi} = \frac{nec\,V}{(hc/e)} = \frac{ne^2V}{h}.$$

Thus the Hall conductance I/V has the quantized form found above. Any localized states present will have their phase shifted by the gauge transformation

3.2.4 Pauli Paramagnetism

Let us now turn our attention to the spin response of a free-electron system. As mentioned at the beginning of this chapter, we can construct the response to an arbitrary static field if we know the response to a single Fourier component. Therefore, let us apply an external field of the form $H \cos(\boldsymbol{q} \cdot \boldsymbol{r})$. The interaction with the electron spins is

$$\mathcal{H} = \sum_i g\mu_B \boldsymbol{s}_i \cdot \boldsymbol{H} \cos(\boldsymbol{q} \cdot \boldsymbol{r}_i) \,. \tag{3.77}$$

According to (1.22), the moment induced by the field may be obtained by computing the change in the energy of the system that is proportional to the field. This we shall do by perturbation theory. Notice that if the quantization axis of our system is chosen to lie along the direction of $\boldsymbol{H}$, then when $\boldsymbol{q} = 0$, (3.77) has first-order diagonal matrix elements. Therefore a perturbation calculation would have to distinguish between $\boldsymbol{q} = 0$ and $\boldsymbol{q} \neq 0$. In order to avoid this difficulty we shall assume that the field lies in the x direction, $\boldsymbol{H} = H\hat{\boldsymbol{x}}$. Then

$$\mathcal{H} = \frac{g\mu_B H}{4} \sum_i (s_i^+ + s_i^-)[\exp(i\boldsymbol{q}\cdot\boldsymbol{r}_i) + \exp(-i\boldsymbol{q}\cdot\boldsymbol{r}_i)] \,. \tag{3.78}$$

Second quantizing this interaction by the prescription given in Chap. 1 gives

$$\mathcal{H} = \frac{g\mu_B H}{4} \sum_k (a^\dagger_{k+q,\uparrow} a_{k\downarrow} + a^\dagger_{k+q,\downarrow} a_{k\uparrow} + a^\dagger_{k-q,\uparrow} a_{k,\downarrow} + a^\dagger_{k-q,\downarrow} a_{k\uparrow}) \,. \tag{3.79}$$

We see immediately that there are no first-order diagonal matrix elements. The second-order correction to the energy is

$$\Delta E = \left(\frac{g\mu_B H}{4}\right)^2 \sum_{k,\sigma} \frac{|\langle a^\dagger_{k+q,-\sigma} a_{k\sigma}\rangle|^2}{\epsilon_{k\sigma} - \epsilon_{k+q,-\sigma}} + \text{(terms involving} -\boldsymbol{q}) \,. \tag{3.80}$$

The sum over $\boldsymbol{k}$ is restricted to those values of $\boldsymbol{k}$ inside the Fermi sphere such that $\boldsymbol{k} + \boldsymbol{q}$ lies outside the Fermi sphere. Since the unperturbed energies are independent of spin, (3.80) reduces to

$$\Delta E = -\frac{g^2\mu_B^2 H^2}{8} \sum_k \left[\frac{f_k(1-f_{k+q})}{\epsilon_{k+q} - \epsilon_k} + \frac{f_k(1-f_{k-q})}{\epsilon_{k-q} - \epsilon_k}\right], \tag{3.81}$$

when the sum now runs over all $\boldsymbol{k}$ and f_k is the Fermi function

$$f_k = \frac{1}{\exp[(\epsilon_k - \epsilon_F)/k_B T] + 1} \,. \tag{3.82}$$

Since the sum does run over all k, we may write $k \to k+q$ in the second term. The two terms then combine to give

$$\sum_k \frac{f_k - f_{k+q}}{\epsilon_{k+q} - \epsilon_k}.$$

Since

$$\Delta E = -\tfrac{1}{2} \int dr\, M(r) \cdot H(r) = -\frac{H}{2} \int dr\, M(r) \cos(q \cdot r), \tag{3.83}$$

the derivative of $-\Delta E$ with respect to H gives $M(q)$ directly. Therefore, with $H(q) = HV/2$, the susceptibility becomes

$$\chi(q) = \frac{g^2 \mu_B^2}{V} \sum_k \frac{f_k - f_{k+q}}{\epsilon_{k+q} - \epsilon_k}. \tag{3.84}$$

The sum in (3.84) may be evaluated by converting it into an integral. At $T = 0\mathrm{K}$ we obtain

$$\chi(q) = \frac{3g^2 \mu_B^2 (N/V)}{8 \epsilon_F} F\!\left(\frac{q}{2k_F}\right), \tag{3.85}$$

where

$$F\!\left(\frac{q}{2k_F}\right) = \frac{1}{2} + \frac{k_F}{2q}\left(1 - \frac{q^2}{4k_F^2}\right) \ln\left|\frac{2k_F + q}{2k_F - q}\right| \tag{3.86}$$

and N is the total number of electrons in the system. Notice that

$$\lim_{q \to 0} F\!\left(\frac{q}{2k_F}\right) = 1. \tag{3.87}$$

With $g = 2$ and $\mu_B = e\hbar/2mc$, (3.85) may be rewritten as

$$\boxed{\chi(q) = \chi_{\text{Pauli}}\, F\!\left(\frac{q}{2k_F}\right)} \tag{3.88}$$

where

$$\chi_{\text{Pauli}} = \frac{3(N/V)e^2}{4mc^2 k_F^2}. \tag{3.89}$$

The *Pauli susceptibility*, given by (3.89), describes the spin response of an electron system to *a uniform field*. To see this, note that for a uniform field $H(q) = H\Delta(q)$. Therefore

$$M(r) = \sum_{q} e^{iq \cdot r} \chi(q) H(q) = \chi_{\text{Pauli}} H. \tag{3.90}$$

The wave-vector dependence of $\chi(q)$ is sketched in Fig. 3.3. Notice that this function has an infinite derivative when $q = 2k_F$. The susceptibility (3.88) was derived under the assumption that the temperature was small in comparison with the Fermi temperature. Since Fermi temperatures are typically of the order of 10,000 K, this is generally a good approximation. However, there are situations, as in metals like palladium and in nondegenerate semiconductors, where the effective Fermi temperature is much lower. It is then possible to reach temperatures where the electrons behave like Boltzman particles and exhibit a Curie-like susceptibility.

It is also interesting to consider how the spin susceptibility is changed if the metal becomes superconducting. In a superconductor electrons with opposite spin are correlated over a distance ξ_0 called the coherence length. This suppresses the magnetic response of such pairs at wave vectors below ξ_0^{-1}. When this effect is combined with the decreasing response beyond $2k_F$ we obtain a maximum in the susceptibility at $q_0 = (2\pi k_F^2/\xi_0)^{1/3}$ as shown in Fig. 3.9.

As with the orbital susceptibility, the periodicity of the lattice also modifies the spin susceptibility. However, in the limit of weak spin–orbit coupling one

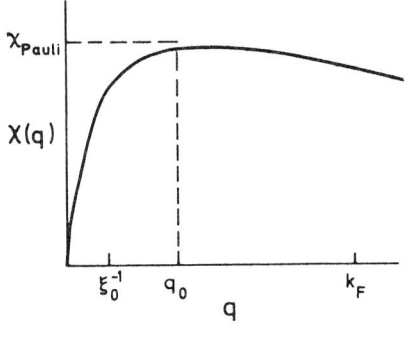

Fig. 3.9. Wave-vector dependence of the spin susceptibility in a superconductor

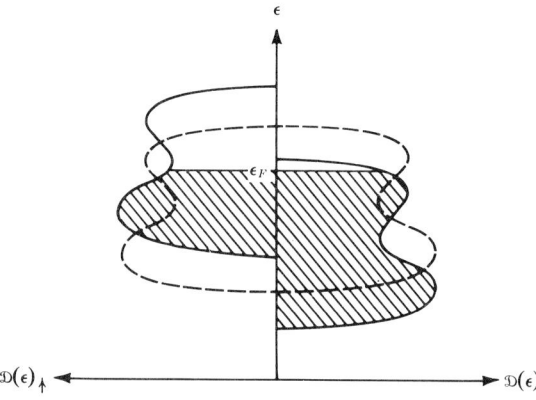

Fig. 3.10. Relative splitting of a d band in the presence of an external field

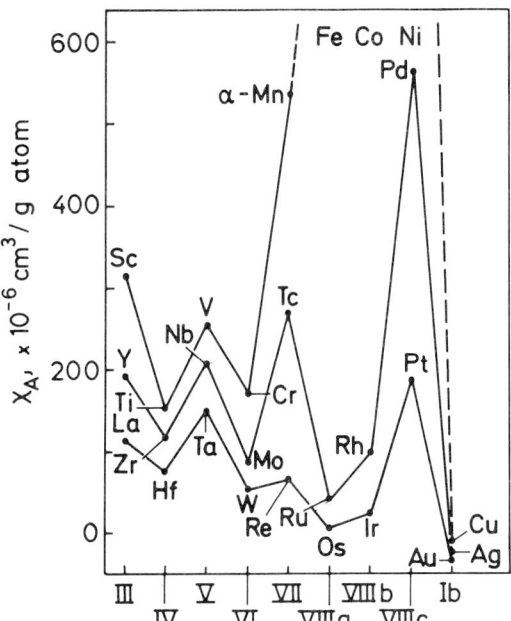

Fig. 3.11. The susceptibility of the transition metals at room temperature [3.19]

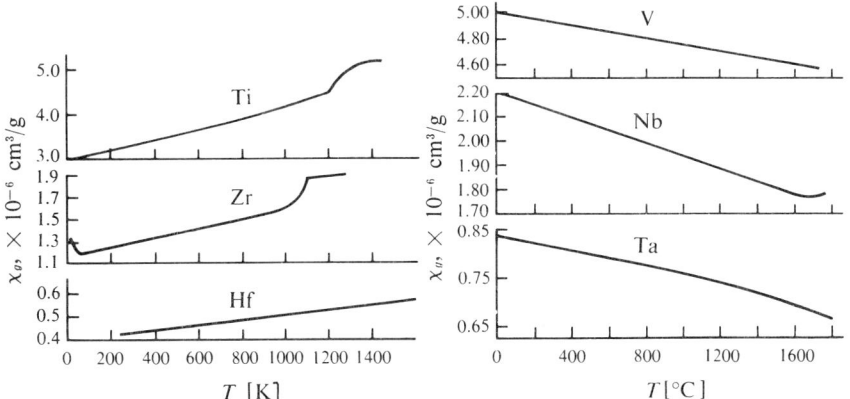

Fig. 3.12. The temperature dependence of the specific susceptibility of selected transition metals [3.20]. The structure in the data for Ti and Zr near 1200 K is associated with structural phase transitions

can talk about spin-up and spin-down bands and the Pauli susceptibility takes a particularily simple and useful form. Let us consider a metal whose band structure gives rise to a density of states, i.e., number of states per unit energy per unit volume, as illustrated in Fig. 3.10. To lowest order, the effect of the applied is to shift the relative energy distributions of the up and down spins. Since the electrons do, in fact, interact with one another, the up- and down-spin distributions will reach equilibrium in such a manner that their Fermi levels are the same.

Therefore, as we see from Fig. 3.10 there will be more *spins* pointing anti-parallel to the field than parallel to it. Since the gyromagnetic ratio of an electron is negative, this leads to a net *moment* in the direction of the field. To find this moment we need only compute the difference in the number of up- and down-spin electrons,

$$M = \tfrac{1}{2} g \mu_B \int_{-\infty}^{\infty} [\mathscr{D}(\epsilon) f(\epsilon - \tfrac{1}{2} g \mu_B H) - \mathscr{D}(\epsilon) f(\epsilon + \tfrac{1}{2} g \mu_B H)] d\epsilon. \qquad (3.91)$$

We expand the Fermi function about $H = 0$ to yield

$$M \simeq \tfrac{1}{2} g^2 \mu_B^2 H \int_0^{\infty} \left(-\frac{\partial f}{\partial \epsilon}\right) \mathscr{D}(\epsilon) d\epsilon. \qquad (3.92)$$

At low temperatures the Fermi function is essentially a step function at the Fermi level. Therefore its derivative is a delta function. Thus (3.92) leads to the susceptibility

$$\boxed{\chi_{\text{Pauli}} = \tfrac{1}{2} g^2 \mu_B^2 \mathscr{D}(\epsilon_F).} \qquad (3.93)$$

For a free-electron system

$$\mathscr{D}(\epsilon_F) = \frac{m k_F}{2 \pi^2 \hbar^2}, \qquad (3.94)$$

and (3.93) reduces to our previous result (3.89).

From (3.93) we would expect the Pauli susceptibility to be roughly ten times larger when the Fermi surface lies within a *d* band than when it lies within an *s* band since *d* bands are generally narrower with a higher density of states. We might also expect this *d*-band contribution to exhibit a temperature dependence. The reason for this is that at a temperature T the susceptibility essentially involves an average of $\mathscr{D}(\epsilon_F)$ over a region of the order of $k_B T$. Therefore, if the width of the structure in the *d*-band density of states is of this order, it will be reflected in the susceptibility. Interactions between the electrons can also lead to a temperature dependence.

Susceptibilities for the various transition metals are shown in Fig. 3.11 First, note that the susceptibility of those metals with unfilled *d* bands is much larger than for the *s*-band metals Cu, Ag, and Au. Second, note the variation in the susceptibility across the periodic table. This might be associated with variations in *d*-band density of states. This rigid-band explanation is consistent with the temperature dependence of these susceptibilities. As shown in Fig. 3.12, the susceptibility of metals such as Ti, Zr, and Hf, which have a relative minimum in Fig. 3.11, increases with temperature, while for those with a maximum it decreases. Those metals with d^1 and d^9 configurations are further complicated by interaction effects which we shall discuss in the next chapter.

In Mn, Fe, Co, and Ni interactions among the electrons are sufficient to produce an ordered magnetic state. Above their ordering temperature these metals exhibit a relatively strong temperature dependence as shown in Figs. 3.13, 14. This behavior is indicative of some sort of local-moment formation. It is not simply Curie–Weiss, however, for $d(\chi^{-1})/dT$ is not temperature independent.

If band effects may be represented by an effective mass, then

$$\chi_{\text{Pauli}} = g^2 \mu_B^2 \frac{m^* k_F}{4\pi^2 \hbar^2}. \tag{3.95}$$

Notice that the mass entering the Bohr magneton is unaffected. However, the Landau–Peierls susceptibility (3.58) becomes

$$\chi_{\text{LP}} = -\frac{e^2 k_F}{12\pi^2 m^* c^2} = -\frac{1}{3}\left(\frac{m}{m^*}\right)^2 \frac{e^2 \hbar^2 m^* k_F}{m^2 c^2 4\pi^2 \hbar^2}. \tag{3.96}$$

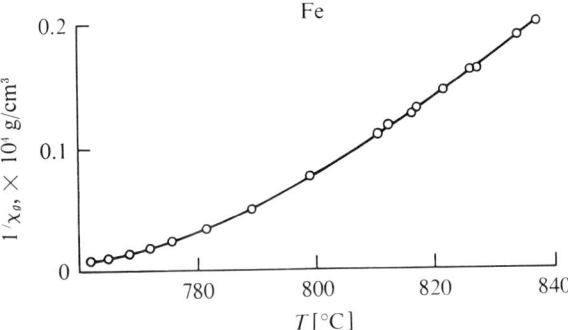

Fig. 3.13. Temperature dependence of the reciprocal susceptibility of Fe [3.19]

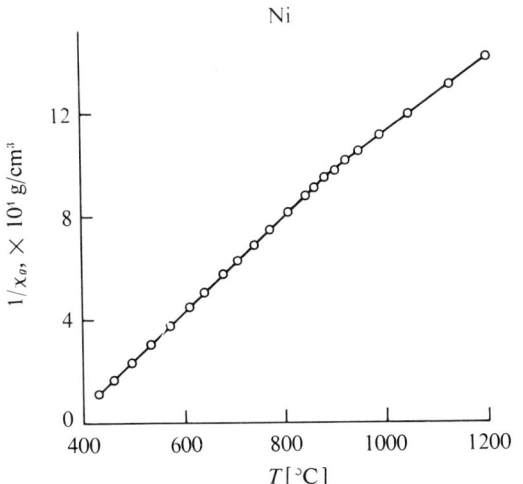

3.14. Temperature dependence of the reciprocal susceptibility of Ni [3.19]

Therefore

$$\chi_{LP} = -\frac{1}{3}\left(\frac{m}{m^*}\right)^2 \chi_{Pauli} . \tag{3.97}$$

Collecting all our results, we may express the total susceptibility of a nonmagnetic metal as

$$\chi(T) = \chi_{core} + \chi_{LP} + \chi_{orb} + \chi(T)_{Pauli} . \tag{3.98}$$

The temperature dependence arises primarily from the d-band contribution to the Pauli susceptibility.

3.3 Measurement of the Susceptibility

At this point it is appropriate to say a little about measuring the susceptibility. A broad class of techniques use the fact that when a sample with a susceptibility χ_g per gram is placed in a magnetic field gradient it experiences a force. In particular, if the field is in the x direction with a gradient in the z direction, this force is

$$\Delta F_z = \left(H_x \frac{dH_x}{dz}\right) \chi_g \cdot \omega t(gr) . \tag{3.99}$$

If the sample is mounted at the end of a pendulum and suspended at a right angle to the field gradient, there will be a torque tending to displace the pendulum. This is the *pendulum galvanometer* [3.21]. The torque due to the sample can be balanced by an opposing torque due to current flowing in a solenoid surrounding the sample. The measurement of the susceptibility is thus reduced to measuring the current flow at zero displacement and consequently has a large dynamic range well suited for ferromagnets and strong paramagnets.

For strongly magnetic materials Foner's *vibrating-sample magnetometer* is particularly convenient (Fig. 3.15a) [3.22]. In this arrangement the sample is placed at the end of a rod which vibrates up and down inside a pickup coil. If the whole assembly is placed between the pole faces of a magnet, a moment is induced in the sample which, by virtue of its vibration, induces a signal in the pickup coil that is proportional to the magnetization of the sample.

For weakly magnetic materials it is convenient to measure the force directly with an electrobalance. The most popular balance technique today is the *Faraday balance* (Fig. 3.15b). For this method the pole pieces of the magnet are carefully machined to produce a fairly large region over which $H_x dH_x/dz$ is constant. Samples small with respect to this region are suspended in the region and the force measured. The quantity $H_x dH_x/dz$ is determined by calibrating the system

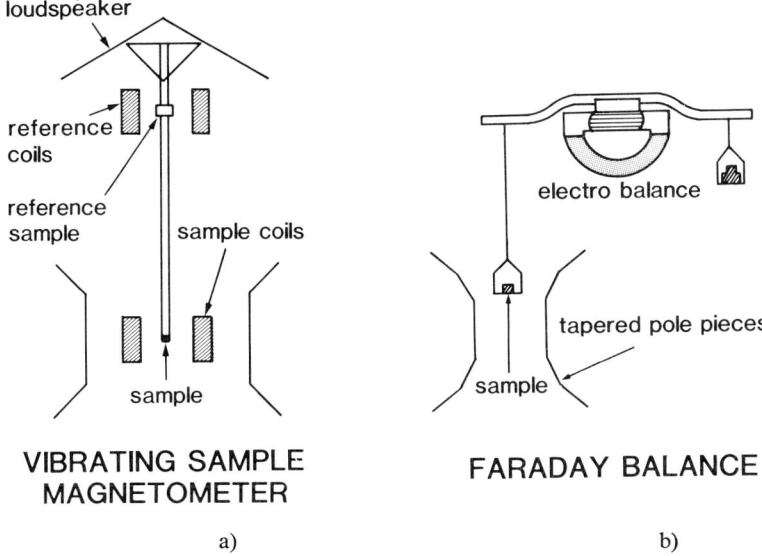

Fig. 3.15a, b. Simplified illustrations of (a) a vibrating-sample magnetometer and (b) a Faraday susceptibility balance

with a standard such as platinum. A typical microbalance can resolve 5 micrograms. In a related technique, known as the *Gouy method*, one employs a long uniform sample, frequently a liquid, which extends over the entire field region. Thus the force is the integral of (3.99).

Another way of measuring the susceptibility of small or magnetically weak samples is by a *mutual-inductance technique* [3.23]. The sample is mounted inside a pair of secondary coils on top of which is wound a primary coil. Measurement of the mutual inductance of this coil system with and without the sample yields an absolute determination of the susceptibility of the sample.

A very different method for obtaining the static susceptibility makes use of the Kramers–Kronig relation

$$\chi(0) = \frac{2}{\pi} \int_0^\infty \frac{\chi''(\omega) d\omega}{\omega}, \tag{3.100}$$

where $\chi(0)$ is measured in relation to $\chi(\infty)$. If the imaginary part of the susceptibility is strongly peaked around a particular frequency ω_0, then

$$\chi(0) \simeq \frac{2}{\pi\omega_0} \int_0^\infty \chi''(\omega) d\omega. \tag{3.101}$$

If, as in most cases, the experiment entails varying the magnetic field at a fixed frequency, this becomes

$$\chi(0) = \frac{g\mu_B}{\pi\hbar\omega} \int_0^\infty \chi''(H) dH . \tag{3.102}$$

Therefore the static susceptibility is simply given by the area under the magnetic resonance. Since this entails an absolute intensity measurement, it is rather difficult. However, *Schumacher* and *Slichter* [3.24], in determining the spin susceptibility of lithium and sodium, neatly solved this problem by comparing spin resonances with the nuclear resonance in the same sample. The nuclear susceptibility χ_n may be computed from the Langevin susceptibility (3.27). The spin susceptibility is determined by measuring the area under the conduction-electron spin resonance A_e and the area under the nuclear resonance A_n and using the relation

$$\chi_{el} = \chi_n \frac{\mu_B A_{el}}{\mu_n A_n} . \tag{3.103}$$

There is another technique involving nuclear magnetic resonance which is very helpful in determining the various contributions to the total susceptibility in a metal. As we saw in Sect. 2.2, s-like conduction electrons produce a contact hyperfine field at the nuclei of a metal. Thus, if the conduction electrons are polarized by an external field, they will, in turn, shift the nuclear magnetic resonance frequency from what it would be in a diamagnetic environment. This is known as the *Knight shift*. Since the ability of the conduction electrons to be polarized is determined by their Pauli susceptibility, the resulting Knight shift K will be proportional to this susceptibility. In particular,

$$K = \left(\frac{8\pi}{3}\right) \chi_P \Omega \langle |\psi_s(0)|^2 \rangle_F , \tag{3.104}$$

where χ_P is the Pauli susceptibility per unit volume, Ω is the volume of the unit cell, and $\langle |\psi_s(0)^2| \rangle_F$ is the average value of the s-electron probability density at the nucleus for electrons near the Fermi surface.

As was also pointed out in Sect. 2.2, d-like electrons do not contribute to the contact hyperfine field. However, their presence can lead to a slight polarization of the otherwise paired core s electrons. This *core polarization* depends on the d-band polarization and will therefore contribute a term to the Knight shift that is proportional to the d-band susceptibility. It turns out that the core polarization generally leads to a negative Knight shift. Further more, since χ_p is temperature dependent, the resulting Knight shift will also be temperature dependent. Thus the Knight shift, together with direct susceptibility measurements, enables us to unravel the various contributions to the total susceptibility given by (3.98). A beautiful example of this is the work of *Clogston* et al. on metallic platinum [3.25].

In some cases, we are interested in the magnetic response of a system at very high fields, such as in de Haas–van Alphen studies or studies of magnetic impuri-

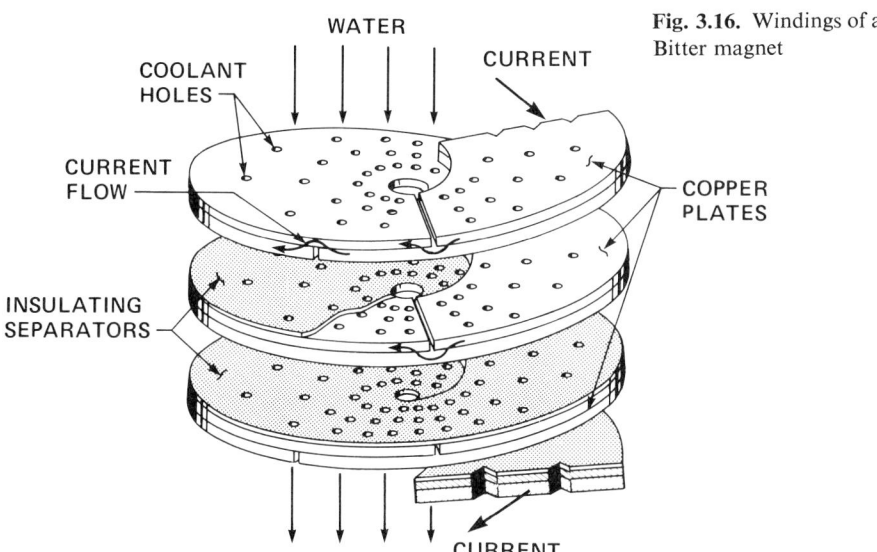

Fig. 3.16. Windings of a Bitter magnet

ty levels in semiconductors. The production of such high fields has become a specialty practiced only within the world's few high-field laboratories, such as the U.S. National Magnet Laboratory at MIT or the joint West German–French high-field laboratory in Grenoble.

The classical approach to generating large continuous magnetic fields is to use water-cooled coils. In the late 1930's, Francis Bitter pioneered the design of such high-field magnets. A typical Bitter magnet is illustrated in Fig. 3.16. In order to produce 15 T in a 5-cm bore, five megawatts of power are required. If the coil is not stress limited, the central field is proportional to the square root of the power dissipated. As a result, the practical limit for such resistive magnets is about 20 T. If one tries to increase the field by reducing the inner diameter, the Lorentz force on the conductor may exceed its tensile strength.

The limitation on the power can be overcome by using superconducting magnets. However, the stress problem remains, and one must insure that the design is stable against the superconductor going normal locally. One must also not exceed the upper critical field of the superconductor. The currently available technical superconductors, Nb–Ti and Nb_3Sn, for example, have upper critical fields of 15 and 23 T, respectively.

To achieve fields in excess of 20 T, hybrid systems making use of both approaches are employed. The resistive magnet is placed in the center where the field is highest and the superconducting section on the outside where the field is low. Although the volume of the superconducting magnet is large, it does not carry the power penalty of a resistive magnet.

4. The Static Susceptibility of Interacting Systems

In this chapter we shall turn our attention to the static response of systems in which interactions among the constituents, either localized moments or itinerant electrons, may lead to long-range order. A thorough treatment of such interactions and the resulting phase transition is an extremely complicated subject. Although very sophisticated techniques have been developed to treat these problems, there are still gaps in our understanding. However, except for the region very close to the critical point, an effective field theory works quite well. Therefore we shall employ this approximation throughout most of this chapter.

4.1 Localized Moments

In Sect. 2.2 we found that the Coulomb interaction between the valence electrons on different ions could be expressed as an effective interaction between the individual electron spins. We also found that under certain conditions this interaction could be expressed in terms of the total ionic spin or, in some cases, the effective spin of the ground-state multiplet. Let us consider a system whose interactions may, in fact, be described by the Heisenberg exchange Hamiltonian

$$\mathscr{H}_{ex} = -\sum_{i}\sum_{j\neq i} J_{ij} \mathbf{S}_i \cdot \mathbf{S}_j \tag{4.1}$$

where J_{ij} is a function of the relative separation $\mathbf{R}_i - \mathbf{R}_j$.

Although many systems do not satisfy our conditions for writing the interaction in this form, it is often found that such systems may, nevertheless, be described surprisingly well by it. However, as we shall see, the treatment of this interaction entails certain approximations. Hence our success may result more from our methods than from the interaction with which we start. Since the exchange interaction is often large in comparison with other interactions, such as the magnetic-dipole-dipole interaction, the gross features of such a system may be described by (4.1) alone.

Let us now see how such a system responds to a static applied field. Since (4.1) is rotationally invariant, the direction of this applied field may be taken to define the z axis. Just as in the preceding chapter, we apply a field of the form

$$\mathbf{H}(\mathbf{r}) = H\hat{\mathbf{z}} \cos(\mathbf{q}\cdot\mathbf{r}) \tag{4.2}$$

where q corresponds to some point within the first Brillouin zone of the lattice. If the g value associated with the effective spin is g, then the total Hamiltonian becomes

$$\mathcal{H} = -\sum_i \sum_{j \neq i} J_{ij} \mathbf{S}_i \cdot \mathbf{S}_j + g\mu_B H \sum_i S_i^z \cos(\mathbf{q} \cdot \mathbf{R}_i). \tag{4.3}$$

As in most cases involving coupling between lattice sites, it is convenient to express the Hamiltonian in Fourier components. Since

$$\mathbf{S}(\mathbf{q}') = \int d\mathbf{r}\, e^{-i\mathbf{q}' \cdot \mathbf{r}} \mathbf{S}(\mathbf{r}) = \sum_i \int d\mathbf{r}\, e^{-i\mathbf{q}' \cdot \mathbf{r}} \mathbf{S}_i \delta(\mathbf{r} - \mathbf{R}_i) = \sum_i e^{-i\mathbf{q}' \cdot \mathbf{R}_i} \mathbf{S}_i, \tag{4.4}$$

we may write (4.3) as

$$\mathcal{H} = -\sum_{q'} J(-\mathbf{q}') \mathbf{S}(\mathbf{q}') \cdot \mathbf{S}(-\mathbf{q}') + \tfrac{1}{2} g\mu_B H[S_z(\mathbf{q}) + S_z(-\mathbf{q})] \tag{4.5}$$

where

$$J(-\mathbf{q}') \equiv \frac{1}{N} \sum_{i \neq j} J(\mathbf{R}_i - \mathbf{R}_j) e^{i\mathbf{q}' \cdot (\mathbf{R}_i - \mathbf{R}_j)}. \tag{4.6}$$

Notice that if the crystal has inversion symmetry, $J(-\mathbf{q}) = J(\mathbf{q})$. The analysis of the Hamiltonian (4.5) is very difficult and has been the subject of much work. The simplest approach is to consider one of the spins in (4.1) and replace its interaction with the other spins by an effective field [4.1]. This is referred to as the *mean-field approximation*. The concept appears to have first been applied by van der Waals in 1873 to develop the liquid–gas equation of state. Its application to ferromagnetism was made by P. Weiss in 1907. The same approximation was employed by Bragg and Williams in 1934 to describe order–disorder transitions such as occur in β-brass when the copper and zinc atoms, through diffusion, order themselves with the copper at the center and the zinc at the corners of the body-centered-cubic lattice. The physics of such order–disorder transitions is identical to that of an Ising magnet. In 1937 Landau developed a generalized mean-field theory applicable to all second-order phase transitions. We shall return to Landau's theory later in this chapter.

Since we have been emphasizing the usefullness of working with Fourier components it is appropriate to apply a variation of the mean-field approximation in which each of the Fourier components in (4.5) is assumed to be independent. This is referred to as the *random-phase approximation*. In this approximation the exchange interaction becomes

$$\mathcal{H}_{\text{ex}} = -\sum_{q'} J(-\mathbf{q}') \mathbf{S}(\mathbf{q}') \cdot \langle \mathbf{S}(-\mathbf{q}') \rangle. \tag{4.7}$$

4.1.1 High Temperatures

If the temperature is higher than that at which long-range order sets in, we may assume that the only components with nonzero average values are those driven by the applied field. Thus $\langle S(-q')\rangle$ is replaced by $\langle S_z(-q')\rangle \hat{z} \Delta(q-q')$. The total Hamiltonian in the random-phase approximation becomes

$$\mathcal{H}_{RPA} = -[J(q)\langle S_z(-q)\rangle S_z(q) + J(q)\langle S_z(q)\rangle S_z(-q)]$$
$$+ \tfrac{1}{2}g\mu_B H[S_z(q) + S_z(-q)], \qquad (4.8)$$

which defines the total effective field seen by the qth component as

$$H(-q)_{\text{eff}} = \frac{J(-q)V}{g\mu_B}\langle S_z(-q)\rangle \hat{z} + \tfrac{1}{2}HV\hat{z}. \qquad (4.9)$$

In discussing noninteracting systems we found that the response of the qth component of the spin at high temperatures was determined by

$$\chi_0(q) = \frac{C}{T} \equiv \chi_0 \qquad (4.10)$$

where

$$C = \frac{N}{V}g^2\mu_B^2 \frac{S(S+1)}{3k_B}.$$

Therefore, employing our definition of the susceptibility, we have

$$M_z(q) = -g\mu_B\langle S_z(q)\rangle = \chi_0(q)H_z(q)_{\text{eff}} = -\frac{J(q)\chi_0(q)V}{g\mu_B}\langle S_z(q)\rangle$$
$$+ \tfrac{1}{2}HV\chi_0(q). \qquad (4.11)$$

Solving for $\langle S_z(q)\rangle$ and using (4.10) gives us the susceptibility of the interacting system at high temperatures,

$$\chi(q) = \frac{\chi_0}{1 - \chi_0 \dfrac{J(q)V}{g^2\mu_B^2}} = \frac{C/T_c}{(T-T_c)/T_c + [1 - J(q)/J(Q)]}. \qquad (4.12)$$

We notice immeciately that as the temperature decreases a divergence appears at a critical temperature

$$T_c = \frac{J(Q)CV}{g^2\mu_B^2} \qquad (4.13)$$

where Q is the wave vector for which $J(Q)$ is a maximum. The fact that the susceptibility diverges for this particular wave vector means that the component of the magnetization with that wave vector would remain nonzero even if the probing field were to go to 0. Thus we have a spontaneous magnetization. The wave vector Q for which this transition occurs depends on the nature of $J(q)$. Let us consider four possible cases.

Ferromagnetism. Suppose that the exchange J_{ij} is *positive* and extends only to nearest neighbors. Then

$$J(q) = \frac{J}{N} \sum_{\delta} \exp(-i q \cdot \delta) \tag{4.14}$$

where δ is a vector to the nearest neighbor. In particular, if we have a simple cubic lattice with a lattice parameter a, then

$$J(q) = \frac{2J}{N} (\cos q_x a + \cos q_y a + \cos q_z a) . \tag{4.15}$$

Figure 4.1 shows $1/\chi(q)$ plotted as a function of temperature for various values of q. We see immediately that as the temperature decreases the first divergence appears at $Q = 0$.

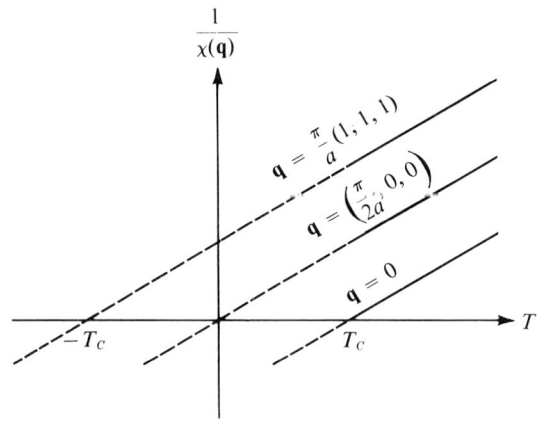

Fig. 4.1. Temperature dependence of the inverse susceptibility $1/\chi(q)$ of a ferromagnet for various values of q

This instability results in a uniformly magnetized system, a *ferromagnet*. The temperature associated with this transition is the Curie temperature, which, for our simple cubic example, is

$$T_c = \frac{J(0)CV}{g^2 \mu_B^2} = \frac{6JS(S+1)}{3k_B} . \tag{4.16}$$

The uniform susceptibility is

$$\boxed{\chi(0)_{\text{ferro}} = \frac{C}{T - T_{\text{C}}}} \,. \tag{4.17}$$

In fact, ferromagnetism is not compatible with cubic symmetry. This has to do with time reversal. In the ferromagnetically ordered state time reversal by itself is not a symmetry operation, but only occurs in conjunction with some other symmetry operator. Consider, for example, a body-centered-cubic (bcc) system. In the paramagnetic state the symmetry operations which transform this structure into itself constitute the octahedral group denoted O_h or m3m. If, in the ordered state, the spins at each site all point in the z direction then those operations of O_h which involve rotations about the x or y axes must be combined with time reversal to preserve the direction of the spins. The resulting allowed operations constitute the so-called *magnetic* point group, which, in this case, is denoted 4/mmm. The important feature of this group is that it is *tetragonal*. When a material develops a spontaneous magnetization, it also generally develops a strain proportional to the magnetization. This magnetostrictive effect, as it is known, reflects the symmetry of the system. Thus, iron, which is bcc above its Curie temperature, develops a tetragonal distortion in the ferromagnetic state. However, since this distortion is very small one often finds ferromagnetic iron characterized as bcc.

Another general result which is obscured by the above mean field approximation has to do with the dimensionality of the system. *Landau* and *Lifshitz* [4.2] showed that one cannot have long-range order in one dimension. Consider, for example, a chain of spins. The long-range order is destroyed by simply "flipping" one spin. Although this costs exchange energy, the entropy gained is proportional to the logarithm of the number of spins. Thus, for an infinitely long chain the entropy term will always dominate the free energy.

In two dimensions the situation depends upon the details of the interaction— a Heisenberg system, for example, does not exhibit ferromagnetism, while an Ising system does.

Antiferromagnetism. From Fig. 4.1 it is obvious that if the exchange constant between nearest neighbors is *negative*, then the first divergence appears for $Q = (\pi/a)(\hat{x} + \hat{y} + \hat{z})$ or one of the other similar wave vectors in a body-diagonal direction. This means that the phase of the magnetization changes by π from one plane of atoms to the next as we proceed along the direction of Q. Thus we have an *antiferromagnetic* system. The temperature at which $\chi(Q)$ diverges is the *Néel temperature*,

$$T_{\text{N}} = \frac{6|J|S(S+1)}{3k_{\text{B}}} \,. \tag{4.18}$$

The uniform susceptibility is given by

$$\boxed{\chi(0)_{\text{antiferro}} = \frac{C}{T + T_{\text{N}}}}\qquad(4.19)$$

which is well behaved at the Néel temperature.

In general the exchange interaction will extend beyond just the nearest neighbors. This can lead to a variety of antiferromagnetic configurations. Antiferromagnetism *is* compatible with cubic symmetry, and several "types" of antiferromagnetic orderings in cubic lattices are illustrated in Fig. 4.2. The oxides MnO, FeO, CoO, and NiO are examples of the fcc type II order.

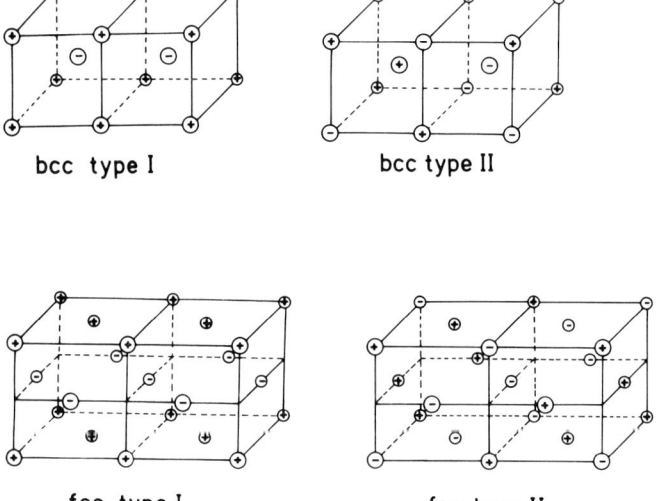

Fig. 4.2. Possible types of antiferromagnetic order in body-centered and face-centered cubic lattices. The symbols ⊕ and ⊖ refer to "up" and "down" magnetic moments, respectively

Helimagnetism. In Chap. 7 we shall discuss the RKKY exchange interaction, an indirect exchange between localized moments which arises through the presence of conduction electrons. For our present purposes it is sufficient to note that this interaction is long range and oscillatory. To see what effect such an exchange may have on our spin configuration, let us consider our simple cubic lattice and take

$$J_{ij} = \begin{cases} J_1 & \text{for the six nearest neighbors.} \\ -J_2 & \text{for the twelve next-nearest neighbors.} \end{cases}$$

Then

$$J(\boldsymbol{q}) = \frac{2J_1}{N}(\cos q_x a + \cos q_y a + \cos q_z a) - \frac{4J_2}{N}[(\cos q_x a)(\cos q_y a)$$
$$+ (\cos q_y a)(\cos q_z a) + (\cos q_z a)(\cos q_x a)]. \tag{4.20}$$

This exchange has an extremum when $a\boldsymbol{Q} = \cos^{-1}(J_1/4J_2)(\hat{\boldsymbol{x}} + \hat{\boldsymbol{y}} + \hat{\boldsymbol{z}})$. If $J_1 < 4J_2$, it will lead to a spin configuration whose wavelength is incommensurate with the lattice spacing. An example of such a system is the rare-earth metal Tm at temperatures between 40 and 56 K. The spin configuration of Tm in this region is illustrated in Fig. 4.3. This particular configuration is sometimes referred to as a *longitudinal wave*.

The general situation is actually more complicated. So far our probing field has had a sinusoidally varying *amplitude* but has been fixed in direction. In principle, we could also imagine a field whose direction also varies in space. Should the response to such a peculiar field diverge, a correspondingly peculiar spin configuration would occur. Such spin configurations are, in fact, found in the heavy rare-earth metals as indicated in Fig. 4.3. The insert illustrates various spin configurations. Such systems are referred to as *helimagnets*. Dealing with the generalized susceptibility is not a very practical way of determining the existence of such spin configurations. A better method is to minimize the exchange energy. This is, in fact, the way these configurations were first obtained [4.4].

For our longitudinal-wave helimagnet, $J(\boldsymbol{Q}) = 3J_1^2/4NJ_2$ and

$$T_C = \frac{S(S+1)J_1^2}{4k_B J_2}. \tag{4.21}$$

Therefore the uniform susceptibility is

$$\boxed{\chi(0)_{\text{heli}} = \frac{C}{T - [8J_2(J_1 - 2J_2)/J_1^2]T_C}.} \tag{4.22}$$

Since $J_1 < 4J_2$, the coefficient of T_C in the denominator is always less than 1. Thus the uniform susceptibility will be finite at the transition temperature. However, its high-temperature behavior will resemble that of a ferromagnet if $J_1 > 2J_2$ and that of an antiferromagnet if $J_1 < 2J_2$.

Ferrimagnetism. In the preceding three cases we have assumed that the spins and their environments were all identical. If some spins or some environments are different, then we have an impurity or alloy problem, which will be discussed in Chap. 7. However, if there are two different sites, each of which is translationally equivalent, i.e., two sublattices, then we can have what Néel called a *ferrimagnet*. As an example of such a system, let us consider an NaCl-like structure with spins S_1, say, at the Na-like sites and S_2 at the Cl-like sites. Thus we have a *basis* of two spins, just as NaCl has a basis of two ions. The Bravais lattice is face-centered cubic.

108   4. The Static Susceptibility of Interacting Systems

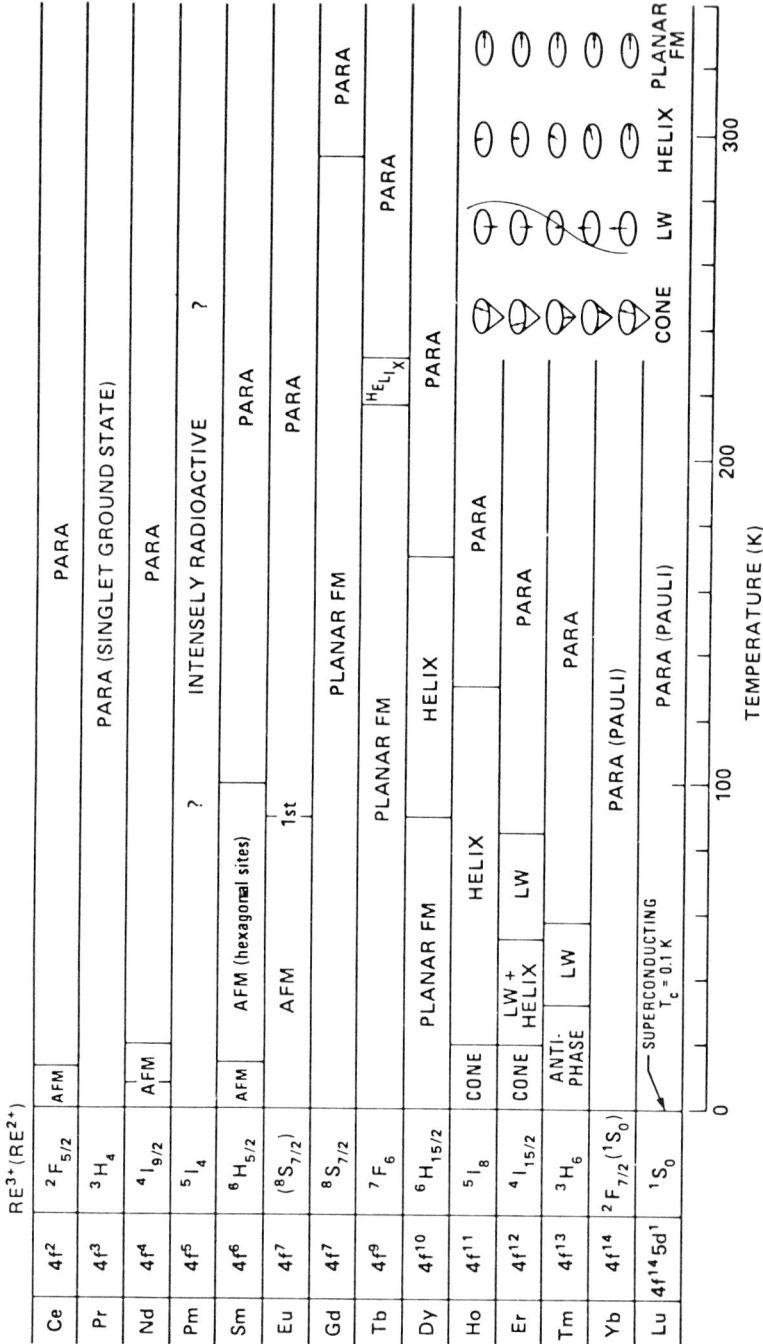

Fig. 4.3. Magnetic properties of the rare-earth metals. Data assembled from [4.3]

If we denote the spin in the ith cell with the basis vector $\boldsymbol{b}$ as $\boldsymbol{S}_{ib}$ and assume a nearest-neighbor exchange interaction, then the exchange Hamiltonian is

$$\mathcal{H}_{\text{ex}} = -J \sum_{i,j} \sum_{b,b'} \boldsymbol{S}_{ib} \cdot \boldsymbol{S}_{jb'} . \tag{4.23}$$

where the sums involve only nearest-neighbor spins. The reciprocal lattice associated with our system is that appropriate to the face-centered-cubic Bravais lattice. Therefore the Fourier expansion of the spins will involve wave vectors whose components are multiples of $2\pi/L$ up to $\pi/2a$, where L is the crystal dimension and a is the ionic spacing. Applying an external field with a spatial variation characterized by one of these wave vectors gives a Zeeman Hamiltonian of the form

$$\mathcal{H}_Z = g\mu_B H \sum_{i,b} S^z_{ib} \cos(\boldsymbol{q} \cdot \boldsymbol{R}_i + \boldsymbol{q} \cdot \boldsymbol{b}) . \tag{4.24}$$

In Fourier components, the total Hamiltonian is

$$\mathcal{H} = -\sum_{q'} \sum_{b \neq b'} J(-\boldsymbol{q}') \boldsymbol{S}_b(\boldsymbol{q}') \cdot \boldsymbol{S}_{b'}(-\boldsymbol{q}') + \tfrac{1}{2} g\mu_B H \\ \times \sum_b [S^z_b(\boldsymbol{q})e^{-i\boldsymbol{q}\cdot\boldsymbol{b}} + S^z_b(-\boldsymbol{q})e^{-i\boldsymbol{q}\cdot\boldsymbol{b}}] \tag{4.25}$$

where

$$J(-\boldsymbol{q}) = \frac{J}{N} {\sum_{\boldsymbol{\delta}}}' e^{i\boldsymbol{q}\cdot\boldsymbol{\delta}} . \tag{4.26}$$

The prime indicates that $\boldsymbol{\delta}$ is restricted to 0 and five vectors to nearest neighbors, which depend on the definition of the basis.

If we take $\boldsymbol{b} = a\hat{\boldsymbol{x}}$, then

$$J(-\boldsymbol{q}) = \frac{2J}{N} (\cos q_x a + \cos q_y a + \cos q_z a) \exp(-iq_x a) . \tag{4.27}$$

The effective field seen by the spin S_b is

$$H_b(-\boldsymbol{q})_{\text{eff}} = -\frac{J(-\boldsymbol{q})V}{g\mu_B} \langle S^z_{b'}(-\boldsymbol{q})\rangle \hat{\boldsymbol{z}} + \tfrac{1}{2} H V e^{-i\boldsymbol{q}\cdot\boldsymbol{b}} \hat{\boldsymbol{z}} . \tag{4.28}$$

Making use of (4.10), we have

$$\langle S^z_b(\boldsymbol{q})\rangle = -\frac{\chi(\boldsymbol{q})_0}{g\mu_B} H_b(\boldsymbol{q})_{\text{eff}} = \frac{C_b J(\boldsymbol{q}) V}{g^2 \mu_B^2 T} \langle S^z_{b'}(\boldsymbol{q})\rangle - \frac{C_b H V}{2g\mu_B T} e^{-i\boldsymbol{q}\cdot\boldsymbol{b}} \tag{4.29}$$

with a similar equation for $\langle S^z_{b'}(\boldsymbol{q})\rangle$. Solving these two equations gives

$$\langle S_b^z(\boldsymbol{q})\rangle = -\frac{HV}{2g\mu_B T}\frac{C_b e^{i\boldsymbol{q}\cdot\boldsymbol{b}} + [C_b C_{b'} J(\boldsymbol{q})V/g^2\mu_B^2 T]e^{i\boldsymbol{q}\cdot\boldsymbol{b}'}}{1 - C_b C_{b'} J(\boldsymbol{q})^2 V^2/g^4\mu_B^4 T^2}. \quad (4.30)$$

Since

$$M_z(\boldsymbol{q}) = -g\mu_B \int d\boldsymbol{r}\, e^{-i\boldsymbol{q}\cdot\boldsymbol{r}} \sum_{i,b} \langle S_{ib}^z\rangle \delta(\boldsymbol{r}-\boldsymbol{R}_i-\boldsymbol{b}),$$
$$= -g\mu_B \sum_b \langle S_b(\boldsymbol{q})\rangle\, e^{-i\boldsymbol{q}\cdot\boldsymbol{b}}, \quad (4.31)$$

we obtain for the susceptibility of our ferrimagnetic system,

$$\chi(\boldsymbol{q}) = \frac{(C_1+C_2)T + 2C_1 C_2 J(\boldsymbol{q})\cos(q_x a)V/g^2\mu_B^2}{T^2 - C_1 C_2 J(\boldsymbol{q})^2 V^2/g^4\mu_B^4}. \quad (4.32)$$

From the nature of $J(\boldsymbol{q})$, this susceptibility diverges for $\boldsymbol{q} = 0$. Furthermore, this divergence is independent of whether J is positive or negative. If it is positive, the two spins within the basis are parallel; if it is negative, they are antiparallel. These two possible configurations are shown in Fig. 4.4. The uniform susceptibility of this ferrimagnetic system is

$$\boxed{\chi(0)_{\text{ferri}} = \frac{(C_1+C_2)T + 2\sqrt{C_1 C_2}\, T_c}{T^2 - T_c^2}}, \quad (4.33)$$

where the transition temperature is

$$T_c = \frac{6J\sqrt{S_1 S_2(S_1+1)(S_2+1)}}{3k_B}. \quad (4.34)$$

The high-temperature susceptibilities associated with the various types of magnetic order are shown in Fig. 4.5.

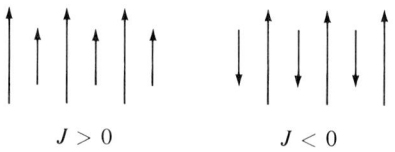

$J > 0 \qquad\qquad J < 0$

Fig. 4.4. Two possible basis configurations for a ferrimagnet

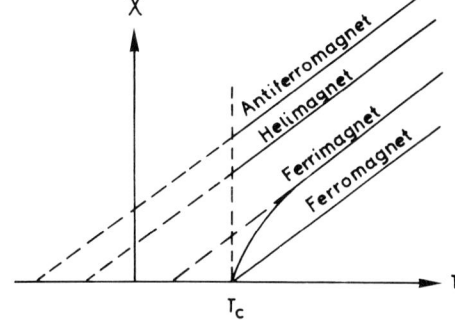

Fig. 4.5. General behavior of the high-temperature inverse susceptibility associated with various systems

The rare-earth iron garnets, $RE_3Fe_5O_{12}$, are typical examples of ferrimagnets.

Although our preceeding discussion has focused on lattices, the existence of such *atomic* long-range order is not required for *magnetic* long-range order. There are a large number of materials that can be prepared in the amorphous state which show magnetic order. In particular, amorphous films of rare-earth and transition metals, such as Gd–Co and Tb–Fe, are ferrimagnetic. Below the ferromagnetic transistion temperature the system is characterized by a spontaneous uniform magnetization. In most ferromagnets the magnetization develops smoothly from zero, identifying such transitions as second order. A typical magnetization curve is shown in Fig. 4.6.

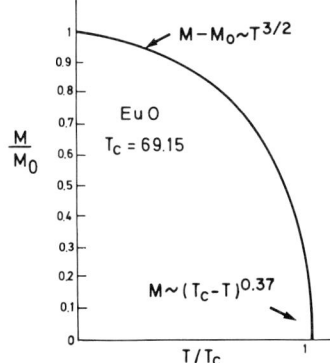

Fig. 4.6. Magnetization of europium oxide as a function of temperature

The magnetization curve for a ferrimagnet is more complex. Figure 4.7 shows the net magnetization associated with a two-component, antiferromagnetically coupled ferrimagnet. Because the temperature dependences of the individual components are different, the net magnetization goes to zero at a point called the *compensation temperature*.

4.1.2 Low Temperatures

Let us now turn briefly to the question of how the systems we have just discussed respond to a static field when they are below their ordering temperature. In the treatment above we considered only an isotropic exchange interaction. If this were the only contribution to the Hamiltonian, then the Hamiltonian, as well as its eigenstates, would be invariant under rotations. This would mean that the isotropic exchange interaction itself could not lead to a macroscopically ordered state. There are, in fact, additional interactions which favor certain crystallographic directions. Nickel, for example, prefers to magnetize along its [1 1 1] direction while iron prefers the [1 0 0] direction. Such magnetic anisotropy has its origin in the spin–orbit interaction as we discussed in Sect. 2.3. In metals this interaction leads to shifts in the energy bands as the magnetization direction is varied. Such shifts are accompanied by an electron redistribution among the bands with a corresponding change in the total electronic energy.

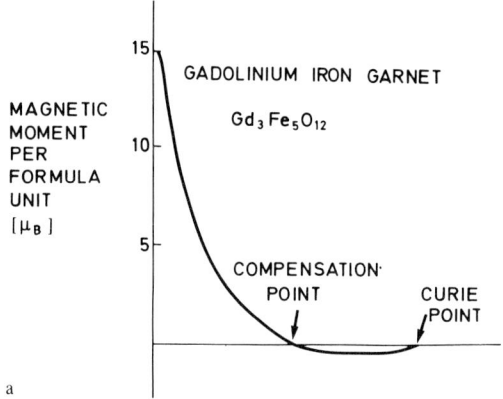

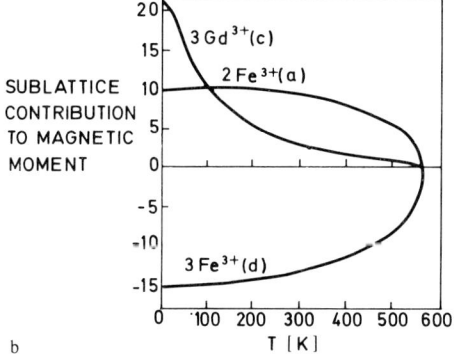

Fig. 4.7a, b. Magnetization of gadolinium iron garnet as a function of temperature (a). The garnets have three inequivalent magnetic sublattices. The "a" ions are arranged on a bcc lattice in octahedral sites; the "c" and "d" ions sit in dodecahedral and tetrahedral sites, respectively. The dominant exchange interaction is an antiferromagnetic coupling between the a and d sites. The calculated contributions to the magnetization from these sublattice is shown in (b)

Macroscopically, the magnetic anisotropy is characterized by an energy which is a function of the direction cosines of the magnetization relative to the crystallographic axes. A uniaxial anisotropy, for example, would be represented by $K_1 \sin^2\theta$. Microscopically, at least for an ionic solid, the anisotropy appears as terms in the Hamiltonian involving products of the components of the spin. In particular, for a uniaxial system,

$$\mathscr{H}_{\text{aniso}} = -D \sum_i (S_i^z)^2 \tag{4.35}$$

where D is the anisotropy constant. The total effective field acting on the ith spin in the molecular field approximation is therefore

$$\boldsymbol{H}_{i,\text{eff}} = -\frac{1}{g\mu_B} \sum_{j \neq i} J_{ij} \langle S_j \rangle - \frac{D}{g\mu_B} \langle S_i^z \rangle \hat{\boldsymbol{z}} + \boldsymbol{H} \cos(\boldsymbol{q} \cdot \boldsymbol{R}_i). \qquad (4.36)$$

Because there is now an inherent direction in the system, the response to an applied field will depend upon the direction of such a field. Let us first consider the case in which the applied field is parallel to the z axis. Then

$$\langle S_j \rangle = \langle S_i^z \rangle_0 \cos[\boldsymbol{Q} \cdot (\boldsymbol{R}_j - \boldsymbol{R}_i)] \hat{\boldsymbol{z}} + \delta S_j^z \hat{\boldsymbol{z}} \qquad (4.37)$$

where $\langle S_i^z \rangle_0$ is the average value of S_i^z in the absence of the applied field. Since the exchange energy exceeds the thermal energy in the ordered state, this is given by the Brillouin function as

$$\langle S_i^z \rangle_0 = -S B_S \left(-\frac{[NSJ(\boldsymbol{Q}) + DS]\langle S_i^z \rangle_0}{k_B T} \right). \qquad (4.38)$$

The term δS_j^z in (4.37) is that part of $\langle S_j^z \rangle$ which is induced by the applied field. This response is what defines the susceptibility. The average value of S_i^z in the presence of the field is given by an expression similar to (4.38), with the applied field added to the argument of the Brillouin function as well as terms involving δS_i^z. However, if all these additional terms are small, we may expand the Brillouin function in powers of these terms. To lowest order, we obtain

$$\langle S_i^z \rangle = \langle S_i^z \rangle_0 + \frac{NS^2 B_S'}{k_B T} \frac{1}{N} \sum_{j \neq i} J_{ij} \delta S_j^z$$
$$+ \frac{DS^2 B_S'}{k_B T} \delta S_i^z - \frac{g\mu_B S^2 H B_S'}{k_B T} \cos(\boldsymbol{q} \cdot \boldsymbol{R}_i) \qquad (4.39)$$

where B_S' is the derivative of the Brillouin function with respect to its argument evaluated at zero field. If we take the Fourier transform of (4.39), the susceptibility becomes

$$\boxed{\chi_\parallel(\boldsymbol{q}) = \frac{3SCB_S'/(S+1)}{T - [3SB_S'/(S+1)][NJ(\boldsymbol{q}) + D]S(S+1)/3k_B}.} \qquad (4.40)$$

By similar arguments, the response to a field applied perpendicular to the z axis is characterized by

$$\boxed{\chi_\perp(\boldsymbol{q}) = \frac{g^2 \mu_B^2 N/V}{N[J(\boldsymbol{Q}) - J(\boldsymbol{q})] + D}.} \qquad (4.41)$$

From these expressions we find that the perpendicular susceptibility is independent of temperature, while the parallel susceptibility decreases to 0 as the temperature does. This leads to a rather interesting phenomenon in antiferromagnets. At zero temperature the application of a parallel field produces no polarization. Thus there is no Zeeman energy associated with this state. A perpendicular field, however, produces a small canting of the spins which does lead to a polarization. The Zeeman energy associated with this state is essentially $-\mathbf{M}\cdot\mathbf{H}$. Thus, if there were no anisotropy energy, the spins in a parallel field would all "flop" over into the lower-energy perpendicular configuration. Because of the anisotropy energy there is a minimum field at which this "spin-flop" transition occurs. By equating the energies in the two states we find, for nearest-neighbor interactions,

$$H_{\text{crit}} = \frac{S}{g\mu_B}\sqrt{4zD|J|} \qquad (4.42)$$

which is the geometrical mean of the exchange field and the anisotropy field. This will not happen in a ferromagnet, even though the parallel response function does go to 0, because of the large dipolar energy associated with the aligned state. Fig. 4.8 shows the phase diagram for an antiferromagnet.

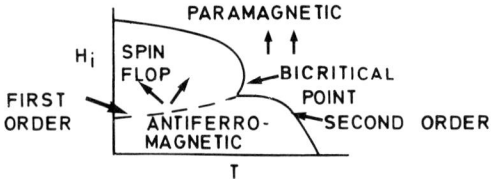

Fig. 4.8. Phase diagram for an antiferromagnet (H_i is the internal field)

Metamagnetism. In a material with strong anisotropy the spin-flop state is suppressed and below a certain temperature the system goes directly into the paramagnetic state via a first-order transition. Such systems are called metamagnets. Classic examples are $FeCl_2$ and $DyPO_4$. A schematic phase diagram is shown in Fig. 4–9a. The point at which the line of second-order transitions changes into a line of first-order transitions is called the *tricritical* point. In an antiferromagnet the field conjugate to the staggered magnetization is the staggered field $H(Q) \equiv H_s$. *Griffiths* [4.5] has pointed out that in the H_i–H_s–T phase diagram for a metamagnet there exist two first-order surfaces extending like wings out into the $H_s \neq 0$ region as shown in Fig.4.9a. Notice that these wings are bounded by two lines of second-order transitions. Thus the tricritical point is also the meeting of three critical lines.

In finite samples the internal field H_i and the applied field H_0 differ due to demagnetizing effects. For an elliptical sample characterized by a demagnetizing factor N, the internal field is given by $H_i = H_0 - 4\pi NM$. Since the magnetization increases discontinuously as we cross the line of first-order transitions, the internal field is discontinuously reduced to a value below that required for the transition. The material resolves this contradiction by breaking into domains of two phases. This mixed phase is analogous to the so-called intermediate state in superconductivity.

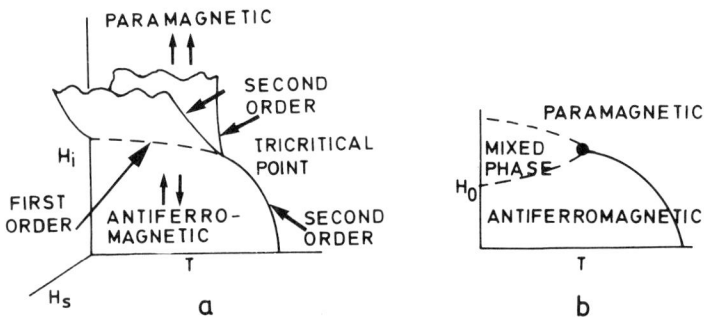

Fig. 4.9a, b. Phase diagram for a metamagnet in terms of the internal field, H_i (a) and the applied field, H_0 (b).

4.1.3 Temperatures near T_c

Let us now turn to the interesting region of the critical point itself. The object of any critical-point theory is to predict accurately the behavior of various physical quantities as the temperature or another variable approaches its critical-point value. Generally this approach is represented by a power law such as $(T - T_c)^\lambda$. Figure 4.6, for example, shows that the magnetization of nickel vanishes as $(T - T_c)^{0.355}$. A successful theory must correctly predict the value of λ, as well as any relationship, if it exists, between the exponents associated with different physical quantities.

There are two basic steps in this problem. First a model must be selected that adequately represents the actual system but is still mathematically tractable. Our discussions so far have been based on the Heisenberg Hamiltonian, $-\sum J_{ij} \mathbf{S}_i \cdot \mathbf{S}_j$. We have also mentioned the *Ising model*, $-\sum J_{ij} S_i^z S_j^z$. Although some materials, such as dysprosium aluminum garnet, may actually be described by this model, its main attraction is that it is much easier to deal with than the Heisenberg model. In fact, the thermodynamic quantities associated with a *two-dimensional* Ising model may be computed exactly [4.6].

Since even the relatively simple Ising model has not yet been solved exactly in three dimensions, we are forced to make certain approximations. Therefore the second step in the critical-point problem is the development of mathematical techniques that enable us to calculate physical properties near the phase transition of our model system. In Sect. 4.1. we employed the mean-field approximation, for low temperatures where each spin was assumed to be statistically independent, that is, $\langle S_i^z S_j^z \rangle_0 = \langle S_i^z \rangle_0 \langle S_j^z \rangle_0$, where the subscript indicates that this average is to be taken in zero field. In the high-temperature region we employed the random-phase approximation, in which only the Fourier components were assumed independent. Notice that use of the molecular field approximation in this region would have implied that the energy of the system, which involves $\langle S_i^z S_j^z \rangle_0$, was 0 above T_C. Thus the specific heat would also have been 0. In actual fact, however, as T_c is approached from above the spins begin to correlate with one another. That is, $\langle S_i^z S_j^z \rangle$ is not 0 for spins that are relatively close together. We speak of this as the appearance of *short-range order*. Owing to the existence of this short-range order, the energy, and consequently the specific heat, are not 0 above the point where the spontaneous magnetization vanishes.

The determination of the correlation function $\langle S_i^z S_j^z \rangle_0$ provides us with a nice application of the fluctuation-dissipation theorem. Introducing Fourier components and requiring that this correlation function depend only upon the relative coordinates of the spins gives

$$\langle S_i^z S_j^z \rangle_0 = \frac{1}{N^2} \sum_q \langle S(q)_z S(-q)_z \rangle_0 \exp[i q \cdot (R_i - R_j)] . \tag{4.43}$$

Since $\mathcal{M}_\nu(q) = -g\mu_B S_\nu(q)$, and since (4.43) is symmetric, the fluctuation-dissipation theorem (1.87) takes the form

$$\langle S(q, t)_z S(-q)_z \rangle_0 = \int \frac{d\omega}{2\pi} \frac{2\hbar V \chi_{zz}''(q, \omega)_S}{g^2 \mu_B^2 (1 - e^{-\beta \hbar \omega})} e^{-i\omega t} \tag{4.44}$$

Setting $t = 0$ and assuming that $k_B T/\hbar$ is larger than any relevant frequencies (we shall discuss the validity of this assumption in Chap. 6), we find

$$\langle S(q)_z S(-q)_z \rangle_0 = \frac{k_B T V}{\pi g^2 \mu_B^2} \int \frac{\chi_{zz}''(q, \omega)_S}{\omega} d\omega . \tag{4.45}$$

By the Kramers–Kronig relation (1.64), the frequency integral is just π times the static susceptibility $\chi_{zz}(q, 0)$. Therefore we finally obtain

$$\langle S_i^z S_j^z \rangle_0 = \frac{k_B T V}{N^2 g^2 \mu_B^2} \sum_q \chi_{zz}(q) \exp[i q \cdot (R_i - R_j)] . \tag{4.46}$$

Notice that since $\langle (S_i^z)^2 \rangle = S(S + 1)/3$, (4.46) leads to the sum rule

$$\boxed{\frac{1}{N} \sum_q \chi(q) = \chi_0} \tag{4.47}$$

where χ_0 is defined as in (4.10).

If we use the random-phase result for $\chi_{zz}(\mathbf{q})$ (4.12) and assume a simple cubic lattice, then in the long-wavelength limit (4.46) becomes

$$\langle S_i^z S_j^z \rangle_0 = \frac{6S(S+1)}{3Na^2} \frac{T}{T_c} \sum_\mathbf{q} \frac{\exp[i\mathbf{q}\cdot(\mathbf{R}_i - \mathbf{R}_j)]}{\kappa^2 + q^2}, \tag{4.48}$$

where

$$\kappa^{-1} = a \sqrt{\frac{T_c}{6(T-T_c)}} \tag{4.49}$$

is referred to as the *correlation length*. Notice that as T becomes very large in comparison with T_c, the correlation length becomes very small and

$$\langle S_i^z S_j^z \rangle_0 \xrightarrow[T > T_c]{} \tfrac{1}{3} S(S+1)\Delta(\mathbf{R}_i - \mathbf{R}_j). \tag{4.50}$$

As T appoaches T_c the correlation function takes the Yukawa-like form

$$\langle S_i^z S_j^z \rangle_0 = \frac{S(S+1)}{16\pi^4 (N/V)a^2} \frac{T}{T_c} \frac{\exp(-\kappa|\mathbf{R}_i - \mathbf{R}_j|)}{|\mathbf{R}_i - \mathbf{R}_j|}. \tag{4.51}$$

Thus as T approaches T_c each spin exerts a polarizing influence on those spins within a sphere of radius κ^{-1}. At $T = T_c$ this sphere encompasses the whole sample, and we say that long-range order has set in.

The specific heat associated with this ordering may also be computed from (4.46). The behavior of the susceptibility and the specific heat in the mean-field and random phase approximations are compared schematically with typical experimetal data in Fig. 4.10. The most notable discrepancy is the fact that the actual transition occurs below the value predicted by this theories. This is the consequence of an inadequate treatment of short-range order.

We can obtain a better approximation for $\chi(\mathbf{q})$ by explicitly taking short-range order in account. The basic idea is that the spins surrounding any given spin will be correlated to the motion of that spin and will, therefore, not contribute to the mean field seen by that spin. This is incorporated into the theory by subtracting a term from the mean field which is proportional to $\langle S_i^z \rangle$ itself,

$$H_i^{\text{eff}} = -\sum_j \frac{J_{ij}}{g\mu_B} \langle S_j^z \rangle + \sum_j \frac{\lambda_{ij} J_{ij}}{g\mu_B} \langle S_i^z \rangle. \tag{4.52}$$

The parameters λ_{ij} are assumed to be temperature dependent and characterize this so-called Onsager reaction field. Proceeding as in the derivation of (4.12) we obtain

$$\chi(\mathbf{q}) = \frac{\chi_0}{1 - \frac{\chi_0 V}{g^2 \mu_B^2}[J(\mathbf{q}) - \lambda]} \tag{4.53}$$

118   4. The Static Susceptibility of Interacting Systems

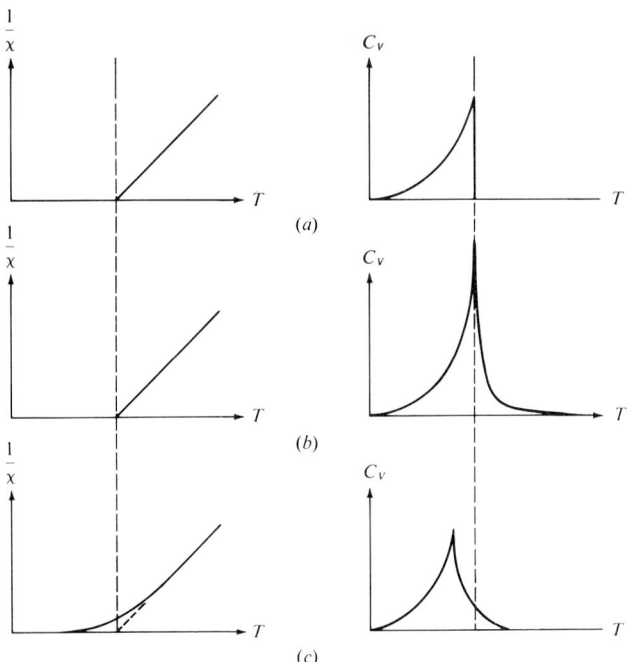

Fig. 4.10a-c. The effect of various approximations on the susceptibility and the specific heat of a Heisenberg ferromagnet: (a) molecular field, (b) random phase, and (c) typical experimental behavior

where $\lambda = (1/N) \sum \lambda_{ij} J_{ij}$. This parameter is obtained by requiring that the susceptibility (4.53) satisfies the sum rule (4.47). The result is

$$\lambda_{ij} = \frac{\langle S_i^z S_j^z \rangle_0}{S(S+1)/3}. \tag{4.54}$$

Thus, (4.46) now becomes a self-consistent equation for the correlation function $\langle S_i^z S_j^z \rangle_0$.

This improved treatment of short-range order also renormalizes the transition temperature relative to the mean-field value T_c. If we sum (4.53) over q and again use the sum rule we obtain the relation

$$G(s) = T_c/T \tag{4.55}$$

where $s = T/T_c + \lambda/J(Q)$ and the sum

$$G(s) = \frac{1}{N} \sum_q \frac{1}{s - J(q)/J(Q)} \tag{4.56}$$

is called the lattice Green's function. From (4.53) we see that $\chi(\mathbf{q})$ diverges at a temperature which corresponds to $s = 1$. Therefore,

$$T_c^* = T_c/G(1) \,. \tag{4.57}$$

For simple cubic, body-centered-cubic, and face-centered-cubic lattices the values of $G(1)$ are 1.517, 1.393, and 1.345, respectively. Thus we see that incorporating the Onsager reaction field predicts a lower transition temperature as we observed in Fig. 4.10.

Just below T_c the quantity of thermodynamic interest is the spontaneous magnetization. As T_c is approached from below, $\langle S_i^z \rangle_0$ becomes very small, which enables us to expand the Brillouin function in (4.38). The resulting magnetization in this molecular field approximation is found to vanish as $(T_c - T)^{1/2}$.

This idea of treating the magnetization as an expansion parameter is a fundamental aspect of Landau's theory of second-order phase transitions. Landau associated second-order phase transitions with transitions in which there is a "broken symmetry." That is, the new ground state of the system does not possess the total symmetry of the Hamiltonian. In the case of ferromagnetism the rotational invariance of the Heisenberg Hamiltonian is broken by the appearance of the spontaneous magnetization $\mathbf{M}$. Landau identified the parameter whose value becomes nonzero in the unsymmetrical state as the *order parameter*. In this case it is obviously the magnetization. The order parameter represents an additional variable that must now be used to specify the state of the system. For the Heisenberg model the order parameter is a vector. We therefore speak of it having the *dimensionality* $n = 3$. Since an Ising model involves only the z component of the spin, the order parameter has the dimensionality $n = 1$.

If the order parameter, call it M, is to grow from zero above T_c to a finite value below T_c then the Landau theory argues that near T_c the free energy has the form

$$F = a(T - T_c)M^2 + BM^4 \,. \tag{4.58}$$

Minimizing this energy with respect to M tells us that $M \sim (T_c - T)^{1/2}$ as we found above. If we add an applied field term, $-MH$, then it can be shown that the susceptibility diverges as $\chi \sim (T - T_c)^{-1}$. Similarly, if we assume that the magnetization varies slowly in space, then there will be an additional term in the free energy. Since it costs virtually no energy to rotate the magnetization uniformly, we expect that the energy associated with a gradual variation may be expressed in terms of derivatives of $\mathbf{M}$. In order to satisfy time reversal, this must be a quadratic expression. There are several such possibilities. In an isotropic medium, for example, there are three: $|\nabla \mathbf{M}|^2$, $(\nabla \cdot \mathbf{M})^2$, and $|\nabla \times \mathbf{M}|^2$. If we assume that all those terms reflecting the spatial symmetry are taken into account through an anisotropy energy, then the exchange energy must be invariant with respect to rotations in *spin* space. For example, a rotation of π about the z axis in spin space takes $M_x \to -M_x$ and $M_y \to -M_y$, but leaves x and y unchanged. Thus, the exchange energy in a cubic crystal must have the form

$$\frac{\delta E}{V} = \frac{A}{M_0^2} |\nabla M|^2 \tag{4.59}$$

where A is referred to as the exchange stiffness. This contribution introduces a characteristic length, the correlation length, which varies as $\xi \sim (T_c - T)^{-1/2}$, also as we have already found.

These critical exponents,

$\beta = \frac{1}{2}$ for the magnetization,
$\gamma = 1$ for the susceptibility, and
$\nu = \frac{1}{2}$ for the correlation length,

do not agree with those observed experimentally. This is not too surprising for the theory itself predicts that as $T \to T_c$ fluctuations in the order parameter become larger than the order parameter itself. Careful measurements of these and other critical exponents in the 1960's stimulated much theoretical activity. It was particularly intriguing that particular critical exponents in different systems seemed to have nearly the same values. Furthermore, certain combinations of exponents seemed to be related.

The first major assault on this problem was Widom's hypothesis that the critical part of free energy, f, was a *homogeneous function* of its independent variables. In the case of magnetism this takes the form

$$f(L^x \epsilon, L^y H) = L^d f(\epsilon, H) \tag{4.60}$$

where $\epsilon = (T - T_c)/T_c$, x and y are two critical indices, d is the dimensionality and L was subsequently shown by Kadanoff to correspond to a scaling length. With this "scaling" hypothesis all the critical exponents now become combinations of x, y, and d. In 1971 K.Wilson introduced the renormalization group which clarified the nonanalytic nature of the phase transition and provided the basis for scaling. Various methods have been developed for performing renormalization-group calculations for the critical exponents themselves. Such calculations show that the critical exponents are only functions of the dimensionality of space and the dimensionality of the order parameter, a situation refered to as *universality* (see [4.30]).

It is difficult to make accurate comparisons with experiment. First of all, impurities and strains tend to smear out the transition, making it difficult to measure close to the transition. And also, weak anisotropies become important near the transition and can lead to "crossover" from one universality class to another. These difficulties are illustrated in Fig. 4.11 where we have plotted the critical exponent β for four representative materials among the constant β con-

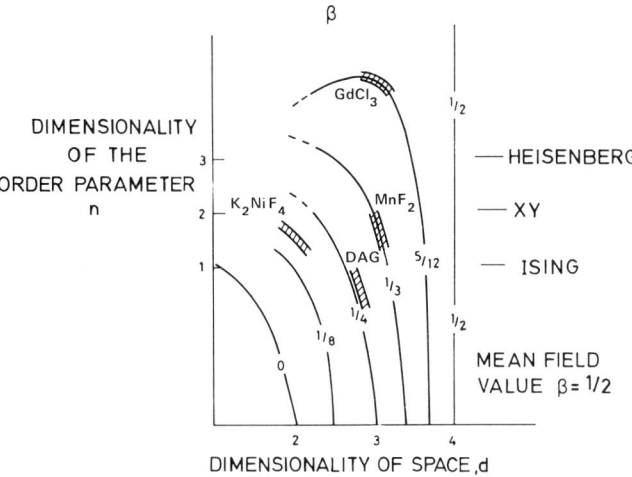

Fig. 4.11. Variation of the critical exponent β (associated with the magnetization) with the dimensionality of space (d) and the order parameter (n). The experimental values for several materials are indicated (see Table 4.1 for references)

tours in the n, d plane obtained from a renormalization-group calculation [4.7]. Dysprosium aluminum garnet (DAG) is considered to be a three-dimensional Ising system, K_2NiF_4 a two-dimensional Heisenberg system, and $GdCl_3$ a three-dimensional Heisenberg system but one in which long-range dipolar interactions are important.

At very low temperatures the molecular field expression (4.38) predicts that the magnetization decreases with increasing temperature as $e^{-2T_c/(S+1)T}$. The decrease is found experimentally to be much less rapid. The reason for this has to do with the existence of spin–wave modes, which we shall discuss in Chap.7.

4.1.4 Topological Long-Range Order

Notice that the theoretical curves shown in Fig. 4.11 do not extend to $d = 2$. In two dimensions, the nature of the long-range order depends critically on the dimensionality of the order parameter. *Mermin* and *Wagner* [4.9] have shown

Table 4.1. The critical exponent β for selected materials

Material	β	$\dfrac{T_c-T}{T_c}$	Reference
$GdCl_3$	$\beta = 0.43$	$0.002 \leq \varepsilon \leq 0.07$	J. Phys. C9, 1291 (1974)
DAG	$\beta = 0.26$	$0.001 \leq \varepsilon \leq 0.056$	Phys. Rev. **186**, 557 (1969)
K_2NiF_4	$\beta = 0.138$	$0.0003 \leq \varepsilon \leq 0.2$	J. Appl. Phys. **41**, 1303 (1970)
MnF_2	$\beta = 0.333$	$0.00006 \leq \varepsilon \leq 0.08$	Phys. Rev. **146**, 403 (1966)

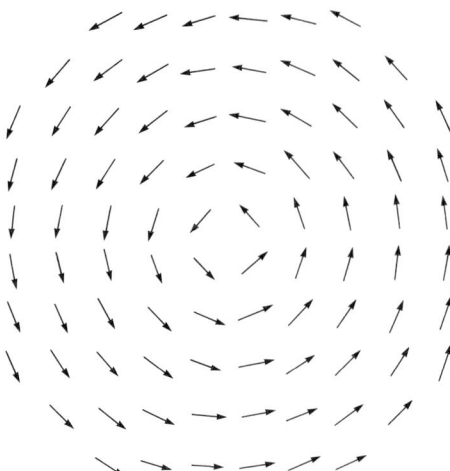

Fig. 4.12. Spin vortex

that two-dimensional systems with a continuous symmetry, such as a Heisenberg system, will not exhibit conventional long-range order as characterized, for example, by a spontenous magnetization or a sublattice magnetization. However, this does not exclude the possibility of a phase transition associated with more complex spin dynamics. The two-dimensional *XY* model is the classic example. *Berezinskii* [4.10] has calculated the spin–spin correlation function $\langle S_i S_j \rangle$ for this model using a low-temperature expansion, and found a power-law behavior

$$\langle S_i S_j \rangle \sim \left(\frac{|R_i - R_j|}{R_0} \right)^{-k_B T / 4\pi J}.$$

This implies that the susceptibility is infinite, i.e., that the system is undergoing large fluctuations which destroy any long-range order. *Kosterlitz* and *Thouless* [4.11] however, have pointed out that, in addition to the long-wavelength spin fluctuations responsible for this power-law behavior, there also exist local "topological defects" in the order parameter which, in this case, correspond to vortices as illustrated in Fig. 4.12. Other examples include dislocations in crystalline solids and vortices in superfluid helium. To calculate the energy of an isolated vortex, we write the *XY* Hamiltonian as

$$\mathcal{H}_{XY} = -2J \sum_{i>j} (S_i^x S_j^x + S_i^y S_j^y) = -2J \sum_{i>j} \cos(\varphi_i - \varphi_j)$$

where φ_i is the angle the *i*th spin makes with some arbitrary axis. Expanding the cosine,

$$E - E_0 \approx J \sum_r [\Delta\varphi(r)]^2$$

where Δ denotes the first difference operator. For a vortex as shown in Fig.4.12, $\Delta\varphi(r) = 2\pi/2\pi r$. Therefore,

$$E - E_0 = 2\pi J \ln(R/a)$$

where R is the radius of the system and a the nearest-neighbor distance. Since this vortex could be centered on any one of the $\pi R^2/a^2$ sites, the entropy is

$$S = 2k_B \ln(R/a)$$

The free energy is $E - E_0 - TS$. Since both the energy and entropy depend upon the size of the system in the same logarithmic way, the energy term will dominate the free energy at low temperatures, and the probability of a vortex appearing is vanishingly small. However, when the temperature exceeds $\pi J/k_B$, the entropy dominates and free vortices are likely to appear. If one now includes interactions between vortices, the critical temperature is given by the solution of [4.11]

$$\frac{\pi J}{k_B T_{KT}} - 1 = 2\pi \exp(-\pi^2 J/k_B T_{KT}).$$

Below T_{KT}, the vortices are bound in pairs of zero total vorticity. This is referred to as topological long-range order. The spin-spin correlation function still has a power-law behavior indicating an infinite suceptibility. However, above T_{KT}, the vortices give the correlation function an exponential form with a susceptibility given by

$$\chi(T) \sim \exp(2.625 t^{-1/2})$$

where $t = (T - T_{KT})/T_{KT}$.

4.2 Metals

The long-range magnetic order in metals is very similar to that observed in insulators as illustrated by the magnetization curve of nickel shown in Fig.4.13. However, the electrons participating in this magnetic state are itinerant; that is, they also have translational degrees of freedom. How such a system of interacting electrons responds to a magnetic field is a many-body problem with all its attendant difficulties.

The many-body corrections to the Landau susceptibility and the Pauli susceptibility must be treated separately. *Kanazawa* and *Matsudaira* [4.12] found that the many-body corrections to the Landau susceptibility are small (less than one per cent) for high electron densities.

124   4. The Static Susceptibility of Interacting Systems

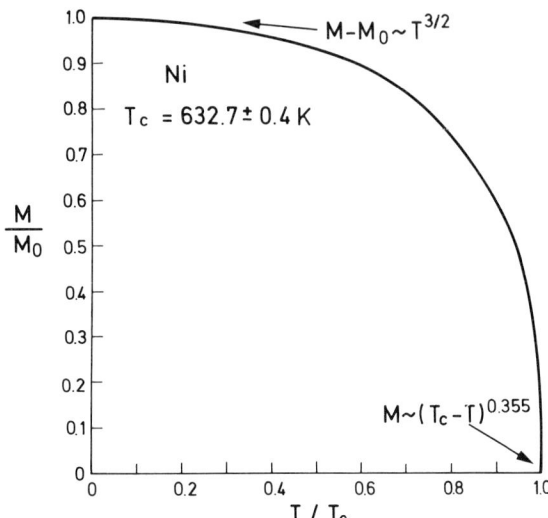

Fig. 4.13. Magnetization of nickel as a function of temperature. The original data of *Weiss* and *Forrer* [4.8] taken at constant pressure has been corrected to constant volume to eliminate the effects of thermal expansion

We shall approach the effect of electron–electron interactions on the spin susceptibility in two ways. The first, Fermi liquid theory, is a phenomenological approach. It involves parameters completely analogous to the parameters entering the spin Hamiltonian. These parameters may be determined experimentally or they may be obtained from the second approach which is to assume a specific microscopic model from which various physical properties can be calculated.

4.2.1 Fermi Liquid Theory

The phenomenological theory of an interacting fermion system was developed by *Landau* in 1956 [4.13]. Although *Landau* was mainly interested in the properties of liquid He^3, his theory may also be applied to metals. Modifications of this theory in terms of the introduction of a magnetic field have been made by *Silin* [4.14].

Let us begin by considering the ground state of a system of N electrons. For a noninteracting system the ground state corresponds to a well-defined Fermi sphere. *Landau* assumed that as the interaction between the electrons is gradually "turned on" the new ground state evolves smoothly out of the original Fermi sphere; if $|0\rangle$ is this new ground state, it is related to the original Fermi sphere $|FS\rangle$ by a unitary transformation,

$$|0\rangle = U|FS\rangle . \tag{4.61}$$

Let us denote the energy associated with $|0\rangle$ as E_0.

Landau also applied this assumption to the excitations of the interacting system. For example, suppose we add one electron, with momentum $\hbar k$, to the noninteracting system. This state has the form $a_{k\sigma}^\dagger|FS\rangle$, where $a_{k\sigma}^\dagger$ is the creation operator for an electron. If the interactions are gradually turned on, let us approximate the new state as

$$|\mathbf{k}\sigma\rangle = U a_{\mathbf{k}\sigma}^\dagger |FS\rangle. \tag{4.62}$$

Because the electron possesses spin, this wave function is a spinor.

Let us define the difference between the energy of $|\mathbf{k}\sigma\rangle$ and $|0\rangle$ as $\epsilon^0(\mathbf{k}, \sigma)$. Since the wave function is a spinor, this energy will be a 2×2 matrix. If the system is isotropic, and in particular if there is no external magnetic field, then this energy is independent of the spin,

$$\epsilon^0(\mathbf{k}, \sigma)_{\alpha\beta} = \epsilon^0(\mathbf{k})\delta_{\alpha\beta}.$$

Because the whole Fermi sphere has readjusted itself as a result of the interactions, the energy $\epsilon^0(\mathbf{k})$ will be quite different from the energy of a free particle. As we do not know this energy, we shall assume that $\mathbf{k}$ is close to k_F and expand in powers of $k - k_F$. Thus we obtain

$$\epsilon^0(\mathbf{k}) = \mu + \frac{\hbar^2 k_F}{m^*}(k - k_F) + \dots \quad \text{where} \tag{4.63}$$

$$\frac{\hbar^2 k_F}{m^*} \equiv \left.\frac{\partial \epsilon^0(\mathbf{k})}{\partial k}\right|_{k=k_F}. \tag{4.64}$$

This electron, "dressed" by all the other electrons, is called a *quasiparticle*. Notice that the energy required to create a quasiparticle at the Fermi surface is μ, the chemical potential. Its increase in energy as it moves away from the Fermi surface is characterized by its effective mass m^*. We restrict ourselves to the region close to the Fermi surface because it is only in this region that quasiparticle lifetimes are long enough to make their description meaningful.

We could just as well have *removed* an electron from some point within the Fermi sphere. This would have created a "hole," which the interactions would convert into a *quasi-hole*. The energy associated with a hole is the energy required to remove an electron at the Fermi surface, $-\mu$, plus the energy it takes to move the electron at $\mathbf{k}$ up to the surface, $(\hbar^2 k_F/m^*)(k_F - k)$. However, if we define the total energy of the system containing a quasi-hole as $E_0 - \epsilon^0(\mathbf{k})$, then

$$\epsilon^0(\mathbf{k}) = \mu + \frac{\hbar^2 k_F}{m^*}|k - k_F|.$$

Thus the excitation spectrum associated with our Fermi liquid has the form shown in Fig. 4.14.

Suppose that other quasiparticles are now introduced into the system. This could occur, for example, as a result of an external field producing electron–hole pairs. Since the energy of a quasiparticle depends on the distribution of all the other quasiparticles, any change in distribution will lead to a change in the

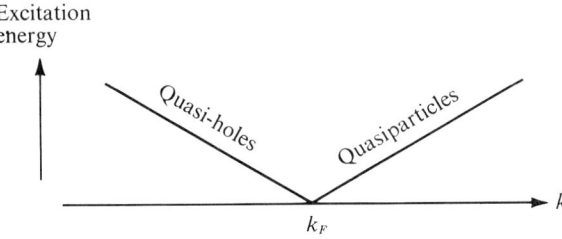

Fig. 4.14. Single-particle excitation spectrum of a Fermi liquid

quasiparticle energy. Let us denote the change in the quasiparticle distribution by $\delta n(\mathbf{k}, \sigma)$.

The quasiparticle distribution is essentially the density matrix associated with the quasiparticle. In particular, it is a 2×2 matrix. For example, $\delta n(\mathbf{k}, \sigma)_{11}$ gives the probability of finding an electron of momentum $\hbar \mathbf{k}$ with spin up. Therefore the quasiparticle energy may be written in phenomenological terms as

$$\epsilon(\mathbf{k}, \sigma) = \epsilon^0(\mathbf{k}, \sigma) + \frac{1}{V} \text{Tr}_{\sigma'} \{\sum_{\mathbf{k}'} f(\mathbf{k}, \sigma; \mathbf{k}', \sigma') \delta n(\mathbf{k}', \sigma')\} \; . \tag{4.65}$$

The quantity $f(\mathbf{k}, \sigma; \mathbf{k}', \sigma')$ is a product of 2×2 matrices analogous to a dyadic vector product. Again, if the system is isotropic, the most general form this quantity can have is

$$f(\mathbf{k}, \sigma; \mathbf{k}', \sigma') = \varphi(\mathbf{k}, \mathbf{k}')\mathbf{1}\mathbf{1}' + \psi(\mathbf{k}, \mathbf{k}')\boldsymbol{\sigma} \cdot \boldsymbol{\sigma}' \tag{4.66}$$

where $\mathbf{1}$ is the 2×2 unit matrix. Furthermore, since this theory is valid only near the Fermi surface, we may take $|\mathbf{k}'| \simeq |\mathbf{k}| = k_\text{F}$. Then φ and ψ depend only upon the angle θ between $\mathbf{k}'$ and $\mathbf{k}$, and we may expand φ and ψ in Legendre polynomials:

$$\varphi(\mathbf{k}, \mathbf{k}') = \frac{\pi^2 \hbar^2}{m^* k_\text{F}} A(\hat{\mathbf{k}} \cdot \hat{\mathbf{k}}') = \frac{\pi^2 \hbar^2}{m^* k_\text{F}} [A_0 + A_1 P_1(\cos \theta) + \ldots] \; , \tag{4.67}$$

$$\psi(\mathbf{k}, \mathbf{k}') = \frac{\pi^2 \hbar^2}{m^* k_\text{F}} B(\hat{\mathbf{k}} \cdot \hat{\mathbf{k}}') = \frac{\pi^2 \hbar^2}{m^* k_\text{F}} [B_0 + B_1 P_1(\cos \theta) + \ldots] \; . \tag{4.68}$$

If we know the quasiparticle distribution function, then we can compute, just as for the electron gas, all the relevant physical quantities. These will involve the parameters A_n and B_n. The beauty of this theory is that some of the same parameters enter different physical quantities. Therefore by measuring certain quantities we can predict others. The difficulty, of course, is in determining the distribution function $\delta n(\mathbf{k}, \sigma)$. For static situations this is relatively easy. However, for dynamic situations, as we shall see in the following chapters, we have to solve a Boltzmann-like equation.

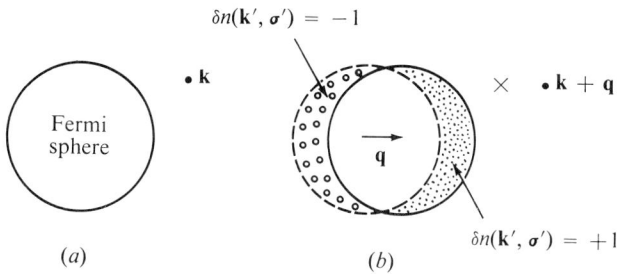

Fig. 4.15a, b. Effect of a uniform translation in momentum space on a state containing one extra particle

Since the **k** dependence of the quasiparticle energy is a result of interactions, there should be a relation betweem m^* and the parameters A_n and B_n. To obtain this relation let us consider the situation at $T = 0$ in which we have one quasi-particle at **k** with spin up, as illustrated in Fig. 4.15a. Now, suppose the momentum of this system is increased by $\hbar \mathbf{q}$, giving us the situation in Fig. 4.15b. This corresponds to placing the whole system on a train moving with a velocity $\hbar \mathbf{q}/m$. To an observer at rest with respect to the train it will appear that the quasiparticle has acquired an additional energy

$$\delta\epsilon(\mathbf{k}, \sigma)_{11} = \hbar^2 \mathbf{k} \cdot \mathbf{q}/m \tag{4.69}$$

for small **q**. However, the quasiparticle itself experiences a change in energy associated with its own motion in momentum space, which, from (4.63), is just $\hbar^2 \mathbf{k} \cdot \mathbf{q} m^*$. In addition, it sees the redistribution of quasiparticles indicated in Fig. 4.15b. Since this momentum displacement does not produce any spin flipping, $\delta n(\mathbf{k}, \sigma)_{\alpha\beta}$ will have the form $\delta n(\mathbf{k}) \delta_{\alpha\beta}$, where $\delta n(\mathbf{k})$ is $+1$ for the quasiparticles and -1 for the quasi-holes. This gives a contribution of

$$\frac{2}{V} \sum_{\mathbf{k}'} \varphi(\mathbf{k}, \mathbf{k}') \delta n(\mathbf{k}')$$

to the 1,1 component of the energy, where the factor 2 arises from the spin trace. Equating these two changes in energy by Galilean invariance and converting the sum over $\mathbf{k}'$ to an integral leads to our desired relation,

$$m^* = m\left(1 + \frac{A_1}{3}\right). \tag{4.70}$$

It can be shown that the specific heat of a Fermi liquid has the same form as that for an ideal Fermi gas, with m replaced by m^*. Thus by measuring the specific heat we can determine the Fermi liquid parameter A_1.

Exchange Enhancement of the Pauli Susceptibility. We are now ready to consider our original question of the response of a Fermi liquid to a magnetic field. In the presence of a magnetic field the noninteracting quasiparticle energy $\epsilon^0(k, \sigma)$ is no longer independent of the spin, but contains a Zeeman contribution,

$$\epsilon^0(k, \sigma) = \epsilon^0(k)\mathbf{1} + \mu_B H \sigma_z . \tag{4.71}$$

We shall assume that any field-induced contributions to the interaction term are small. Therefore the total quasiparticle energy is

$$\epsilon(k, \sigma) = \epsilon^0(k)\mathbf{1} + \mu_B H \sigma_z + \frac{1}{V} \operatorname{Tr}_{\sigma'} \{ \sum_{k'} f(k, \sigma; k', \sigma') \delta n(k' \sigma') \} . \tag{4.72}$$

It is energetically more favorable for the quasiparticles to align themselves opposite to the field, since their gyromagnetic ratio is negative. However, each time a quasiparticle flips over it changes the distribution, thereby bringing in contributions from the last term in (4.72). Thus, if we start with two equal spin distributions, as shown in Fig. 4.16a, an equilibrium situation will eventually be reached, as illustrated in Fig. 4.16b, in which the energy of a quasiparticle on the up-spin Fermi surface is equal to that of a quasiparticle on the down-spin surface; that is,

$$\epsilon(k_F + \delta k_F, \sigma)_{22} = \epsilon(k_F - \delta k_F, \sigma)_{11} . \tag{4.73}$$

From (4.63, 72) this condition becomes

$$\frac{\hbar^2 k_F}{m^*} \delta k_F - \mu_B H + \frac{1}{V} \operatorname{Tr}_{\sigma'} \{ \sum_{k'} [\varphi(\hat{k} \cdot \hat{k}') \mathbf{1}' - \psi(\hat{k} \cdot \hat{k}') \sigma'_z] \delta n(k', \sigma') \}$$
$$= -\frac{\hbar^2 k_F}{m^*} \delta k_F + \mu_B H + \frac{1}{V} \operatorname{Tr}_{\sigma'} \{ \sum_{k'} [\varphi(\hat{k} \cdot \hat{k}') \mathbf{1}' + \psi(\hat{k} \cdot \hat{k}') \sigma'_z] \delta n(k', \sigma') \} . \tag{4.74}$$

The change in quasiparticle distribution shown in Fig. 4.16b is characterized by

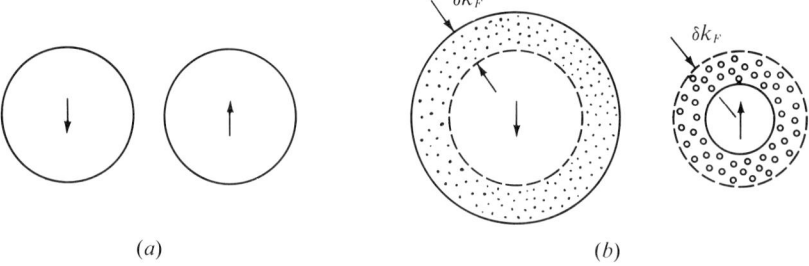

(a) (b)

Fig. 4.16a, b. Effect of a dc magnetic field on the spin-up and the spin-down Fermi spheres

$$\delta n(\bm{k}, \bm{\sigma}) = \begin{cases} \begin{bmatrix} 0 & 0 \\ 0 & 1 \end{bmatrix} & k_F < |\bm{k}| < k_F + \delta k_F \\ \begin{bmatrix} -1 & 0 \\ 0 & 0 \end{bmatrix} & k_F - \delta k_F < |\bm{k}| < k_F \end{cases}. \tag{4.75}$$

Therefore

$$\text{Tr}\left\{\sum_{\sigma'}\sum_{\bm{k}'} \varphi(\hat{\bm{k}}\cdot\hat{\bm{k}}')\delta n(\bm{k}', \bm{\sigma}')\right\} = 0 \tag{4.76}$$

while

$$\text{Tr}\left\{\sum_{\sigma'}\sum_{\bm{k}'} \psi(\hat{\bm{k}}\cdot\hat{\bm{k}}')\sigma_z'\delta n(\bm{k}', \bm{\sigma}')\right\} = -\frac{4\pi V}{(2\pi)^3} k_F^2 \delta k_F \int_{-1}^{+1} d(\cos\theta)\psi(\hat{\bm{k}}\cdot\hat{\bm{k}}'). \tag{4.77}$$

Equation (4.74) then reduces to

$$\frac{2\hbar^2 k_F}{m^*}\delta k_F - 2\mu_B H + \frac{2\hbar^2 k_F \delta k_F}{m^*} B_0 = 0. \tag{4.78}$$

Since the magentization is

$$M_z = -\frac{\mu_B}{V}\text{Tr}\left\{\sum_{\sigma}\sum_{\bm{k}}\sigma_z \delta n(\bm{k}, \bm{\sigma})\right\} = 2\mu_B \frac{4\pi k_F^2 \delta k_B}{(2\pi)^3}, \tag{4.79}$$

the uniform susceptibility of a Fermi liquid at $T = 0$ is

$$\boxed{\chi(0) = \frac{1 + \tfrac{1}{3}A_1}{1 + B_0}\chi_{\text{Pauli}}}. \tag{4.80}$$

Thus we find that in addition to the appearance of the effective mass in place of the bare mass, the susceptibility is also modified by the factor $(1 + B_0)^{-1}$. In the Hartree–Fock approximation

$$B_0 = -\frac{me^2}{\pi\hbar^2 k_F} = -0.166 r_s$$

and we speak of the susceptibility as being exchange enhanced. As the electron density decreases and $r_s \to 6.03$ the susceptibility diverges. This is usually taken to imply that such a material will be ferromagnetic. There has been a great deal of discussion [4.15] about the magnetic state of an interacting electron system, and it is generally agreed that such a system will not become ferromagnetic at *any* electron density. That is, the Hartree–Fock approximation favors ferromagnetism. The reason is that in this approximation parallel spins are kept apart by the exclusion principle while antiparallel spins are spatially uncorrelated.

Thus the antiparallel spins have a relatively large Coulomb energy to gain by becoming parallel. In an exact treatment one would expect the antiparallel spins to be somewhat correlated, thereby reducing this Coulomb difference. The differences between the exact properties of an interacting electron system and those obtained in the Hartree–Fock approximation are referred to as *correlation* effects. Estimates of these correlation corrections indicate that the nonmagnetic ground state of the electron gas has a lower energy than the ferromagnetic one.

The appearance of ferromagnetism in real metals is related to the presence of the ionic cores which tend to localize the itinerant electrons and introduce structure in the electronic density of states. We shall now consider two models that incorporate these features.

4.2.2 The Stoner Model

In the 1930's both *Slater* [4.16] and *Stoner* [4.17] combined Fermi statistics with the molecular field concept to explain itinerant ferromagnetism. This one-electron approach is now generally referred to as the Stoner model. It bears similarities to Landau's Fermi liquid theory in that the effect of the electron–electron interactions is to produce a spin-dependent potential that simply shifts the original Bloch states.

Stoner's result is contained within the generalized susceptibility $\chi(\boldsymbol{q})$ of an interacting electron system. As a model Hamiltonian we take a form similar to (1.132) where ϵ_k now refers to the Bloch band energy. Since the Coulomb interaction in a metal is screened, let us take a delta-function interaction of the form $I\delta(\boldsymbol{r}_i - \boldsymbol{r}_j)$. In this case one need not add a compensating background charge density and (1.132) becomes

$$\mathcal{H}_0 = \sum_{k,\sigma} \epsilon_k a^\dagger_{k\sigma} a_{k\sigma} + \frac{I}{V} \sum_k \sum_{k'} \sum_q \sum_\sigma a^\dagger_{k-q,\sigma} a^\dagger_{k'+q,-\sigma} a_{k',-\sigma} a_{k,\sigma} \tag{4.81}$$

If we now add a spatially varying field $H\hat{z}\cos(\boldsymbol{q}\cdot\boldsymbol{r})$, the Zeeman Hamiltonian becomes

$$\mathcal{H}_Z = -\frac{H}{2}[\mathcal{M}_z(\boldsymbol{q}) + \mathcal{M}_z(-\boldsymbol{q})] \tag{4.82}$$

where

$$\mathcal{M}_z(\boldsymbol{q}) = \tfrac{1}{2}g\mu_B \sum_k (a^\dagger_{k-q,\uparrow} a_{k\uparrow} - a^\dagger_{k-q,\downarrow} a_{k\downarrow}) \ . \tag{4.83}$$

Susceptibility. The susceptibility is obtained by calculating the average value of $\mathcal{M}_z(\boldsymbol{q})$ to lowest order in H. In particular, we must calculate the average of

$$m_{k,q} = a^\dagger_{k-q,\uparrow} a_{k,\uparrow} - a^\dagger_{k-q,\downarrow} a_{k,\downarrow} \ .$$

Following *Wolff* [4.18] we shall do this by writing the equation of motion for $m_{k,q}$ and using the fact that in equlibrium $\partial \langle m_{k,q} \rangle / \partial t = 0$, Thus,

$$\langle [m_{k,q}, \mathscr{H}_0 + \mathscr{H}_z] \rangle = 0. \tag{4.84}$$

This commutator involves a variety of twofold and fourfold products of electron operators. These are simplified by making a random-phase approximation in which we retain only those pairs which are diagonal or have the forms appearing in $m_{k,q}$ itself. Furthermore, the diagonal pairs are replaced by their average in the noninteracting ground state. This is equivalent to a Hartree–Fock approximation. The result is

$$(\epsilon_k - \epsilon_{k+q})\langle m_{k,q}\rangle - \frac{2I}{V}(n_{k+q} - n_k)\sum_{k'}\langle m_{k',q}\rangle - g\mu_B H(n_{k+q} - n_k) = 0. \tag{4.85}$$

Dividing by $(\epsilon_{k+q} - \epsilon_k)$ and summing over k gives

$$\sum_k \langle m_{k,q} \rangle = \frac{g\mu_B H \sum_k \frac{n_k - n_{k+q}}{\epsilon_{k+q} - \epsilon_k}}{1 - \frac{2I}{V}\sum_k \frac{n_k - n_{k+q}}{\epsilon_{k+q} - \epsilon_k}}. \tag{4.86}$$

Since the equilibrium occupation number n_k is just the Fermi function f_k, we recognize the sum appearing in this expression as the same as that which appears in the noninteracting susceptibility (3.84), which we shall denote as $\chi_0(q)$. Thus,

$$\boxed{\chi(q) = \frac{\chi_0(q)}{1 - \frac{2I}{g^2\mu_B^2}\chi_0(q)}.} \tag{4.87}$$

We could also have calculated $\langle m_{k,q} \rangle$ directly by the formulation indicated in (1.80). However, the appearance of the interaction term in the exponentials which enter $m(t)_{k,q}$ requires many-body perturbation techniques which are beyond the scope of this monograph. We shall use the diagrams introduced in Chap. 1 to make the results of such a treatment plausible. The interaction term in (4.81) has the diagrammatic form

$$\begin{array}{c}
k - q, \sigma \diagdown \qquad \diagup k', -\sigma \\
\qquad \diagup\!\!\!\!\!- - -\!\!\!\!\!\diagdown \\
k, \sigma \diagup \qquad \diagdown k' + q, -\sigma
\end{array}$$

There are two ways one can close the electron and hole lines:

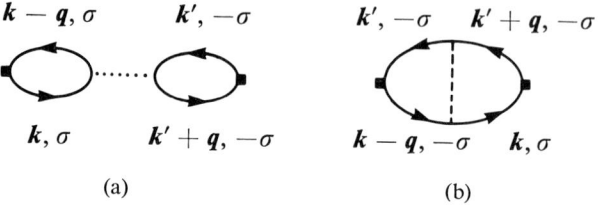

In (a) the vertices indicated by the small squares involve a momentum change q but no spin flip. This *longitudinal* spin fluctuation represents the first-order correction to χ_{zz}. The susceptibility (4.87) corresponds to the summation of all such diagrams:

$$\chi_{zz}(q) = \text{(diagrams)} \qquad (4.88)$$

In diagram (b) the vertices also involve a spin flip. This diagram is the first-order correction to the transverse susceptibility $\chi_{-+}(q)$ whose noninteracting from is given by (3.88). Summing all such *transverse* spin fluctuations gives the total susceptibility.

$$\chi_{-+}(q) = \text{(diagrams)} \qquad (4.89)$$

In the paramagnetic state rotational invariance requires $\chi_{-+}(q) = 2\chi_{zz}(q)$.

Returning to (4.87), since $\lim_{q \to 0} \chi_0(q) = \tfrac{1}{3} g^2 \mu_B^2 \mathscr{D}(\epsilon_F)$,

we see that

$$\chi(0) = \frac{\chi_0(0)}{1 - I\mathscr{D}(\epsilon_F)}. \qquad (4.90)$$

This has the same form as our Fermi liquid result (4.80). Thus we have an expression, at least in the random-phase approximation, for the Fermi liquid parameter B_0 in terms of the intraatomic exchange integral I and the density of states.

Notice that the criterion for the appearance of ferromagnetism is that $I\mathscr{D}(\epsilon_F) \geq 1$. This is referred to as the *Stoner criterion*.

One might wonder how ferromagnetism can occur with only an intraatomic Coulomb interaction. To see the physical origin of this, suppose we have a spin up at some site α. If the spin on a neighboring site α' is also up, it is forbidden by the exclusion principle from hopping onto site α. Therefore the two electrons do not interact, and we might define the energy of such a configuration as 0. However, if the spin of site α' is down, it has a nonzero probability of hopping onto site α. Thus the energy of this configuration is higher than that of the "ferromagnetic" one. However, hopping around can lower the kinetic energy of the electrons. This is reflected in the appearance of the density of states in the Stoner criterion. The occurrence of ferromagnetism depends, therefore, on the relative values of the Coulomb interaction and the kinetic energy.

Equation (4.90) has interesting consequences for metals which are paramagnetic but have a large enhancement factor. For example, an impurity spin placed in such a host produces a large polarization of the conduction electrons in its vicinity. Such *giant moments* have been observed in palladium and certain of its alloys. This seems reasonable, since palladium falls just below nickel in the periodic table and has similar electronic properties. Thus we might suspect that although palladium is not ferromagnetic like nickel, it at least possesses a large exchange enhancement.

Spin-Density Waves. Just as in the case of localized moments, a divergence of $\chi(q)$ for $q \neq 0$ would imply a transition to a state of nonuniform magnetization. In fact, *Overhauser* has shown that the $\chi(q)$ associated with an unscreened Coulomb interaction in the Hartree–Fock appoximation does [4.19] diverge as $q \to 2k_F$, as shown in Fig. 4.17. This would lead to a ground state characterized by a periodic spin density, called a *spin-density wave*. The effects of screening and electron correlations, however, tend to suppress this divergence. Consequently, a spin-density wave can form only under rather special conditions. We get some feeling for these conditions by considering the behavior of the non-interacting susceptibility $\chi_0(q)$ in one and two dimensions as shown in Fig.4.18. We see that lower dimensional systems are more likely to become unstable with respect to spin–density wave formation. The reason for this has to do with the

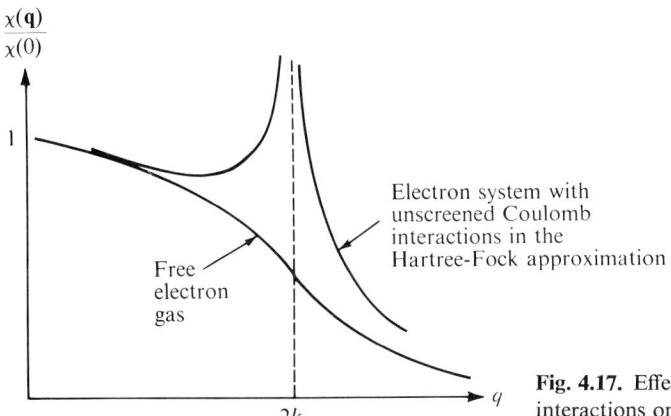

Fig. 4.17. Effect of electron–electron interactions on the susceptibility

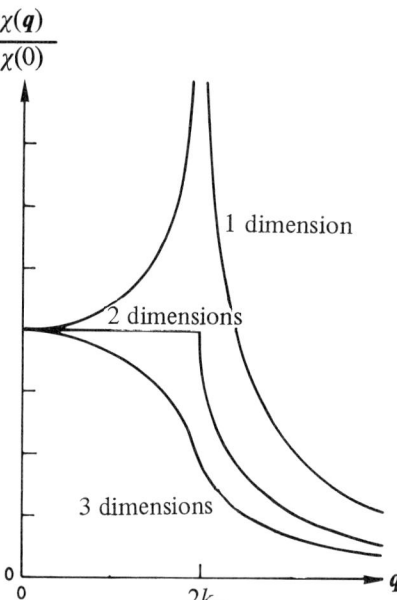

Fig. 4.18. Effect of dimensionality on the free electron susceptibility

fact that in a state characterized by a wave vector q electrons with wave vectors which differ by q become correlated. This effectively removes them from the Fermi sea. This is often described as a "nesting" of the corresponding states. In one and two dimensions Fermi surfaces are geometrically simpler, which means that nesting will have more dramatic effects. These same considerations, however, also apply to *charge*-density instabilities. And, in fact, most of the materials which do show such Fermi-surface-related instabilities show charge-density waves.

To date, the only example of a material possessing a spin-density-wave ground state is chromium. The reason for this has to do with the band structure of chromium.

If the susceptibility does not actually diverge at some nonzero wave vector, but nevertheless becomes very large, the system may be said to exhibit *antiferromagnetic* exchange enhancement. Experiments on dilute alloys of Sc:Gd indicate that scandium metal may be an example of such a type [4.20].

Exchange Splitting. If the system *is* ferromagnetic, then $\langle m_{k,q=0} \rangle = (n_{k\uparrow} - n_{k\downarrow}) \neq 0$ even in the absence of an applied field. In this case the Hamiltonian (4.81) may be written in a particularly revealing form by considering only the diagonal terms in (4.81):

$$\tfrac{1}{2} n_{k'\downarrow} a^\dagger_{k\uparrow} a_{k\uparrow} + \tfrac{1}{2} n_{k'\uparrow} a^\dagger_{k\downarrow} a_{k\downarrow} \; . \quad \text{Since} \tag{4.91}$$

$$M = \frac{g\mu_B}{2V} \sum_k (n_{k\uparrow} - n_{k\downarrow}) , \qquad (4.92)$$

the Hamiltonian becomes

$$\mathscr{H}'_0 = E_0 + \sum_{k,\sigma} \epsilon_{k\sigma} a^\dagger_{k\sigma} a_{k\sigma} \qquad (4.93)$$

where

$$\epsilon_{k\sigma} = \epsilon_k + \frac{NI}{4V} - \frac{IM}{2g\mu_B} \sigma . \qquad (4.94)$$

Thus the spin-up and spin-down energy bands are split by an amount proprotional to the magnetization. This was the basic idea in Stoner's original theory of ferromagnetism. More generally, this splitting arises from the difference between the spin-up and spin-down exchange-correlation potentials seen by the electrons. Since the Stoner model was first proposed there has been a great deal of progress in specifying these potentials.

In 1951 *Slater* [4.21] suggested that we approximate the effect of exchange by the potential

$$V_x(\mathbf{r}) = -6[(3/8\pi)\rho(\mathbf{r})]^{1/3} .$$

This approximate form has been used extensively both in atomic and solid-state calculations. Physically, this density to the one-third power arises from the fact that in the Hartree–Fock approximation parallel spins are kept farther apart than antiparallel spins. Therefore, there is an "exchange hole" around any particular spin associated with a deficiency of similar spins. The radius of this hole must be such that $(\frac{3}{4})\pi r^3 \rho = 1$. Since the potential associated with this deficiency is proportional to $1/r$, we obtain an exchange potential proportional to $\rho^{1/3}$.

In 1965 *Kohn* and *Sham* [4.22] rederived the exchange potential by a different method and obtained a value two thirds that of Slater's. This led workers to multiply (4.95) by an adjustable constant, α, which can be determined for each atom by requiring that the so-called $X\alpha$ energy be equal to the Hartree–Fock energy for that atom. An interesting application of the $X\alpha$ method has been made by *Hattox* et al. [4.23]. They calculated the magnetic moment of bcc vanadium as a function of lattice spacing. The result is shown in Fig. 4.19. We see that the moment falls suddenly to zero at a spacing 20% larger than the actual observed spacing. This decrease is due to the broadening of the 3d band as the lattice spacing decreases. This is consistent with the fact that bcc vanadium, where the vanadium–vanadium distance is $\sqrt{3}\ a_0/2 = 2.49$ Å, is observed to be nonmagnetic. In Au₄V, however, the vanadium–vanadium distance has increased to 3.78 Å and the vanadium has a moment near one Bohr magneton.

If one uses this approach to find α for the magnetic transition metals one finds that the resulting magnetic moments are not in agreement with those observ-

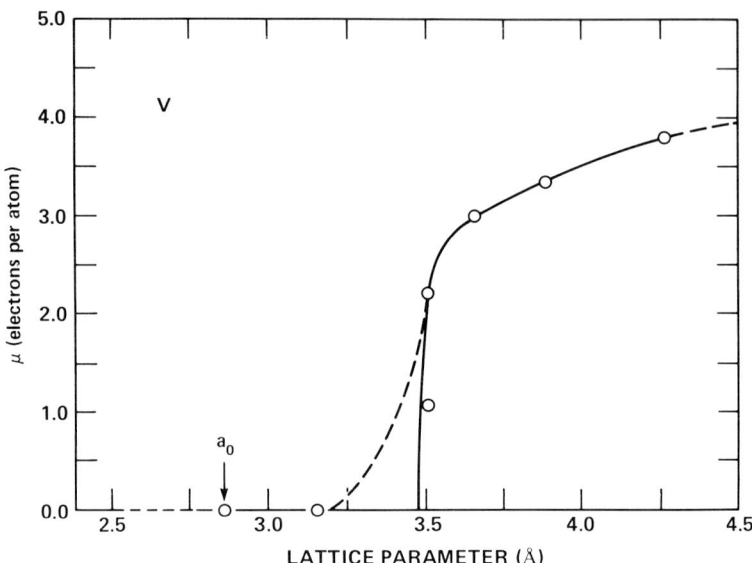

Fig. 4.19. Calculated magnetic moment of vanadium metal as a function of lattice parameter. The two points for $a = 3.5$ Å correspond to two distinct self-consistent solutions associated with different starting potentials. These two solutions are the result of a double minimum in the total energy versus magnetization curve. At 4.25 Å and 3.15 Å the calculations converged to unique values [4.23].

ed. This is not surprising when we consider that we have replaced a nonlocal potential (the Hartree–Fock potential) by a local potential (the Slater $\rho^{1/3}$). Furthermore, the fact that these are different means that the Slater potential must, by definition, include some correlation. However, we do not know if it is in the right direction.

The arbitrariness inherent in the $X\alpha$ method may be avoided by using the "spin-density functional" formalism. This essentially enables one to utilize the results of many-body calculations for the homogeneous electron gas in determining the exchange and correlation potentials in transition metals. This approach is based on a theorem by *Hohenberg* and *Kohn* [4.24] which states that the ground-state energy of an inhomogeneous electron gas is a functional of the electron density $\rho(\mathbf{r})$ and the spin density $m(\mathbf{r})$. The effective potential, for example, is given by

$$V_{\text{eff}} = v(\mathbf{r}) + e^2 \int \frac{\rho(\mathbf{r'})d\mathbf{r'}}{|\mathbf{r} - \mathbf{r'}|} + \frac{\delta E_{xc}\{\rho\}}{\delta \rho(\mathbf{r})}, \qquad (4.95)$$

where $v(\mathbf{r})$ is the one-electron potential, the second term is the Hartree term, and $E_{xc}\{\rho\}$ is the exchange and correlation energy. The importance of this theorem is that it reduces the many-body problem to a set of one-body problems for the one-electron wave functions $\phi_{k\sigma}(\mathbf{r})$ which make up the density,

$$\rho(\mathbf{r}) = \sum_{k,\sigma} \phi_{k\sigma}^*(\mathbf{r})\phi_{k\sigma}(\mathbf{r}) \ . \tag{4.96}$$

The *local* density approximation consists of replacing the unknown functional $E_{xc}\{\rho\}$ by $\int d^3 r \rho(\mathbf{r})\epsilon_{xc}^h[\rho(\mathbf{r})]$ where $\epsilon_{xc}^h[\rho(\mathbf{r})]$ is the exchange and correlation contibution to the energy of a *homogeneous* interacting electron gas of density $\rho(\mathbf{r})$. Although one might question the use of such an approximation in transition metals where there are rapid variations in the charge density, the fact that it is only the spherical average of the exchange-correlation hole which enters makes $E_{xc}\{\rho\}$ fairly insensitive to the details of this hole.

The density functional approach leads to a set of self-consistent Hartree equations. It is tempting to identify the eigenvalues as effective single particle energies. However, detailed studies of the energy bands indicate that this identification may not be appropriate in certain cases. Photoemission has become a powerful technique for probing the electronic states of solids. In this technique, light (actually ultraviolet or x rays) is absorbed by a solid. Those electrons excited above the vacuum level leave the solid and are collected. Initially, all the electrons emitted were collected and analyzed. However, with the availability of high-intensity synchrotron x-ray sources, it became possible to resolve the direction of the emitted electrons. This enables one to reconstruct the energy-momentum relation of the electrons in the solid. Figure 4.20 compares the results of such an experiment on copper [4.25] with the calculated band structure. The agreement is remarkable. This technique has now been extended to determine the spin polarization of the photoemitted electrons. This extension has been pioneered by H. C. Siegmann in Zurich. The photoemitted electrons are accelerated to relativistic velocities and then scattered from a gold foil. Any initial spin polarization is reflected in asymmetric, or Mott, scattering. Figure 4.20 also shows the results of such measurements on Ni [4.26]. Although the overall features are described by the theory, the detailed fit is not nearly as good as in Cu.

Despite this problem with the identifiction of single-particle energies, one can calculate ground-state properties such as the bulk modulus and the magnetic moment. The results are in very good agreement with the experimental values. What makes this agreement all the more impressive is that the only input to such calculations are the atomic numbers and the crystal structures.

Although the Stoner theory works reasonably well for magnetic properties at $T=0$, it fails when applied to finite temperature properties. For example, the only place that temperature enters the susceptibility (4.90) is in the agruments of the Fermi functions. The Curie temperature is calculated from the relation

$$M = \frac{g\mu_B}{2V}\sum_k (f_{k\uparrow} - f_{k\downarrow}) \tag{4.97}$$

where $f_{k\sigma}$ is the Fermi function with $\epsilon_{k\sigma} = \epsilon_k + NI/4V - (IM/2g\mu_B)\sigma$. Since at T_c M is very small, the Fermi functions may be expanded about $M=0$. The condition for T_c then becomes

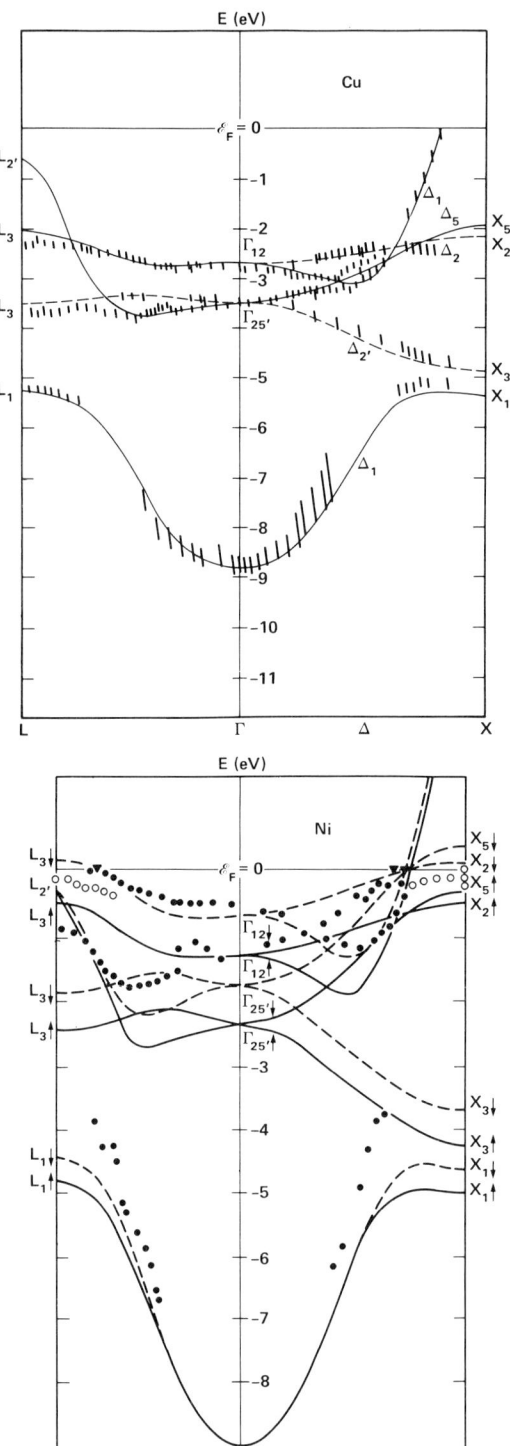

Fig. 4.20a, b. Comparison of angle-resolved photoemission data with calculated band structures for (a) copper and (b) nickel (open data points corraspond to spin up, solid points to spin down) The trianglas are de Haas van Alphen data

$$I \int d\epsilon \frac{\partial f(T_c)}{\partial \epsilon} \mathscr{D}(\epsilon) + 1 = 0. \tag{4.98}$$

Using the values of I that give the correct moments at $T = 0$ for Fe, Co, and Ni, we obtain T_c's from (4.98) that are about 5 times larger than the observed values.

The problem with this application of the Stoner model is that the introduction of "up" and "down" spin directions destroys the rotational symmetry. That is, at nonzero temperatures the *direction* of the effective field arising from the electron–electron interactions as well as its magnitude will vary from site to site as illustrated in Fig. 4.21. Independent calculations by *Hubbard* [4.27] and *Heine* and collaborators [4.28] show that the energy associated with such local changes in the direction of the magnetization is much less than that associated with changes in the *magnitude* of the magnetization. That is, the exchange stiffness A is less than the Stoner parameter I. The problem remains, however, to relate this observation to the thermodynamic properties.

4.2.3 The Hubbard Model

The one-electron Stoner model described above is expected to apply to systems with fairly broad bands. As the bands become narrower intraionic correlation effects become more important. In this case it is convenient to work in the Wannier representation which emphasizes the atomic aspect of the problem. In this representation the one-electron terms become

$$\sum_{\alpha\alpha'\sigma} t_{\alpha\alpha'} a^{\dagger}_{\alpha'\sigma} a_{\alpha\sigma}, \tag{4.99}$$

where

$$t_{\alpha\alpha'} = \frac{1}{V} \sum_k \epsilon_k \exp\left[i\mathbf{k} \cdot (\mathbf{R}_\alpha - \mathbf{R}_{\alpha'})\right] \tag{4.100}$$

describes the hopping of an electron from sits α to site α'. The interaction terms are given by (2.79). Since we are now dealing, with an itinerant situation, we

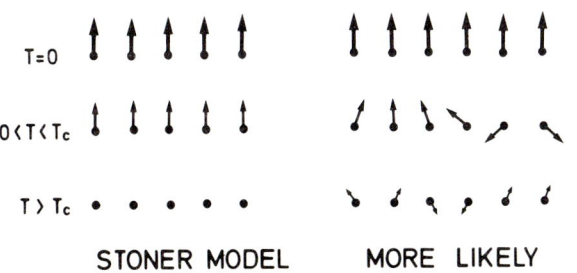

Fig. 4.21 Pictorial comparison of the exchange fields seen by an electron in the Stoner model and what is more likely the case

shall assume that screening effects restrict the interaction to one site. If there is a single nondegenerate orbital $\varphi_0(\mathbf{r} - \mathbf{R}_\alpha)$ associated with each site, then the interaction becomes

$$\sum_{\alpha,\sigma} U_0 n_{\alpha\sigma} n_{\alpha,-\sigma} \tag{4.101}$$

where

$$U_0 = \langle 00|V|00\rangle = \iint d\mathbf{r}\, d\mathbf{r}'\, \varphi_0(\mathbf{r} - \mathbf{R}_\alpha)\varphi_0(\mathbf{r}' - \mathbf{R}_\alpha) \times V$$
$$\varphi_0(\mathbf{r}' - \mathbf{R}_\alpha)\varphi_0(\mathbf{r} - \mathbf{R}_\alpha). \tag{4.102}$$

Because of the screening, which we can think of as s-d correlation effects, the value of U_0 is of the order of 2eV.

Equations (4.99, 101) constitute the Hubbard Hamiltonian. It contains the same physics as the Stoner model. In fact, it gives the same susceptibility (4.87) in the mean-field approximation with U_0 in place of I. In a series of papers Hubbard [4.29] investigated the effects of correlation within this model. He found, for example, that for large correlation the electronic band is split into two subbands separated by U_0. The transition-metal oxides such as NiO and CoO are generally cited as examples of materials where such correlation effects are very important.

In order to understand the variation in magnetic properties as one moves across the transition-metal series, it is necessary to generalize the model above to include orbital degeneracy. This obviously introduces many more Coulomb and exchange integrals. The first simplification is to neglect interactions involving more than two orbitals. One then assumes that all the off-diagonal Coulomb and exchange integrals are the same, i.e.,

$$\left.\begin{array}{l}\langle mm'|V|mm'\rangle = U \\ \langle mm'|V|m'm\rangle = J\end{array}\right\} \quad m' \neq m. \tag{4.103}$$

The exchange integral J is smaller than U, of the order of 0.5 eV. We also take all the diagonal integrals to have the same value. If we require that our model Hamiltonian preserve the rotational invariance of the original Hamiltonian, then the diagonal integral is related to the off-diagonal integrals by

$$\langle mm|V|mm\rangle = U + J. \tag{4.104}$$

The resulting generalized Hubbard Hamiltonian becomes

$$\mathcal{H}' = \tfrac{1}{2}(U+J)\sum_{\alpha,m,\sigma} n_{\alpha m\sigma}n_{\alpha m,-\sigma}$$
$$+ \tfrac{1}{2}\sum_{\alpha,m,m'\sigma}(1 - \delta_{mm'})[Un_{\alpha m\sigma}n_{\alpha m',-\sigma} + (U-J)n_{\alpha m\sigma}n_{\alpha m'\sigma}$$
$$- Ja^\dagger_{\alpha m\sigma}a_{\alpha m-\sigma}a^\dagger_{\alpha m'-\sigma}a_{\alpha m'\sigma}]. \tag{4.105}$$

4.2 Metals

By making use of the spin representation (2.85–87) the last term may be written

$$(U - \tfrac{1}{2}J) \sum_{\alpha, m<m', \sigma, \sigma'} n_{\alpha m \sigma} n_{\alpha m' \sigma'} - 2J \sum_{\alpha, m<m'} \mathbf{S}_{\alpha m} \cdot \mathbf{S}_{\alpha m'}, \qquad (4.106)$$

which clearly reveals the Hund's rule exchange coupling. Let us now consider how these exchange interactions modify the generalized susceptibility. As in the derivation of (4.87) we shall employ the Hartree–Fock approximation. This amounts to writing

$$n_{\alpha m \sigma} = \langle n_{\alpha m \sigma} \rangle + (n_{\alpha m \sigma} - \langle n_{\alpha m \sigma} \rangle) \qquad (4.107)$$

and assuming the term in parenthesis is small. Thus,

$$n_{\alpha m \sigma} n_{\alpha m' \sigma'} = \langle n_{\alpha m \sigma} \rangle n_{\alpha m' \sigma'} + \langle n_{\alpha m' \sigma'} \rangle n_{\alpha m \sigma} - \langle n_{\alpha m \sigma} \rangle \langle n_{\alpha m' \sigma'} \rangle, \qquad (4.108)$$

and the Hartree–Fock Hamiltonian becomes

$$\mathcal{H}_{\mathrm{HF}} = \sum_{\alpha, \alpha' m, \sigma} t_{\alpha \alpha'} a^\dagger_{\alpha m \sigma} a_{\alpha' m \sigma} + (U + J) \sum_{\alpha, m \sigma} \langle n_{\alpha m \sigma} \rangle n_{\alpha m, -\sigma}$$

$$+ U \sum_{\alpha, m<m', \sigma} (\langle n_{\alpha m \sigma} \rangle n_{\alpha m', -\sigma} + \langle n_{\alpha m', -\sigma} \rangle n_{\alpha m \sigma}) \qquad (4.109)$$

$$+ (U - J) \sum_{\alpha, m<m', \sigma} (\langle n_{\alpha m \sigma} \rangle n_{\alpha m' \sigma} + \langle n_{\alpha m' \sigma} \rangle n_{\alpha m \sigma}).$$

The magnetic moment per ion is

$$m_\alpha = -\frac{g \mu_\mathrm{B}}{2} \sum_m (\langle n_{\alpha m \uparrow} \rangle - \langle n_{\alpha m \downarrow} \rangle). \qquad (4.110)$$

If the average number of electrons per ion is n, then

$$n = \sum_m (\langle n_{\alpha m \uparrow} \rangle + \langle n_{\alpha m \downarrow} \rangle). \qquad (4.111)$$

If we take the case of a transition metal where $m = 1, \ldots 5$, then

$$\langle n_{\alpha m \uparrow} \rangle = \frac{1}{10}\left(n - \frac{2m_\alpha}{g \mu_\mathrm{B}}\right),$$

$$\langle n_{\alpha m \downarrow} \rangle = \frac{1}{10}\left(n + \frac{2m_\alpha}{g \mu_\mathrm{B}}\right), \qquad (4.112)$$

and the Hartree–Fock Hamiltonian becomes

$$\mathcal{H}_{\mathrm{HF}} = \sum_{\alpha, \alpha' m, \sigma} t_{\alpha \alpha'} a^\dagger_{\alpha m \sigma} a_{\alpha' m \sigma} + \frac{1}{10} N(9U - 3J) n^2$$

$$+ \frac{U + 5J}{g \mu_\mathrm{B}} \sum_{\alpha, m} m_\alpha (n_{\alpha m \uparrow} - n_{\alpha m \downarrow}). \qquad (4.113)$$

142   4. The Static Susceptibility of Interacting Systems

In the presence of an external field $H \cos(\boldsymbol{q}\cdot\boldsymbol{r})\hat{\boldsymbol{z}}$ the Zeeman interaction has the form

$$\mathcal{H}_z = \frac{g\mu_B H}{2} \sum_{\alpha,m} (n_{\alpha m\uparrow} - n_{\alpha m\downarrow}) \cos \boldsymbol{q}\cdot\boldsymbol{R}_\alpha \,. \tag{4.114}$$

If we assume that the induced moments m_α have the same spatial variation as the applied field, i.e.,

$$m_\alpha = m \cos(\boldsymbol{q}\cdot\boldsymbol{R}_\alpha) \,,$$

then by comparing the last term in (4.113) with (4.114) we see that the effect of the interactions is to give an effective field

$$H(\boldsymbol{r})_{\text{eff}} = \frac{2m}{g\mu_B} (U + 5J) \cos(\boldsymbol{q}\cdot\boldsymbol{r}) \,. \tag{4.115}$$

Taking the Fourier transform of the total effective field and using the fact that

$$M(\boldsymbol{q}) = \chi_0(\boldsymbol{q}) H(\boldsymbol{q})_{\text{eff}} \tag{4.116}$$

where $\chi_0(\boldsymbol{q})$ is the susceptibility of the noninteracting electron system, we find for the susceptibility of this Hubbard model

$$\boxed{\chi(\boldsymbol{q}) = \frac{\chi_0(\boldsymbol{q})}{1 - \dfrac{2(U+5J)}{g^2\mu_B^2} \chi_0(\boldsymbol{q})}} \,. \tag{4.117}$$

This has the same form as the Stoner susceptibility with an effective Stoner parameter $I_{\text{eff}} = U + 5J$, which is of the order of 0.9 eV. The corresponding Stoner criterion becomes $I_{\text{eff}} \mathscr{D}(\epsilon_F) > 1$. Thus the presence of intraatomic exchange favors ferromagnetism. However, again, we expect correlation effects to be very important. There have been many calculations of correlation effects within the Hubbard model but their descripion is beyond the scope of this monograph. The general result of including correlation is to reduce the region of parameter space where ferromagnetism, or antiferromagnetism, is expected.

5. The Dynamic Susceptibility of Weakly Interacting Systems

We now turn to the response of magnetic systems to time-dependent excitations. We shall restrict our consideration to systems in which the interactions among the constitutents are not strong enough to produce a spontaneous magnetization. The study of the frequency response of such systems is essentially the study of paramagnetic resonance and relaxation phenomena. This is obviously an enormous subject, and we shall not be able to go into it in great detail. However, we shall examine some of the basic ideas within the framework of our generalized susceptibility.

5.1 Localized Moments

Let us begin by considering a system of identical localized spins characterized by a noninteracting Hamiltonian $\mathcal{H}_0$. In Sect. 3.1 we found that in the presence of a uniform static field $H_0\hat{z}$ such a system develops a magnetization, which we shall denote by $M_0\hat{z}$. Let us now apply an additional time-dependent field $\boldsymbol{H}_1 \cos \omega t$ and investigate the response to this field.

Since the applied field is uniform, we shall drop explicit reference to spatial considerations. Thus we may write the magnetization as

$$\boldsymbol{M} = \frac{1}{V}\operatorname{Tr}\{\rho \mathcal{M}\} \ . \tag{5.1}$$

The presence of the volume in this relation and not in (1.49) is due to our definition of the space-dependent operator in (1.48). Because the Zeeman Hamiltonian is time dependent, the density matrix, and hence the magnetization, will be time dependent. Differentiating (5.1) with respect to time and making use of (1.47), which implies $d\rho/dt = 0$, we obtain

$$\frac{d\boldsymbol{M}}{dt} = -\frac{i}{\hbar V}\operatorname{Tr}\{\rho[\mathcal{M},\mathcal{H}]\} \ . \tag{5.2}$$

For the present let us assume that the Hamiltonian consists of a part which commutes with $\mathcal{M}$ plus a Zeeman part

$$\mathcal{H}_Z = -\mathcal{M}\cdot\boldsymbol{H} \ . \tag{5.3}$$

Equation (5.2) then becomes

$$\frac{d\boldsymbol{M}}{dt} = -\gamma \boldsymbol{M} \times \boldsymbol{H} \qquad (5.4)$$

where $\gamma\hbar = g\mu_B$. The minus sign arises because we are explicitly dealing with electrons which have a negative gyromagnetic ratio.

If the dynamic field $\boldsymbol{H}_1(t) = \boldsymbol{H}_1 \cos \omega t$ is applied in the z direction, then, according to (5.4), it exerts no torque on the equilibrium magnetization $M_0\hat{\boldsymbol{z}}$. Therefore the response to such a field is 0. This raises an interesting point. If the frequency ω goes to 0, (5.4) tells us that there is no response to such a static field. But in Chap. 3 we found that the magnetization does respond to a static field, with the resulting magnetization given by Curie's law. The answer to this paradox lies in the fact that in Chap. 3 we assumed that the spin system was always in equilibrium. This implies that there is a coupling between the individual spins and their environment which enables them to reach equilibrium. The time it might take the spin system to do this is not important in the static case, since we can always keep the field on until equilibrium has been achieved. In the dynamic case, however, this assumption is not valid, since we may want the response at a frequency that is much faster than this relaxation frequency. In fact, this is generally the experimental situation. In the dynamic case we must actually solve for the nonequilibrium density matrix. We shall see later that the only way to get a response in the z direction is to introduce a relaxation mechanism.

Let us now consider a system which has come to equilibrium in an applied dc field, H_0, and ask what the response will be to a time-varying transverse field $\boldsymbol{H}_1$. In particular, let $\boldsymbol{H}_1 = H_1\hat{\boldsymbol{x}}$. If we write the magnetization as

$$\boldsymbol{M} = m_x\hat{\boldsymbol{x}} + m_y\hat{\boldsymbol{y}} + (M_0 - m_z)\hat{\boldsymbol{z}}, \qquad (5.5)$$

(5.4) becomes, in component form,

$$\frac{dm_x}{dt} = -\gamma H_0 m_y, \qquad (5.6a)$$

$$\frac{dm_y}{dt} = -\gamma H_1(t)(M_0 - m_z) + \gamma H_0 m_x, \qquad (5.6b)$$

$$\frac{dm_z}{dt} = -\gamma H_1(t) m_y. \qquad (5.6c)$$

These equations are nonlinear. As a consequence, magnetic systems exhibit a number of interesting features, such as "spin echoes", which we shall discuss later. For the present, let us linearize (5.6) by neglecting terms which are quadratic in H_1 or the components of $\boldsymbol{m}$. The result is

$$\frac{dm_x}{dt} = -\omega_0 m_y, \tag{5.7a}$$

$$\frac{dm_y}{dt} = -\gamma H_1(t) M_0 + \omega_0 m_x, \tag{5.7b}$$

$$\frac{dm_z}{dt} = 0, \tag{5.7c}$$

where $\omega_0 \equiv \gamma H_0$.

We see from (5.7c) than m_z is a constant. Since $[\mathcal{M}\cdot\mathcal{M}, \mathcal{M}\cdot\mathbf{H}] = 0$, this means that the magnitude of the magnetization is a conserved quantity. Therefore, once m_x and m_y are known, m_z may be obtained from the condition that

$$\mathcal{M}\cdot\mathcal{M} = M_0^2. \tag{5.8}$$

The components m_x and m_y may be determined from (5.7a, b). Differentiating (5.7a) with respect to time and using (5.7b), we obtain the equation for m_x,

$$\frac{d^2 m_x}{dt^2} + \omega_0^2 m_x = \gamma H_1 \omega_0 M_0 \cos \omega t. \tag{5.9}$$

To find the susceptibility we must solve this differential equation. We may write it in a more general form as

$$\mathcal{L} m_x(t) = F(t), \tag{5.10}$$

where $\mathcal{L}$ is the differential operator, in this case $d^2/dt^2 + \omega_0^2$. The solution is then given symbolically by

$$m_x(t) = \mathcal{L}^{-1} F(t), \tag{5.11}$$

where $\mathcal{L}^{-1}$ is the operator inverse of $\mathcal{L}$, *provided* that it exists. It is customary to rewrite this as

$$m_x(t) = \int_{-\infty}^{\infty} dt' \, \mathcal{L}^{-1} F(t') \delta(t - t') = \int_{-\infty}^{\infty} dt' \, \mathcal{G}(t, t') F(t'), \tag{5.12}$$

where $\mathcal{G}(t, t')$ is the *Green's function* of the differential operator $\mathcal{L}$, defined as $\mathcal{L}^{-1}$ operating on the delta function.

A very important relationship between the Green's function and the susceptibility is easily obtained from (5.12). Taking the Fourier transform, we have

$$m_x(\omega) = \int_{-\infty}^{\infty} dt \int_{-\infty}^{\infty} dt' \, \mathcal{G}(t, t') F(t') e^{i\omega t}. \tag{5.13}$$

If the medium is stationary, it can be shown that the Green's function depends only on the relative time $t - t'$. Inserting the factor $\exp(-i\omega t') \exp(i\omega t')$ in (5.13), we obtain

$$m_x(\omega) = \int_{-\infty}^{\infty} dt \int_{-\infty}^{\infty} dt' \, \mathscr{G}(t - t') e^{i\omega(t-t')} F(t') e^{i\omega t'} . \tag{5.14}$$

Converting the integral over t into the integral over $\tau \equiv t - t'$ and recalling our definition of the susceptibility (1.58), we obtain

$$\boxed{\chi_{xx}(\omega) = \gamma M_0 \omega_0 \mathscr{G}(\omega) .} \tag{5.15}$$

The factor $\gamma M_0 \omega_0$ arises because the forcing function $F(t)$ differs from the applied field by this quantity. The important result is that the Fourier transform of the Green's function is, essentially, the frequency-dependent susceptibility. Notice that if the so-called double-time Green's function defined on page 18 is introduced into that equation, we obtain (5.15) directly, aside from the proportionality factor. This should not be surprising, for both Green's functions represent the linear response to a time-dependent magnetic field.

Let us now evaluate the Green's function. This is easily done by using the integral representation for the delta function.

$$\delta(t - t') = \frac{1}{2\pi} \int_{-\infty}^{\infty} d\omega \, e^{-i\omega(t-t')} .$$

Thus

$$\mathscr{G}(t - t') = \mathscr{L}^{-1} \delta(t - t') = \frac{1}{2\pi} \int_{-\infty}^{\infty} d\omega \, \frac{e^{i\omega(t-t')}}{-\omega^2 + \omega_0^2} . \tag{5.16}$$

This integral is not defined until we specify how to treat the singularity at $\omega = \omega_0$. To resolve this difficulty we invoke *causality* and require that $\mathscr{G}(t - t') = 0$ for $t < t'$. This enables us to evaluate (5.16) unambiguously. The result is

$$\mathscr{G}(t - t') = \frac{\sin \omega_0(t - t')}{\omega_0} \theta(t - t') \tag{5.17}$$

where $\theta(t - t')$ is 0 for $t < t'$ and 1 for $t > t'$. Taking the Fourier transform of (5.17) and multiplying the result by $\gamma M_0 \omega_0$, we finally obtain

$$\boxed{\chi_{xx}(\omega) = \frac{\gamma M_0 \omega_0}{\omega_0^2 - \omega^2} + i \frac{\pi \gamma M_0}{2} [\delta(\omega - \omega_0) - \delta(\omega + \omega_0)] .} \tag{5.18}$$

It is interesting to note that the real part of this susceptibility comes from the *particular* solution to (5.9), while the imaginary part, which is necessary to satisfy causality, comes from the solutions of the homogeneous version of (5.9). It is easily demonstrated that (5.18) satisfies the Kramers–Kronig relations (1.64, 65).

The power absorbed by the magnetic system from this time-varying source is given by

$$P = \overline{-\mathbf{M} \cdot \frac{d H_1(t)}{dt}} \tag{5.19}$$

where the bar indicates a time average. From (1.55),

$$M_x(t) = H_1[\chi'_{xx}(\omega) \cos \omega t + \chi''_{xx}(\omega) \sin \omega t], \tag{5.20}$$

and so (5.19) becomes

$$\boxed{P = \tfrac{1}{2} \omega H_1^2 \chi''_{xx}(\omega).} \tag{5.21}$$

From (5.18) we see that the power absorption of our noninteracting magnetic system occurs only at the frequency $\omega_0 = \gamma H_0$. Furthermore, this response will be infinite. In reality, of course, interactions within the system make this response finite and spread it over a distribution of frequencies. It is the determination of this response function to which we now address ourselves.

The poles in the response function define the excitation spectrum associated with the system. The real part of the pole gives the frequency of the excitation, and the imaginary part gives its damping. If the poles move too far from the real axis, the concept of an excitation is less well defined. Since we are dealing here with noninteracting localized moments, these poles are independent of wave vector. Therefore the excitation spectrum is as indicated in Fig. 5.1.

5.1.1 The Bloch Equations

If the magnetization is excited to a nonequilibrium value under the influence of an external field, when the field is suddenly removed the magnetization will relax back to its equilibrium value. The details of the relaxation depend on the nature of the interactions in the system. However, if we assume that this re-

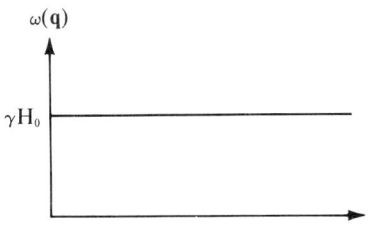

Fig. 5.1. The excitation spectrum of a system of uncoupled spins

laxation, whatever its origin, has an exponential form, then we can develop a phenomenological description of the response function. In general the longitudinal and transverse components of the magnetization may relax with different rates. The equations of motion, called the *Bloch equations*, are [5.1]

$$\frac{dm_z}{dt} = \gamma(M \times H)_z - \frac{m_z}{T_1}. \tag{5.22a}$$

$$\frac{dm_{x,y}}{dt} = \gamma(M \times H)_{x,y} - \frac{m_{x,y}}{T_2} \tag{5.22b}$$

If we again assume a driving field of the form $H_1(t) = H_1 \cos(\omega t)\hat{x}$, the linearized equation for m_x is

$$\frac{d^2 m_x}{dt^2} + \frac{2}{T_2}\frac{dm_x}{dt} + \left(\omega_0^2 + \frac{1}{T_2^2}\right) m_x = \gamma H_1 \omega_0 M_0 \cos \omega t. \tag{5.23}$$

By assuming that m_y relaxes to 0 we have essentially incorporated causality into our solution, for the poles which enter the expression for the Green's function associated with this system, analogous to (5.16), are now displaced off the real axis, as shown in Fig. 5.2. Thus we speak of the susceptibility as being analytic in the upper half of the complex plane. The Green's function is then

$$\mathcal{G}(t - t') = \frac{\sin \omega_0 (t - t')}{\omega_0} e^{-(t-t')/T_2} \theta(t - t'). \tag{5.24}$$

Taking the Fourier transform leads us to

$$\chi_{xx}(\omega) = \chi'_{xx}(\omega) + i\chi''_{xx}(\omega) \tag{5.25}$$

where

$$\chi'_{xx}(\omega) = \tfrac{1}{2}\gamma M_0 \left[\frac{\omega_0 - \omega}{(\omega_0 - \omega)^2 + (1/T_2)^2} + \frac{\omega + \omega_0}{(\omega + \omega_0)^2 + (1/T_2)^2}\right] \tag{5.26}$$

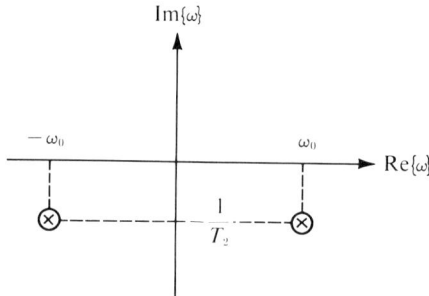

Fig. 5.2. Location of the poles in the imaginary part of the susceptibility

and

$$\chi''_{xx}(\omega) = \frac{\gamma M_0}{2T_2}\left[\frac{1}{(\omega_0 - \omega)^2 + (1/T_2)^2} - \frac{1}{(\omega_0 + \omega)^2 + (1/T_2)^2}\right] \quad (5.27)$$

As $1/T_2$ becomes very small the functions in (5.27) become higher and narrower in such a way as to maintain their area. Therefore, in the limit $1/T_2 \to 0$, they may be considered as representations of delta functions; that is,

$$\lim_{\epsilon \to 0} \frac{\epsilon}{x^2 + \epsilon^2} = \pi\delta(x) \quad (5.28)$$

and (5.26, 27) reduce to our previous result (5.18). The real and imaginary part of the susceptibility are shown in Fig. 5.3.

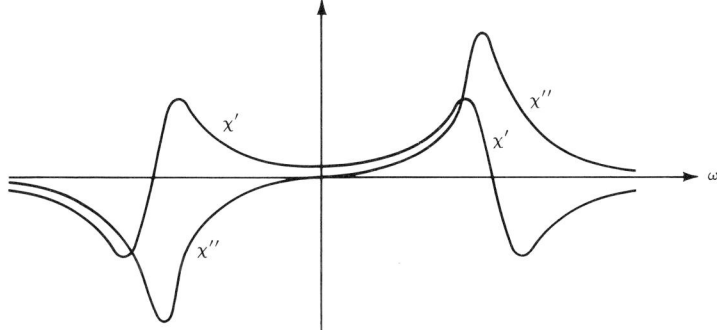

Fig. 5.3. Plot of the real and imaginary parts of the susceptibility $\chi_{xx}(\omega)$

Let us keep in mind that we are dealing with the response to a linearly polarized driving field. Some experiments employ a circularly polarized field. In such cases the response involves the other components of the susceptibility. For example, consider the circularly polarized field

$$\boldsymbol{H}_1(t) = H_1 \cos(\omega t)\hat{\boldsymbol{x}} + H_1 \sin(\omega t)\hat{\boldsymbol{y}} \; .$$

From (1.50), the responses in the x and y directions to such a field may be written as

$$m_x(t) = H_1[(\chi'_{xx} - \chi''_{xy}) \cos \omega t + (\chi''_{xx} + \chi'_{xy}) \sin \omega t] \; ,$$
$$m_y(t) = H_1[(\chi'_{yx} - \chi''_{yy}) \cos \omega t + (\chi''_{yx} + \chi'_{yy}) \sin \omega t] \; .$$

Solving (5.22) for m_y leads to the results

$$\chi'_{yx}(\omega) = -\frac{\gamma M_0}{2T_2}\left[\frac{1}{(\omega_0-\omega)^2+(1/T_2)^2} + \frac{1}{(\omega+\omega_0)^2+(1/T_2)^2}\right],$$

$$\chi''_{yx}(\omega) = \frac{\gamma M_0}{2}\left[\frac{\omega_0-\omega}{(\omega_0-\omega)^2+(1/T_2)^2} - \frac{\omega_0+\omega}{(\omega+\omega_0)^2+(1/T_2)^2}\right].$$

If a field is applied in the y direction, the equation for m_y is identical to (5.23). Therefore $\chi_{yy} = \chi_{xx}$. However, $\chi_{xy} = -\chi_{yx}$. Combining these results, we find that the rotational component $m_+ = m_x + im_y$ has the solution

$$m_+(t) = (\chi'_+ + i\chi''_+)H_1\,e^{i\omega t}$$

where χ''_+, for example, is

$$\chi''_+(\omega) = -\frac{\gamma M_0}{T_2}\frac{1}{(\omega-\omega_0)^2+(1/T_2)^2} \equiv -\pi\gamma M_0 f_L(\omega). \tag{5.29}$$

The line shape $f_L(\omega)$ defined by (5.29) is the familiar *Lorentzian* curve. This shape reflects the fact that it is the lifetime of the quantum states participating in the transition, in this case the Zeeman levels, that govern the profile. It is often found experimentally that the shape of the absorption more nearly resembles the *Gaussian function*

$$f_G(\omega) = \frac{1}{\Delta\sqrt{2\pi}}e^{-(\omega-\omega_0)^2/2\Delta^2} \tag{5.30}$$

where Δ characterizes the width of the Gaussian. The appearance of such a shape is the result of *inhomogeneous broadening*. In general, the absorption may have some arbitrary shape $f(\omega)$. This shape tells us a great deal about the dynamics of the magnetic system.

5.1.2 Resonance Line Shape

There are two approaches that have been very fruitful in understanding the nature of the magnetic-resonance line shape. One is the *method of moments*, which was employed so successfully by *Van Vleck* [5.2]. The other is what we might call the *relaxation-function method*, developed by *Kubo* and *Tomita* [5.3].

The Method of Moments. If the resonance curve is described by a normalized shape function $f(\omega)$ centered at ω_0, then the nth moment with respect to ω_0 is defined by

$$M_n = \int_{-\infty}^{\infty}(\omega-\omega_0)^n f(\omega)\,d\omega. \tag{5.31}$$

If the function $f(\omega)$ is symmetric about ω_0, it is convenient to introduce the function $F(\Omega) = f(\omega_0 + \Omega)$. In terms of this function, the nth moment is

$$M_n = \int_{-\infty}^{\infty} (\Omega)^n F(\Omega) d\Omega . \tag{5.32}$$

If $\Gamma(t)$ is the Fourier transform of $F(\Omega)$,

$$\Gamma(t) = \int_{-\infty}^{\infty} F(\Omega) e^{-i\Omega t} d\Omega = 2 \int_0^{\infty} F(\Omega) \cos(\Omega t) d\Omega , \tag{5.33}$$

then the $2n$th moment may be written as

$$M_{2n} = (-1)^n \left. \frac{d^{2n} \Gamma(t)}{dt^{2n}} \right|_{t=0} . \tag{5.34}$$

In order to obtain an explicit expression for the $2n$th moment let us assume that the width of the absorption at ω_0 is much less than ω_0 itself. We may then approximate the shape function by $\chi''(\omega)/\omega$. Let us also make the high-temperature approximation defined by $g\mu_B H \ll k_B T$. With this approximation the correlation function defined on page 18 becomes [5.4]

$$\langle \mathcal{M}_x(t) \mathcal{M}_x \rangle \to \frac{\text{Tr}\{\mathcal{M}_x(t)\mathcal{M}_x\}}{\text{Tr}\{1\}} \equiv G(t) . \tag{5.35}$$

Under these conditions the fluctuation-dissipation theorem (1.87) reduces to

$$\langle \mathcal{M}_x(t) \mathcal{M}_x \rangle = \frac{k_B T}{\pi} \int_{-\infty}^{\infty} d\omega \, \frac{\chi''_{xx}(\omega)}{\omega} e^{-i\omega t} . \tag{5.36}$$

Therefore, as a result of the meaning we have ascribed to $f(\omega)$,

$$G(t) = C \int_0^{\infty} d\omega \, f(\omega) \cos \omega t . \tag{5.37}$$

where C is the coefficient of proportionality. This may be rewritten as

$$G(t) = C \int_{-\omega_0}^{\infty} d\Omega \, F(\Omega) \cos(\omega_0 + \Omega)t . \tag{5.38}$$

If $F(\Omega)$ falls to 0 by the time Ω reaches $-\omega_0$, so that the lower limit may be extended to $-\infty$, then

$$G(t) = \left[2C \int_0^{\infty} d\Omega \, F(\Omega) \cos \Omega t \right] \cos \omega_0 t = G_1(t) \cos \omega_0 t . \tag{5.39}$$

Therefore

$$M_{2n} = \frac{(-1)^n}{G_1(0)} \left. \frac{d^{2n} G_1(t)}{dt^{2n}} \right|_{t=0} . \tag{5.40}$$

Differentiating (5.35, 39), we obtain the explicit expressions for the second and fourth moments,

$$M_2 = -\frac{\text{Tr}\{[\mathcal{H}, \mathcal{M}_x]^2\}}{\hbar^2 \text{Tr}\{\mathcal{M}_x^2\}} - \omega_0^2, \tag{5.41}$$

$$M_4 = \frac{\text{Tr}\{[\mathcal{H}, [\mathcal{H}, \mathcal{M}_x]]^2\}}{\hbar^4 \text{Tr}\{\mathcal{M}_x^2\}} - 6\omega_0^2 M_2 - \omega_0^4. \tag{5.42}$$

Let us consider what these moments would be for Lorentizian and Gaussian lines. Consider first the normalized Gaussian given by (5.30). From definition (5.31) we find

$$M_2 = \Delta^2 \quad \text{and} \quad M_4 = 3\Delta^4. \tag{5.43}$$

The halfwidth $\Delta\omega$, as defined by $f_G(\omega_0 + \Delta\omega) = \frac{1}{2} f_G(\omega_0)$, is

$$\Delta\omega = \sqrt{(2 \ln 2)}\, \Delta = 1.18\Delta. \tag{5.44}$$

Therefore the square root of the second moment gives a good approximation of the width of a Gaussian line.

Now consider the normalized Lorentzian,

$$f_L(\omega) = \frac{1}{\pi T_2} \frac{1}{(\omega - \omega_0)^2 + (1/T_2)^2}. \tag{5.45}$$

The integrals for the second and higher moments diverge for this function. Therefore a cutoff is usually introduced at $|\omega - \omega_0| = \omega_m$. The moments are then given by

$$M_2 = \frac{2\omega_m}{\pi T_2} \quad \text{and} \quad M_4 = \frac{2\omega_m^3}{3\pi T_2}. \tag{5.46}$$

Notice that

$$\frac{M_4}{M_2^2} = \begin{cases} 3 & \text{for a Gaussian} \\ \dfrac{\pi \omega_m T_2}{6} & \text{for a cutoff Lorentzian} \end{cases} \tag{5.47}$$

If we compute this ratio for some relaxation process and find that the result is very large, we suspect the line shape will be Lorentzian, whereas if the result is near 3, the lines shape will be closer to a Gaussian.

The problem to which Van Vleck addressed himself was that of a system of magnetic moments located in an external field and interacting weakly through both the dipole–dipole interaction (2.50) and the exchange interaction (2.89). Thus the system is characterized by the Hamiltonian

$$\mathcal{H} = \mathcal{H}_z + \mathcal{H}_{dip} + \mathcal{H}_{ex}. \tag{5.48}$$

Notice that the last three terms in (2.50) have matrix elements between states of the system in which the total magnetic quantum number is changed by ± 1 and ± 2. These terms lead to transitions which are separated from the main resonance by multiples of $g\mu_B H_0$ and appear as sidebands on the main Zeeman line at $g\mu_B H_0$. As long as the sidebands do not overlap our main line, we may neglect the last three terms in (2.50), in so far as we are concerned only with the main line. The truncated Hamiltonian is thus

$$\mathcal{H} = \mathcal{H}_z + \mathcal{H}'_{dip} + \mathcal{H}_{ex}, \tag{5.49}$$

where

$$\mathcal{H}'_{dip} = g^2 \mu_B^2 \sum_{\substack{i,j \\ j \neq i}} \left[-\frac{3\cos^2\theta_{ij} - 1}{r_{ij}^3} S_i^z S_j^z \right.$$

$$\left. + \frac{3\cos^2\theta_{ij} - 1}{4r_{ij}^3} (S_i^+ S_j^- + S_i^- S_j^+) \right]. \tag{5.50}$$

Since

$$\mathcal{M}_x = -g\mu_B \sum_i S_i^x \quad \text{and} \quad \mathcal{H}_z = g\mu_B H \sum_i S_i^z,$$

we find that

$$[\mathcal{H}_z, \mathcal{M}_x] = i\hbar\omega_0 \mathcal{M}_y. \tag{5.51}$$

It is also easy to show that $[\mathcal{H}_{ex}, \mathcal{M}_x] = 0$. Therefore the second moment reduces to

$$M_2 = -\frac{\text{Tr}\{[\mathcal{H}'_{dip}, \mathcal{M}_x]^2\}}{\hbar^2 \text{Tr}\{\mathcal{M}_x^2\}}. \tag{5.52}$$

To evaluate this trace Van Vleck used the fact that it is independent of the basis. Therefore, even though this is truly a many-body system, we may use a basis consisting of products of individual spin states, $|M_{S_1}, M_{S_2}, \dots\rangle$. We can then readily carry out the trace and find that

$$M_2 = \frac{3}{4} \frac{g^4 \mu_B^4}{\hbar^2} S(S+1) \sum_i \frac{(1 - 3\cos^2\theta_{ij})^2}{r_{ij}^6}. \tag{5.53}$$

If the sample consists of randomly oriented crystallites, then the angular part of

(5.5) averages to $\frac{4}{3}$. Furthermore, if we have a moment at every site in a simple cubic lattice with a lattice parameter a, then

$$\sum_i \frac{1}{r_{ij}^6} = \frac{8.32}{a^6} \quad \text{and} \quad M_2 = \frac{5\mu^4 S(S+1)}{\hbar^2 a^6}.$$

For *nuclear* moments separated by, say, 3 Å, the corresponding line width $\Delta H = \sqrt{M_2}/\gamma_N$ is about 6 gauss, which is in reasonable agreement with observations in nuclear magnetic-resonance experiments. For *electronic* moments, however, this line width is about 5 kG. This is enormous. But when we have electronic moments as close as 3 Å, the exchange interaction becomes very important, and the second moment, which does not contain exchange effects, is not sufficient to describe the situation.

To study the effects of exchange Van Vleck also computed the fourth moment. What he found was that when the exchange interaction exceeds the dipole–dipole interaction, the fourth moment exceeds the square of the second moment. equation (5.47) then tells us that as the exchange interaction increases, the shape of the absorption line changes from Gaussian to Lorentzian. Furthermore, if the fourth moment increases while the second moment remains unchanged, the intensity in the wings of the absorption curve *increases*. If the total integrated intensity is to remain the same, there must be a decrease in intensity closer to the center of the line. This results in what is called *exchange narrowing*, illustrated in Fig. 5.4. The origin of this narrowing may be better understood by considering the effective field acting on a particular spin. In the absence of any exchange, this spin experiences not only the applied field H_0, but also a dipolar field arising from the other spins in its environment. Since each spin experiences a slightly different environment, this leads to a distribution of resonant frequencies which manifests itself as an *inhomogeneously broadened absorption*. The effect of the exchange interaction is to modulate the dipolar field. As this modulation increases the average dipolar field decreases, with the result that the distribution of resonant frequencies peaks more strongly near ω_0.

The Relaxation-Function Method. Guided by this interpretation of the results of the moment calculations, let us now outline the relaxation-function approach

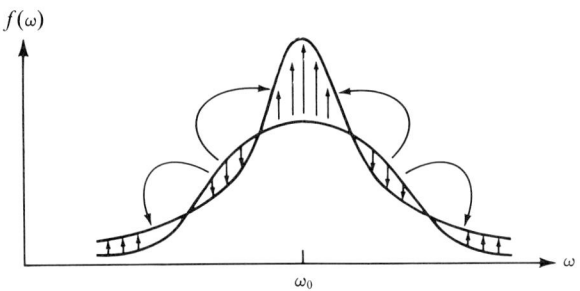

Fig. 5.4. Schematic representation of exchange narrowing. The arrows indicate the transfer of intensity

to obtain the detailed shape of the resonance curve. Our discussion follows that of *Abragam* [5.5]. Since we are assuming that the absorption spectrum is proportional to $\chi''(\omega)/\omega$, we must compute the high-temperature correlation function introduced in (5.35),

$$G(t) = \frac{\text{Tr}\{\mathcal{M}_x(t)\mathcal{M}_x\}}{\text{Tr}\{1\}}. \tag{5.54}$$

We shall not include the exchange interaction explicitly, but shall merely assume that it leads to a time-dependent dipolar interaction. The total Hamiltonian is then $\mathcal{H} = \mathcal{H}_z + \mathcal{H}_{\text{dip}}$. The dipole–dipole interaction is assumed to be small in comparison with the Zeeman interaction. Therefore we look for a solution to $\mathcal{M}_x(t)$ which is an expansion in powers of the dipole–dipole interaction. This is very much different from the method of moments, which is essentially an expansion of $G(t)$ in powers of t.

In order to develop an expansion of $\mathcal{M}_x(t)$ in the dipole interaction it is convenient to remove the Zeeman interaction by the transformation

$$\widetilde{\mathcal{M}}_x \equiv \exp\left(\frac{-i\mathcal{H}_z t}{\hbar}\right) \mathcal{M}_x(t) \exp\left(\frac{i\mathcal{H}_z t}{\hbar}\right). \tag{5.55}$$

The equation of motion for this operator is

$$-i\hbar \frac{d\widetilde{\mathcal{M}}_x}{dt} = \left[\exp\left(\frac{-i\mathcal{H}_z t}{\hbar}\right) \mathcal{H}_{\text{dip}} \exp\left(\frac{i\mathcal{H}_z t}{\hbar}\right), \widetilde{\mathcal{M}}_x\right]. \tag{5.56}$$

The corresponding equation for the matrix element of $\widetilde{\mathcal{M}}_x$ between eigenstates of $\mathcal{H}_z$ is

$$-i\hbar \frac{d}{dt}(\widetilde{\mathcal{M}}_x)_{nn'} = \sum_{n''} \{(\mathcal{H}_{\text{dip}})_{nn''}(\widetilde{\mathcal{M}}_x)_{n''n'} \exp[i(E_{n''} - E_n)t/\hbar] \\ - (\widetilde{\mathcal{M}}_x)_{nn''}(\mathcal{H}_{\text{dip}})_{n''n'} \exp[-i(E_{n''} - E_{n'})t/\hbar]. \tag{5.57}$$

Since the transformation (5.56) has removed the rapid Zeeman precession from $\widetilde{\mathcal{M}}_x$, we may assume that remaining time dependence is relatively slow. Then, since $E_{n''} - E_n$ and $E_{n''} - E_{n'}$ are multiples of $\hbar\omega_0$, those terms in (5.57) for which these differences are not 0 oscillate rapidly, giving a very small contribution which we shall neglect. This leaves us with matrix elements of the form $(\mathcal{H}_{\text{dip}})_{nn''}$ between degenerate states. We shall assume that these states are chosen such that these matrix elements vanish unless $n'' = n$. Thus (5.57) reduces to

$$-i\hbar \frac{d}{dt}(\widetilde{\mathcal{M}}_x)_{nn'} = \hbar\Delta\omega(t)_{nn'}(\widetilde{\mathcal{M}}_x)_{nn'}, \tag{5.58}$$

where

$$\hbar\Delta\omega(t)_{nn'} \equiv (\mathcal{H}_{\text{dip}})_{nn} - (\mathcal{H}_{\text{dip}})_{n'n'} . \tag{5.59}$$

The time dependence of this quantity reflects the exchange modulation. Equation (5.58) has the solution

$$(\tilde{\mathcal{M}}_x)_{nn'} = (\mathcal{M}_x)_{nn'}^0 \exp\left[i\int_0^t \Delta\omega(t')_{nn'} dt'\right]. \tag{5.60}$$

If we invert (5.55), the correlation function becomes

$$G(t) = \sum_{n,n'} \exp\left[i(E_n - E_{n'})t/\hbar\right] |\langle n|\mathcal{M}_x|n'\rangle|^2 \exp\left[i\int_0^t \Delta\omega(t')_{nn'} dt'\right]. \tag{5.61}$$

Let us consider only the main absorption line, for which $E_n - E_{n'} = \hbar\omega_0$. This restricts the sum over n and n' to those pairs whose eigenvalues differ by $\hbar\omega_0$. If we define

$$\int_0^t \Delta\omega(t')_{nn'} dt' \equiv x(t), \tag{5.62}$$

then for those states n and n' which give a particular $x(t)$ we also have a corresponding value for $|\langle n|\mathcal{M}_x|n'\rangle|^2$, which we may write as $P[x(t)]$. Thus the correlation function may be thought of as the average of $\exp[ix(t)]$ with the probability distribution $P[x(t)]$; that is,

$$G(t) = e^{i\omega_0 t} \int_{-\infty}^{\infty} dx P(x) e^{ix} = e^{i\omega_0 t} \langle e^{ix} \rangle \tag{5.63}$$

where $P(x)$ includes a normalizing factor as well as the matrix element appearing in (5.61).

The assumption is now made that $\Delta\omega(t')_{nn'}$ is a Gaussian function of time, with a mean value given by the second moment M_2, which we computed earlier. Thus we take

$$\langle \Delta\omega(t)_{nn'}\Delta\omega(t+\tau)_{nn'} \rangle = M_2 \psi(\tau) \tag{5.64}$$

where $\psi(\tau)$ characterizes the fluctuations in the local field. There is a theorem, called the *central-limit theorem*, which says, in effect, that if $\Delta\omega(t')_{nn'}$ is a Gaussian function, then $x(t)$, as defined by (5.62), is also a Gaussian function. This means that $P[x(t)]$ has the Gaussian form

$$P[x(t)] = \frac{1}{\sqrt{2\pi\langle x^2\rangle}} \exp(-x^2/2\langle x^2\rangle) . \tag{5.65}$$

Thus

$$\langle e^{ix} \rangle = e^{-\langle x^2 \rangle/2}$$

where

$$\langle x^2 \rangle = \left\langle \int_0^t dt' \int_0^t dt'' \, \Delta\omega(t')_{nn'} \Delta\omega(t'')_{nn'} \right\rangle = 2M_2 \int_0^t d\tau (t - \tau) \, \psi(\tau) \, . \tag{5.66}$$

Therefore we finally obtain

$$G(t) = e^{i\omega_0 t} \varphi(t) \tag{5.67}$$

with

$$\varphi(t) = \exp\left[-M_2 \int_0^t (t - \tau) \, \psi(\tau) \, d\tau \right] . \tag{5.68}$$

The narrowing process may be seen from this expression. If the correlation time that characterizes $\psi(\tau)$ is long, then $\psi(\tau) \simeq 1$ over the range of the integral in (5.66) and

$$\varphi(t) \rightarrow \exp(-M_2 t^2/2) \, . \tag{5.69}$$

This leads to a Gaussian line with the Van Vleck second moment. If, on the other hand, $\psi(\tau)$ decays before we reach the upper limit of the integral, then

$$M_2 \int_0^t (t - \tau)\psi(\tau) d\tau \simeq M_2 t \int_0^\infty \psi(\tau) d\tau \equiv M_2 t/\omega_{\text{ex}} \tag{5.70}$$

and

$$\varphi(t) \rightarrow \exp(-M_2 t/\omega_{\text{ex}}) \, . \tag{5.71}$$

This leads to a Lorentzian line with a halfwidth proportional to M_2/ω_{ex}. Thus as the exchange increases, the dipolar broadened line changes from Gaussian to narrow Lorentzian.

An analytic form for $\psi(\tau)$ which implies these Gaussian and Lorentzian limits is

$$\psi(\tau) = \exp(-\pi\omega_{\text{ex}}^2 \tau^2/4) \, . \tag{5.72}$$

However, detailed comparison with measured correlation functions shows that $\psi(\tau)$ actually decays more slowly than such a Gaussian. To obtain a better description of the relaxation at long times we recognize that if we perturb a given spin in the system this perturbation will propagate away by means of the exchange interaction. At long times this process will involve many spins and may therefore be described as a diffusion process in a spin continuum. The correlation function $\psi(\mathbf{r}, t)$ then obeys the diffusion equation,

$$\Lambda \nabla^2 \psi = \frac{\partial \psi}{\partial t}, \tag{5.73}$$

where the diffusion coefficient Λ is related to the correlation time ω_{ex}^{-1}. The difference between Gaussian behavior and diffusive behavior is illustrated in Fig. 5.5.

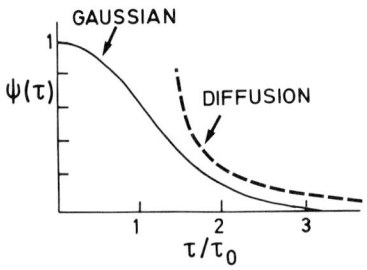

Fig. 5.5. Comparison of the Gaussian and diffusion approximations to the correlation function $\psi(\tau)$. The time is in units of $(1/\tau_0)^2 = 2z\,(J/\hbar)^2 S(S+1)/3$

It is interesting to note that dimensionally $\psi(\tau)$ must vary as $\tau^{-d/2}$ where d is the dimensionality. Thus, in one dimension the correlation time ω_{ex}^{-1} defined by (5.70) diverges and we expect an anomalous line shape. $(CH_3)_4NMnCl_3$, abbreviated TMMC, is a system in which the Mn^{2+} spins are exchange coupled only along chains approximating a one-dimensional spin system. Magnetic resonance studies indeed show [5.6] that the line shape is the Fourier transform of $\exp(-\Lambda t^{3/2})$, intermediate between a Gaussian and a Lorentzian.

Although we have considered here the case of exchange-narrowed dipolar-broadened lines, the relaxation-function approach may also be applied in a variety of other situations, such as exchange-narrowed hyperfine-broadened nuclear magnetic resonance lines [5.7] or motionally narrowed, dipolar-broadened nuclear resonance lines [5.8].

5.1.3 Measurement of T_1

Notice that the longitudinal relaxation time T_1 does not enter the *linear* response function (5.27). Had we included the nonlinear contributions to this response function, the additional term $\gamma^2 H_1^2 T_1/T_2$ would have appeared in the denominator. As a consequence, when the amplitude of the driving field becomes comparable to $(\gamma^2 T_1 T_2)^{-1/2}$ the resonance begins to saturate. The onset of this saturation provides a measure of T_1.

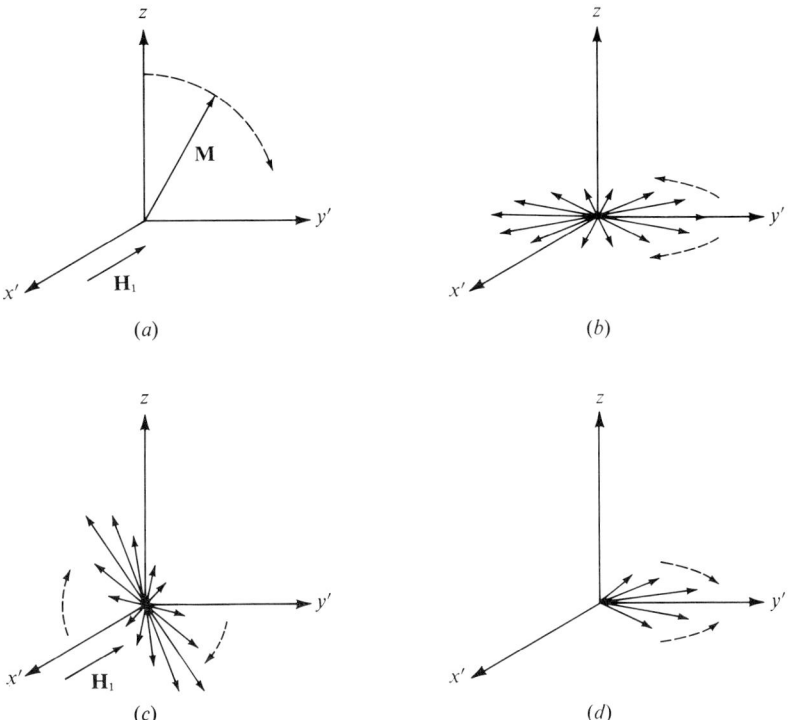

Fig. 5.6a-d. Description of the origin of a spin echo in a coordinate frame rotating about the z axis with the Zeeman frequency (see text for details)

The same nonlinearity is also employed to produce *spin echoes*. This phenomenon is easily understood with the help of Fig. 5.6. A circularily polarized field H_1 at frequency ω_0 is applied transverse to the dc field H_0. We know from the Bloch equations that in a coordinate system rotating about the z axis at a rate ω_0 the magnetization will precess away from the z axis, as illustrated in Fig. 5.6a. If the driving field is turned off just as the magnetization reaches the xy plane, a 90° pulse is produced. Leaving the driving field on for twice as long results in a 180° pulse, which just inverts the magnetization.

Suppose that at time $t = 0$ we apply a 90° pulse. After this pulse is turned off the spins precess at their own resonant frequencies, which, because of the dipolar field, are distributed about ω_0 After some dephasing time T_2^* the spins produce a "pancake" in the rotating system, as shown in Fig. 5.6b. A 180° pulse applied at some time τ to the system in this configuration will cause this pancake to rotate 180° about the x' axis, as shown in Fig. 5.5c. Now, as the spins precess they begin to rephase and at a time τ after the 180° pulse they again become coherent, producing an "echo." If we continue to apply 180° pulses at times 3τ, 5τ, etc., we shall observe echoes at times 4τ, 6τ, etc. If there were no transverse relaxation, these echoes would all have the same height. However, because of this relaxation the height of the nth echo decreases as $\exp(-2n\tau/T_2)$, providing us with a measure of T_2. This scheme is shown in Fig. 5.7.

If $T_1 \gg T_2$, then the longitudinal relaxation time T_1 may be obtained as follows. First we apply a 90° pulse, which takes M_z to zero. After a time τ, M_z has returned to $M_0[1 - \exp(-\tau/T_1)]$. If we now apply a 90° pulse followed by a 180° pulse, the height of the resulting echo will be proportional to $M_z(\tau)$. By repeating this sequence we can follow M_z back to M_0.

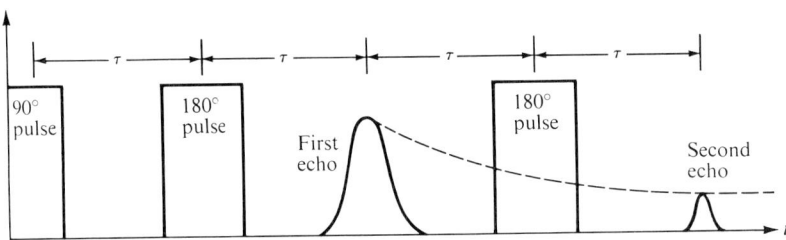

Fig. 5.7. Spin echo scheme for measuring T_2

Let us consider briefly still another way of determining T_1. This consists of measuring the response of the system to a *low-frequency longitudinal* driving field. In general, the magnetization relaxes toward the *instantaneous* field. Therefore the Bloch equation (5.22a) for a field $\boldsymbol{H} = H_0\hat{\boldsymbol{z}} + H_1 \cos \omega t \hat{\boldsymbol{z}}$ should be written as

$$\frac{dm_z}{dt} = \gamma(\boldsymbol{M} \times \boldsymbol{H})_z + \frac{\chi_0 H_1 \cos \omega t - m_z}{T_1}. \tag{5.74}$$

This is only an approximation, in the sense that we have used the static susceptibility χ_0. Solving (5.74) leads directly to the response functions

$$\chi'_{zz}(\omega) = \frac{\chi_0}{1 + \omega^2 T_1^2}, \tag{5.75a}$$

$$\chi''_{zz}(\omega) = \frac{\chi_0 \omega T_1}{1 + \omega^2 T_1^2}. \tag{5.75b}$$

These functions are plotted in Fig 5.8. For electronic spins this broad longitudinal absorption is in the radio-frequency part of the spectrum. This is far below the transverse resonance, which is in the microwave region. Microwave sources were unavailable until World War II, and consequently the relaxation time T_1 was investigated much earlier than T_2. In fact, the response function (5.75b) was derived from thermodynamic considerations long before the development of Bloch's equations.

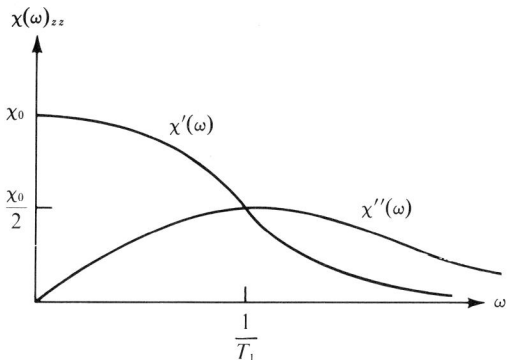

Fig. 5.8. Longitudinal susceptibility of a weakly interacting spin system

5.1.4 Calculation of T_1

Let us now derive an expression for the phenomenological relaxation time T_1. According to the Bloch equation (5.22a), if the z component of the magnetizaton is reduced from its equilibrium value M_0, it will return to this value according to

$$M_z = M_0(1 - e^{-t/T_1}) . \tag{5.76}$$

To obtain an explicit expression for T_1 let us compute M_z for a simple two-level spin system that has been displaced from equilibrium. If the number of spins in the state $|-\tfrac{1}{2}\rangle$ is N_- and the number in $|+\tfrac{1}{2}\rangle$ is N_+, then the total magnetization is

$$M_z = \frac{\tfrac{1}{2}g\mu_B(N_- - N_+)}{V} \equiv \tfrac{1}{2}g\mu_B n . \tag{5.77}$$

Thus, since the z component of the magnetization depends on the relative population of the Zeeman states, any change in this population will result in a change in M_z.

Let us define $W(- \to +)$ as the probability per unit time that a spin in the state $|-\tfrac{1}{2}\rangle$ is "flipped" to the state $|+\tfrac{1}{2}\rangle$. Then the change in the populations may be described by the rate equations

$$\frac{dN_+}{dt} = N_- W(- \to +) - N_+ W(+ \to -) , \tag{5.78a}$$

$$\frac{dN_-}{dt} = N_+ W(+ \to -) - N_- W(- \to +) . \tag{5.78b}$$

The transition probabilities $W(+ \to -)$ and $W(- \to +)$ are not independent, for in equilibrium

162   5. The Dynamic Susceptibility of Weakly Interacting Systems

$$N_-^0 W(- \to +) - N_+^0 W(+ \to -) = 0. \tag{5.79}$$

Therefore, if the levels are separated by $g\mu_B H$ and we define $W(+ \to -) \equiv W$, then

$$W(- \to +) = W \exp(-g\mu_B H/k_B T). \tag{5.80}$$

The equation for n becomes

$$\frac{dn}{dt} = \frac{2N_+ W}{V} - \frac{2N_- W \exp(g\mu_B H/k_B T)}{V}. \tag{5.81}$$

In the high-temperature limit, valid for most experimental situations, the exponential in (5.81) may be expanded. If we replace the occupation numbers in (5.77) by their equilibrium values, we see that n_0 is approximately $(N_-/V)(g\mu_B H/k_B T)$ and

$$\frac{dn}{dt} = -2W(n - n_0). \tag{5.82}$$

Solving this equation and using (5.77), we obtain

$$M_z = M_0(1 - e^{-2Wt}). \tag{5.83}$$

Comparing this with (5.76), we have the result

$$\boxed{\frac{1}{T_1} = 2W.} \tag{5.84}$$

There are various mechanisms that might contribute to the transition probability W. For example, in a paramagnet the spins may be coupled to the lattice vibrations. In this case the spin flips are accompanied by the emission or absorption of phonons. The resulting T_1 is called the *spin-lattice relaxation time*.

Another example is that of nuclear spins in a ferromagnet. In this case the nuclear spins are coupled to the electron spins by the hyperfine interaction. Fluctuations in the electron-spin system can then flip the nuclear spins, leading to a *nuclear-spin relaxation time* T_1. Let us consider this particular mechanism as an example of the calculation of T_1.

Suppose our unperturbed system is a single nuclear spin of $\frac{1}{2}$ plus the exchange-coupled electron system. Thus

$$\mathcal{H}_0 = \mathcal{H}_N + \mathcal{H}_{el} \tag{5.85}$$

where $\mathcal{H}_N = -g_N \mu_N H I_z$ and $\mathcal{H}_{el}$ is the electronic Hamiltonian. The eigenfunctions of $\mathcal{H}_0$ have the form $|m_I, n\rangle$, where n characterizes the state of the electron system. The perturbing Hamiltonian is the hyperfine interaction

$$\mathcal{H}_1 = A\mathbf{I}\cdot\mathbf{S} \tag{5.86}$$

where $\mathbf{S}$ is the spin of the electron associated with the particular nucleus we are considering. If, at time $t = 0$, the system is in the state $|m_I, n\rangle$, then we want to know the probability of finding the system in the state $|m_I', n'\rangle$ at some later time t. To find this probability we must solve the Schrödinger equation

$$i\hbar \frac{\partial \psi}{\partial t} = (\mathcal{H}_0 + \mathcal{H}_1)\psi. \tag{5.87}$$

We shall see in a moment that it is convenient to introduce

$$\psi = \exp\left(\frac{-i\mathcal{H}_0 t}{\hbar}\right)\phi \tag{5.88}$$

where ϕ satisfies

$$i\hbar \frac{\partial \phi}{\partial t} = \mathcal{H}_1(t)\phi \tag{5.89}$$

and

$$\mathcal{H}_1(t) \equiv \exp\left(\frac{i\mathcal{H}_0 t}{\hbar}\right)\mathcal{H}_1 \exp\left(\frac{-i\mathcal{H}_0 t}{\hbar}\right) \tag{5.90}$$

Integrating this equation with the condition that $\phi(t = 0) = |m_I, n\rangle$, we obtain

$$\phi = \exp\left[-\frac{i}{\hbar}\int_0^t dt'\, \mathcal{H}_1(t')\right]|m_I, n\rangle. \tag{5.91}$$

The reason for introducing ϕ is that (5.91) may be expanded in powers of the perturbation $\mathcal{H}_1$,

$$\phi = |m_I, n\rangle - \frac{i}{\hbar}\int_0^t dt'\, \mathcal{H}_1(t')|m_I, n\rangle + \ldots. \tag{5.92}$$

Therefore the probability of finding the system in the state $|m_I', n'\rangle$ is

$$|\langle m_I', n'|\psi\rangle|^2 = \frac{1}{\hbar^2}\int_0^t dt' \int_0^t dt'' \langle m_I, n|\mathcal{H}_1(t')|m_I', n'\rangle$$
$$\times \langle m_I', n'|\mathcal{H}_1(t'')|m_I, n\rangle + \ldots. \tag{5.93}$$

The total probability per unit time that the system will make a transition is obtained by dividing (5.93) by t and letting $t \to \infty$. To find W we set $m_I = -\frac{1}{2}$ and $m'_I = +\frac{1}{2}$. We also average over all the electronic degrees of freedom. That is, we multiply (5.93) by the probability that the electronic system will be in the initial state n and then we sum over n as well as over all the final states n'. The sum over n' gives unity by closure, leaving us with just the *thermal average* of $\mathcal{H}_1(t')\mathcal{H}_1(t'')$, which we denote by $\langle \cdots \rangle$. This is not to be confused with the bras and kets above. Thus

$$W = \lim_{t \to \infty} \frac{A^2}{4\hbar^2 t} \int_0^t dt' \int_0^t dt'' \exp[i\omega_N(t' - t'')]\langle S^-(t')S^+(t'')\rangle \tag{5.94}$$

where $\omega_N = g_N \mu_N H$ is the nuclear magnetic resonance frequency. Changing variables leads to the result

$$\boxed{\frac{1}{T_1} = \frac{A^2}{2\hbar^2} \int_{-\infty}^{\infty} dt \, \langle S^-(t)S^+\rangle \exp(i\omega_N t) .} \tag{5.95}$$

Thus we see that the longitudinal relaxation frequency in this case is the Fourier component at the nuclear magnetic resonance frequency of the electron-spin correlation function. It the correlation function relaxes exponentially, i.e.,

$$\langle S^-(t)S^+\rangle = \langle S_\perp^2\rangle \, e^{-t/\tau},$$

then

$$\frac{1}{T_1} = \frac{A^2 \langle S_\perp^2\rangle}{\hbar^2} \frac{\tau}{1 + \omega_N^2 \tau^2}. \tag{5.96}$$

The variation of T_1 with τ is shown in Fig. 5.9. Thus as τ varies, due to a variation in temperature for example, T_1 goes through a minimum when $\omega_N \tau \sim 1$.

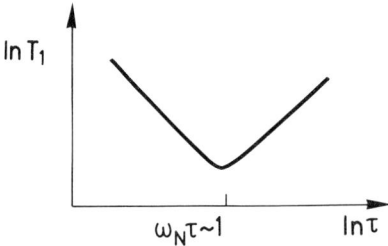

Fig. 5.9. Variation of T_1 with the correlation time τ according to the BPP model

This characteristic behavior was first derived by *Bloembergen* et al. [5.9] and is often simply referred to as the BPP result.

5.2 Metals

Let us now investigate the response of a metal to a time-dependent magnetic field. Experimentally, such investigations are very difficult because of problems associated with the fact that the driving field penetrates only a short distance into the metal. The first resonance experiments involving conduction-electron spins were carried out in the early 1950s. These experiments entailed measuring the surface impedance of samples located in a microwave cavity.

In 1964 electron spin resonance was observed in transmission. The analysis of both resonance techniques requires a knowledge of the nonlocal susceptibility function $\chi(z, z', \omega)$. This may be obtained from the generalized susceptibility $\chi(\mathbf{q}, \omega)$ along with a boundary condition on $\mathbf{M}$. If we assume that the boundaries do not relax the magnetization then the magnetization current $\nabla \mathbf{M}$ must be conserved at the boundary. Let us now consider the generalized susceptibility.

There are a number of ways we might calculate the dynamic susceptibility. One that is particularly instructive makes use of the fluctuation-dissipation theorem. As we saw earlier, the presence of a time-dependent field leads to a coupling between the x and y components of the magnetization. Consequently, it is more convenient to work with the rotational components $M_\pm(\mathbf{q}, \omega)$. In terms of these components, the susceptibility $\chi''_{xx} + \chi''_{yy}$ is proportional to the correlation function $\langle \mathcal{M}_-(\mathbf{q}, t) \mathcal{M}_+(-\mathbf{q}) \rangle$. This is just the susceptibility by $\chi_{-+}(\mathbf{q},\omega)$ we have already encountered.

The magnetization operator associated with the spin of a system of itinerant electrons is

$$\mathcal{M}(\mathbf{r}) = -\sum_i \mu_B \boldsymbol{\sigma}_i \delta(\mathbf{r} - \mathbf{r}_i) \,. \tag{5.97}$$

Second quantizing this operator in terms of the plane-waves states (1.123) gives

$$\mathcal{M}(\mathbf{r}) = -\frac{\mu_B}{V} \sum_{k,q} \sum_{\alpha,\beta} \boldsymbol{\sigma}_{\alpha\beta} e^{i\mathbf{q}\cdot\mathbf{r}} a^\dagger_{k-q,\alpha} a_{k,\beta} \,. \tag{5.98}$$

Therefore

$$\mathcal{M}_+(-\mathbf{q}) = -2\mu_B \sum_k a^\dagger_{k+q,\uparrow} a_{k\downarrow} \tag{5.99}$$

and

$$\mathcal{M}_-(\mathbf{q}) = -2\mu_B \sum_k a^\dagger_{k-q,\downarrow} a_{k\uparrow} \,. \tag{5.100}$$

The desired correlation function is then

166   5. The Dynamic Susceptibility of Weakly Interacting Systems

$$\langle \mathcal{M}_-(\mathbf{q},t)\mathcal{M}_+(-\mathbf{q})\rangle = 4\mu_B^2 \sum_k \sum_{k'} \Big\langle \exp\Big(\frac{i\mathcal{H}t}{\hbar}\Big) a_{k'-q,\downarrow}^\dagger a_{k'\uparrow}$$
$$\times \exp\Big(\frac{-i\mathcal{H}t}{\hbar}\Big) a_{k+q,\uparrow}^\dagger a_{k\downarrow}\Big\rangle \quad (5.101)$$

where

$$\mathcal{H} = \sum_k \sum_\sigma \epsilon_{k\sigma} a_{k\sigma}^\dagger a_{k\sigma}. \quad (5.102)$$

The eigenvalue $\epsilon_{k\sigma}$ is $\hbar^2 k^2/2m + \mu_B H_0 \sigma$, where σ is $+1$ for an up spin and -1 for a down spin.

At $T = 0\,K$ the ground state consists of up- and down-spin Fermi spheres. As in our calculation of the static susceptibility in Chap. 3, the wave vector $\mathbf{k}$ is restricted to those values within the down-spin Fermi sphere such that $\mathbf{k} + \mathbf{q}$ lies outside the up-spin Fermi sphere. The correlation function at $T = 0\,K$ is then

$$\langle \mathcal{M}_-(\mathbf{q},t)\mathcal{M}_+(-\mathbf{q})\rangle = 4\mu_B^2 \sum_k f_{k\downarrow}(1-f_{k+q,\uparrow}) \exp\frac{-i(\epsilon_{k+q,\uparrow}-\epsilon_{k\downarrow})t}{\hbar}. \quad (5.103)$$

Taking the Fourier transform and applying (1.87) at $T = 0\,K$ leads to the susceptibility,

$$\chi''_{-+}(\mathbf{q},\omega)_s = \frac{8\pi\mu_B^2}{V}\sum_k f_{k\downarrow}(1-f_{k+q,\uparrow})\delta(\hbar\omega - \epsilon_{k+q,\uparrow} + \epsilon_{k\downarrow}). \quad (5.104)$$

The real part of the susceptibility is obtained from the Kramers–Kronig relation (1.64),

$$\boxed{\chi'_{-+}(\mathbf{q},\omega)_s = \frac{8\mu_B^2}{V}\sum_k \frac{f_{k\downarrow}(1-f_{k+q,\uparrow})}{\epsilon_{k+q,\uparrow}-\epsilon_{k\downarrow}-\hbar\omega}.} \quad (5.105)$$

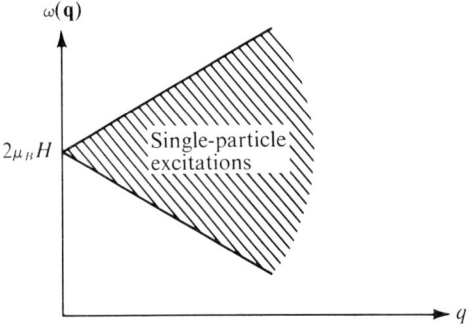

Fig. 5.10. The magnetic-excitation spectrum of a noninteracting electron gas

Notice that χ_{-+} involves different spin indices whereas χ_{zz}, as given in (3.84), for example, involves the same spin indices. In zero field $\epsilon_{k\uparrow} = \epsilon_{k\downarrow} = \epsilon_k$, and in the limit $\omega \to 0$ (5.105) correctly reduces to twice the static result.

The single-particle energies entering the delta function of (5.104) have the value

$$\epsilon_{k+q,\uparrow} - \epsilon_{k\uparrow} = 2\mu_B H_0 + \frac{\hbar^2}{m} \mathbf{k} \cdot \mathbf{q} + \frac{\hbar^2}{2m} q^2 \,. \tag{5.106}$$

Therefore the excitation spectrum has the form shown in Fig. 5.10. The $\mathbf{q} = 0$ mode is the conduction-electron spin resonance at $\omega = \omega_s = g\mu_B H$.

5.2.1 Paramagnons

The result for $\chi_{-+}(\mathbf{q}, \omega)$ in the presence of the delta-function interaction introduced in (4.81) is readily obtained by the equation-of-motion approach used in that section. In order to calculate $\chi_{-+}(\mathbf{q}, \omega)$, as opposed to $\chi_{zz}(\mathbf{q}, \omega)$, we must determine the response

$$M_+(\mathbf{q}) = -2\mu_B \sum_k \langle a^\dagger_{k-q,\uparrow} a_{k\downarrow} \rangle \tag{5.107}$$

to a transverse field $H[\cos(\omega t)\hat{\mathbf{x}} + \sin(\omega t)\hat{\mathbf{y}}]\cos(\mathbf{q} \cdot \mathbf{r})$. The corresponding Zeeman interaction is

$$\mathcal{H}_Z = -\tfrac{1}{4} H \{ [\mathcal{M}_-(\mathbf{q}) + \mathcal{M}_-(-\mathbf{q})] e^{i\omega t} + [\mathcal{M}_+(\mathbf{q}) + \mathcal{M}_+(-\mathbf{q})] e^{-i\omega t} \} \,. \tag{5.108}$$

Adding this interaction to (4.81) the equation of motion for $\langle a^\dagger_{k-q,\uparrow} a_{k\downarrow} \rangle$ in the random-phase approximation becomes

$$\begin{aligned} i\hbar \frac{d}{dt} \langle a^\dagger_{k-q,\uparrow} a_{k\downarrow} \rangle = &(\epsilon_k - \epsilon_{k-q}) \langle a^\dagger_{k-q,\uparrow} a_{k\downarrow} \rangle \\ &+ \frac{2I}{V} \sum_{k'} (n_{k'\uparrow} - n_{k'\downarrow}) \langle a^\dagger_{k-q,\uparrow} a_{k\downarrow} \rangle \\ &- \frac{2I}{V} (n_{k-q,\uparrow} - n_{k\downarrow}) \sum_{q'} \langle a^\dagger_{k-q'-q,\uparrow} a_{k-q',\downarrow} \rangle \\ &+ \tfrac{1}{2}\mu_B H (n_{k-q,\uparrow} - n_{k\downarrow}) e^{i\omega t} \,. \end{aligned} \tag{5.109}$$

We see that the "spin fluctuation" $\langle a^\dagger_{k-q,\uparrow} a_{k\downarrow} \rangle$ is coupled to many other fluctuations through the third term on the right-hand side. If we assume that these spin fluctuations all have a time dependence of the form $\exp(i\omega t)$, then

$$\langle a^\dagger_{k-q,\uparrow} a_{k\downarrow} \rangle = \frac{(2I/V)(n_{k-q,\uparrow} - n_{k\downarrow}) \sum_{q'} \langle a^\dagger_{k-q'-q,\uparrow} a_{k-q',\downarrow} \rangle}{\hbar\omega - (\tilde{\epsilon}_{k-q,\uparrow} - \tilde{\epsilon}_{k\downarrow})}$$
$$- \frac{\frac{1}{2}\mu_B H(n_{k-q,\uparrow} - n_{k\downarrow})}{\hbar\omega - (\tilde{\epsilon}_{k-q,\uparrow} - \tilde{\epsilon}_{k\downarrow})} \tag{5.110}$$

where

$$\tilde{\epsilon}_{k\sigma} = \epsilon_k + \frac{I}{2V} \sum_{k'} n_{k'\sigma}. \tag{5.111}$$

Since were interested in the magnetization, we must sum over **k**. The index **q'** then becomes a *dummy index*, which enables us to solve for the magnetization. From definition (1.55) of the susceptibility we obtain for the real part

$$\boxed{\chi'_{-+}(\mathbf{q}, \omega) = \frac{(2\mu_B^2/V)\Gamma(\mathbf{q}, \omega)}{1 - (2I/V)\Gamma(\mathbf{q}, \omega)}} \tag{5.112}$$

where

$$\Gamma(\mathbf{q}, \omega) = \sum_k \frac{n_{k-q,\uparrow} - n_{k\downarrow}}{\hbar\omega - (\tilde{\epsilon}_{k-q,\uparrow} - \tilde{\epsilon}_{k\downarrow})}. \tag{5.113}$$

In zero field and in the paramagnetic state $\epsilon_{k\uparrow} = \epsilon_{k\downarrow}$. In this case we can evaluate the real and imaginary parts of the susceptibility for simple parabolic bands. The imaginary part is shown in Fig. 5.11. When the interaction $I = 0$, we obtain a broad response. However, as I increases toward the Stoner critical value, the response develops a sharp peak at low frequency whose position varies linearly with the wave vector $\mathbf{q}$. The presence of this low-frequency peak, or *paramagnon*, enhances the effective mass m^*. In particular, it is found [5.11] that the mass contains a contribution which varies as the *logarithm* of the Stoner enhancement factor.

The above treatment does not include the interactions between the electrons and their host lattice. These interactions are primarily responsible for the resistivity and are characterized by a relaxation time τ. If this relaxation time is much smaller than the inverse cyclotron frequency ω_c, i.e., $\omega_c\tau \ll 1$, then the electron's flight between collisions is approximately a straight line. The electron's overall motion will then be that of a random walk with a diffusion constant $D = \frac{1}{3} v_F^2 \tau$. We shall see below that this behavior, while not shifting the conduction-electron spin resonance frequency, has a dramatic effect on the line shape. It is possible to incorporate collision effects into the susceptibility (5.105) in analogy with *Mermin's* [5.12] treatment of the Lindhard dielectric function. We shall, however, do this within the framework of Fermi liquid theory.

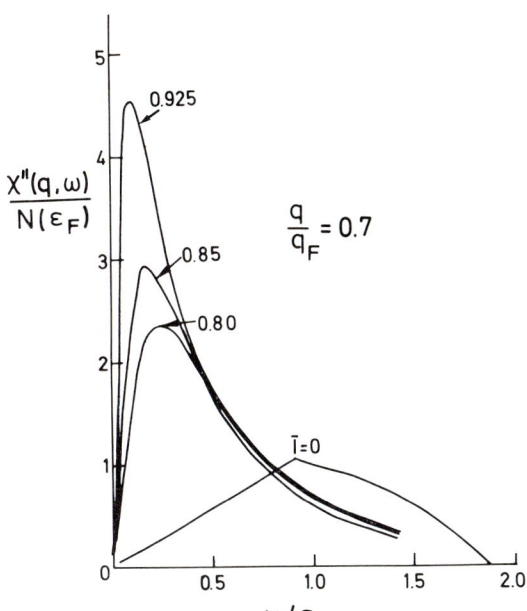

Fig. 5.11. Frequency response of the imaginary part of the susceptibility, showing the paramagnon peak [5.10]

5.2.2 Fermi Liquid Theory

As we saw in Sect. 4.2, the important quantity in the Fermi liquid approach is the distribution of quasiparticles $n(k, \sigma)$. Since we shall eventually be interested in the response to space-varying excitations, let us consider a region of space which is small in comparison with the wavelength of such excitations but large enough to contain many quasiparticles. Then $n(k, \sigma)$ becomes a function of position $n(r, k, \sigma)$. In Chap. 4 we were able to guess correctly how the distribution function was changed by a constant velocity displacement or by the presence of a uniform magnetic field. In the case of a time-dependent perturbation, however, it is not obvious what the resulting change will be.

In the absence of any magnetic fields the quasiparticle energy is $\epsilon^0(k)\,\mathbf{1}$. When magnetic fields are applied the energy changes for two reasons. First, there are orbital and spin Zeeman terms. Second, these Zeeman terms alter the distribution of quasiparticles, thereby bringing interaction terms into the energy. If we write the change in the quasiparticle distribution in the general form

$$\delta n(r, k, \sigma) = n_1(r, k)\,\mathbf{1} + n_2(r, k)\cdot\sigma \,, \tag{5.114}$$

then the change in the quasiparticle energy may be written as

$$\delta\epsilon(r, k, \sigma) = \epsilon_1(r, k)\,\mathbf{1} + \epsilon_2(r, k)\cdot\sigma \tag{5.115}$$

where

$$\epsilon_1(r, k) = -\frac{e\hbar}{2m^*c} k \times H_0 \cdot r + \frac{2}{V} \sum_{k'} \varphi(\hat{k} \cdot \hat{k}') n_1(r, k') \qquad (5.116)$$

and

$$\epsilon_2(r, k) = \mu_B H_0 \hat{z} + \mu_B H_1(r, t) + \frac{2}{V} \sum_{k'} \psi(\hat{k} \cdot \hat{k}') n_2(r, k') \qquad (5.117)$$

where $H_1(r, t)$ is the probing magnetic field. The first term in $\epsilon_1(r,k)$ is the orbital Zeeman contribution.

The magnetization at r arising from the quasiparticles in state k is

$$m(r, k) = -\frac{\mu_B}{V} \mathrm{Tr}_{\sigma} \{[n(r, k, \sigma)\sigma]\} = -\frac{2\mu_B}{V} n_2(r, k). \qquad (5.118)$$

To determine the magnetization we must therefore determine the spin-dependent part of the change in the quasiparticle distribution function. This may be found by solving the equation of motion associated with $n(r, k, \sigma)$

To set up this equation we ask for the different ways in which $n(r,k,\sigma)$ may change with time. First of all, it might depend explicitly on time. This rate of change has the form $\partial n/\partial t$. Second, quasiparticles may drift into and out of the region in real space defined by r. The rate of change of $n(r,k,\sigma)$ associated with this process is

$$\frac{1}{2\hbar} \left(\frac{\partial n}{\partial r} \frac{\partial \epsilon}{\partial k} + \frac{\partial \epsilon}{\partial k} \frac{\partial n}{\partial r} \right). \qquad (5.119)$$

Since $\epsilon(r, k, \sigma)$ and $n(r, k, \sigma)$ are matrices, this is written in the symmetrized form. A third way in which the quasiparticle distribution might change is for quasiparticles to change their momentum $\hbar k$. Using Hamilton's equation $\partial p/\partial t = -\partial \epsilon/\partial r$, we have for this contribution

$$-\frac{1}{2\hbar} \left(\frac{\partial n}{\partial k} \frac{\partial \epsilon}{\partial k} + \frac{\partial \epsilon}{\partial r} \frac{\partial n}{\partial k} \right). \qquad (5.120)$$

Another way for $n(k, r, \sigma)$ to change is for the quasiparticle to change its spin state. This change is determined by the commutator $[\epsilon, n]$. Finally, the interaction with the lattice will also produce a change in the distribution. Therefore the total equation of motion becomes

$$\frac{\partial n}{\partial t} + \frac{1}{\hbar} \left\{ \frac{\partial n}{\partial r} \frac{\partial \epsilon}{\partial k} \right\} - \frac{1}{\hbar} \left\{ \frac{\partial n}{\partial k} \frac{\partial \epsilon}{\partial k} \right\} - \frac{i}{\hbar} [\epsilon, n] = \frac{\partial n}{\partial t} \bigg|_{\text{collisions}} \qquad (5.121)$$

where the braces indicate the symmetrized product. Notice that the second and third terms may be written as a symmetrized Poisson bracket.

The equation of motion for the magnetization is obtained by multiplying (5.121) by $-(\mu_B/V)\boldsymbol{\sigma}$ and taking the spin trace. Let us consider the second term. Since the unperturbed distribution function is spatially independent, the spatial derivative becomes

$$\frac{\partial n}{\partial r} = \frac{\partial n_1}{\partial r} + \frac{\partial}{\partial r}(\boldsymbol{n}_2 \cdot \boldsymbol{\sigma}). \tag{5.122}$$

The derivative of the energy with respect to the wave vector is

$$\frac{\partial \epsilon}{\partial \boldsymbol{k}} = \frac{\partial \epsilon^0}{\partial \boldsymbol{k}} + \frac{\partial \epsilon_1}{\partial \boldsymbol{k}} + \frac{\partial}{\partial \boldsymbol{k}}(\boldsymbol{\epsilon}_2 \cdot \boldsymbol{\sigma}) = \hbar v + \frac{\partial}{\partial \boldsymbol{k}}(\boldsymbol{\epsilon}_2 \cdot \boldsymbol{\sigma}). \tag{5.123}$$

Since $\text{Tr}\{\boldsymbol{\sigma}\} = 0$, $\text{Tr}\{(\boldsymbol{A}\cdot\boldsymbol{\sigma})\boldsymbol{\sigma}\} = 2\boldsymbol{A}$, and $\text{Tr}\{(\boldsymbol{A}\cdot\boldsymbol{\sigma})(\boldsymbol{B}\cdot\boldsymbol{\sigma})\boldsymbol{\sigma}\} = 2i\boldsymbol{A}\times\boldsymbol{B}$, we obtain

$$-\frac{\mu_B}{V}\text{Tr}\left\{\frac{\partial n}{\partial r}\frac{\partial \epsilon}{\partial \boldsymbol{k}}\right\}\boldsymbol{\sigma} = -\frac{2\mu_B}{V}\left(\frac{\partial n_1}{\partial r}\frac{\partial}{\partial \boldsymbol{k}}\right)\boldsymbol{\epsilon}_2 + \hbar(v\cdot\boldsymbol{\nabla})\boldsymbol{m}. \tag{5.124}$$

But

$$\frac{\partial n_1}{\partial r} = \frac{\partial n_0}{\partial \epsilon^0}\frac{\partial \epsilon_1}{\partial r} = -\frac{\partial n_0}{\partial \epsilon^0}\left(\frac{e\hbar}{2m^*c}\right)\boldsymbol{k}\times\boldsymbol{H}_0. \tag{5.125}$$

Therefore

$$-\frac{\mu_B}{\hbar V}\text{Tr}\left\{\frac{\partial n}{\partial r}\frac{\partial \epsilon}{\partial \boldsymbol{k}}\right\}\boldsymbol{\sigma} = \frac{\mu_B}{V}\frac{\partial n_0}{\partial \epsilon^0}\frac{e}{m^*c}\left[(\boldsymbol{k}\times\boldsymbol{H}_0)\cdot\frac{\partial}{\partial \boldsymbol{k}}\right]\boldsymbol{\epsilon}_2 + (v\cdot\boldsymbol{\nabla})\boldsymbol{m}. \tag{5.126}$$

The third term in (5.121) is evaluated in the same manner, giving

$$\frac{\mu_B}{\hbar V}\text{Tr}\left\{\frac{\partial n}{\partial \boldsymbol{k}}\frac{\partial \epsilon}{\partial r}\right\}\boldsymbol{\sigma} = \frac{2\mu_B}{V}\frac{\partial n_0}{\partial \epsilon^0}(v\cdot\boldsymbol{\nabla})\boldsymbol{\epsilon}_2 + \frac{e}{2m^*c}\left[(\boldsymbol{k}\times\boldsymbol{H}_0)\cdot\frac{\partial}{\partial \boldsymbol{k}}\right]\boldsymbol{m}. \tag{5.127}$$

Notice that the terms in (5.126, 127) which do not explicitly involve $\boldsymbol{m}$ contain the factor $\partial n_0/\partial \epsilon^0$. This implies that $\boldsymbol{m}$ is proportional to $\partial n_0/\partial \epsilon^0$. Since $\partial n_0/\partial \epsilon^0$ is essentially nonzero only at the Fermi surface, the magnetization arises only from those quasiparticles at the Fermi surface. This, together with the fact that we are interested in the transverse response of the system, suggests that we introduce the quantity $g(\boldsymbol{r}, \boldsymbol{k})$, defined by

$$m_+(\boldsymbol{r}, \boldsymbol{k}) = m_x + im_y = -\frac{\partial n_0}{\partial \epsilon^0}g(\boldsymbol{r}, \boldsymbol{k}). \tag{5.128}$$

Since $\boldsymbol{k} \simeq k_F\hat{\boldsymbol{k}}$, the dependence of $g(\boldsymbol{r}, \boldsymbol{k})$ upon $\boldsymbol{k}$ involves only the polar angles of $\boldsymbol{k}$ with respect to the z axis.

172   5. The Dynamic Susceptibility of Weakly Interacting Systems

The contribution to the equation of motion for m_+ from the fourth term involving the commutator is

$$i\frac{\mu_B}{\hbar V}\operatorname{Tr}\{[\epsilon, n]\sigma_+\} = i\left[\left(\frac{\mu_B H_0}{\hbar} + \frac{V}{2\hbar\mu_B}G_z\right)m_+ + \frac{V}{2\hbar\mu_B}m_z G - \frac{\mu_B}{\hbar}m_z H_+\right] \quad (5.129)$$

where

$$G(r, k) = \frac{2}{V}\sum_{k'}\psi(\hat{k}\cdot\hat{k}')m_+(r, k') \quad (5.130)$$

and G_z involves m_z in place of m_+. To proceed further with this expression we must determine $m_z(r, k)$. In the last chapter we found that the magnetization resulting from a uniform field arises from a spherically symmetric shell of states at the Fermi surface. Therefore we take

$$m_z(r, k) = -\frac{\partial n_0}{\partial \epsilon^0}g(r)_z . \quad (5.131)$$

To find $g(r)_z$ we note that the magnetization at r, which is given by

$$m_z(r) = \sum_k m_z(r, k) , \quad (5.132)$$

is also related to the static susceptibility by

$$m_z(r) = \frac{1}{V}\sum_q \chi(q)H(q) = \chi(0)H_0 . \quad (5.133)$$

We use the fact that

$$\frac{\partial n_0}{\partial \epsilon^0} = -\delta(\mu - \epsilon^0(k)) = -\frac{m^*}{\hbar^2 k_F}\delta(k - k_F) \quad (5.134)$$

along with our result (4.80) for the static susceptibility, which may be written as

$$\chi(0) = \frac{\mu_B^2 m^* k_F}{\pi^2 \hbar^2 (1 + B_0)} . \quad (5.135)$$

We then obtain

$$m_z(r, k) = -\frac{\partial n_0}{\partial \epsilon^0}\frac{2\mu_B^2 H_0}{V(1 + B_0)} . \quad (5.136)$$

The collision term on the right-hand side of (5.121) is handled by assuming that the distribution function $n(r, k, \sigma)$ relaxes to its local average value $\bar{n}(r, \sigma)$, defined by

$$\bar{n}(r, \sigma) = \frac{1}{\frac{4}{3}\pi k_F^2} \int dk\, n(r, k, \sigma). \tag{5.137}$$

If this relaxation is characterized by a time τ, then

$$\left.\frac{\partial g}{\partial t}\right|_{\text{collisions}} = -\frac{1}{\tau}\left(g - \frac{1}{4\pi}\int d\Omega g\right). \tag{5.138}$$

Combining all these results in the equation of motion for $g(r, k)$ yields

$$\frac{\partial g}{\partial t} + \left[v\cdot\nabla + \frac{e}{m^*c}(k \times H_0)\cdot\frac{\partial}{\partial k} + i\Omega_0\right](g + G)$$
$$= -\frac{1}{\tau}\left(g - \frac{1}{4\pi}\int d\Omega g\right) + \frac{2\mu_B^2}{V}(v\cdot\nabla + i\Omega_0)H_+ \tag{5.139}$$

where

$$\Omega_0 = \frac{2\mu_B H_0}{\hbar(1 + B_0)}. \tag{5.140}$$

Equation (5.139) is a linear integral equation that is difficult to solve in general. Therefore certain simplifying approximations are made. For example, *Silin* has considered the spatially uniform normal modes of a collisionless system [5.13]. This corresponds to solving (5.139) without the $v\cdot\nabla$ term on the left and with the right side set to 0. The resulting solutions for $g(\theta,\varphi)$ are spherical harmonics, $Y_n^m(\theta, \varphi)$, with the corresponding eigenfrequencies.

$$\omega_{nm} + (\Omega_0 + m\omega_c)\left(1 + \frac{B_n}{2n + 1}\right) \tag{5.141}$$

where $\omega_c = eH_0/m^*c$ is the cyclotron frequency. The mode $\omega_{00} \equiv \omega_s \equiv 2\mu_B H_0/\hbar$ is the conduction-electron spin resonance. We see that as a result of the interactions there are also modes involving combined spin and orbital excitations.

Platzman and *Wolff* have investigated the q dependence of the ω_{00} mode [5.14]. However, they also retained the collision term. Their result for the susceptibility, including also a spin-relaxation term, is

$$\chi(q, \omega) = \frac{m^*}{m}\frac{\chi_{\text{Pauli}}}{1 + B_0}\frac{\omega_s}{\omega_s - \omega - iD^*q^2 - i/\bar{T}_1}, \tag{5.142}$$

where

$$iD^* = \tfrac{1}{3}v_F^2(1 + B_0)(1 + B_1)(\bar{\omega}_s - \bar{\Omega}_0)$$
$$\times \left[\frac{\sin^2\Delta}{\bar{\omega}_c^2 - (\bar{\omega}_s - \bar{\Omega}_0)^2} - \frac{\cos^2\Delta}{(\bar{\omega}_s - \bar{\Omega}_0)^2}\right]. \tag{5.143}$$

Here Δ is the angle that q makes with the z axis, and

174   5. The Dynamic Susceptibility of Weakly Interacting Systems

$$\bar{\omega}_s = \omega_s + \frac{i}{\tau}(1 + B_1), \tag{5.144}$$

$$\bar{\omega}_c = \frac{eH_0}{m^*c}(1 + B_1), \tag{5.145}$$

$$\bar{\Omega}_0 = \Omega_0(1 + B_1), \tag{5.146}$$

$$\bar{T}_1 = T_1(1 + B_0)^{-1}.$$

Notice that as ω and $q \to 0$ and $\tau \to \infty$, (5.142) reduces to the result we obtained in Chap. 4.

Conduction-Electron Spin Resonance. An interesting limit of (5.142) is when there are no quasiparticle interactions, that is, $B_0 = B_1 = 0$. In this case $\bar{\omega}_s - \bar{\Omega}_0 = i/\tau$ and D^* is a real number. The susceptibility (5.142) then contains a pole at the conduction-electron spin resonance frequency ω_s. The fact that the lifetime of this mode is partially governed by the diffusion term has an interesting consequence for the resonance spectrum. The point is that in a metal the electromagnetic field that excites the resonance is not uniform across the sample. To find the field distribution we must solve Maxwell's equations with the appropriate boundary conditions.

Let us consider the geometry shown in Fig. 5.12 where the electromagnetic wave is incident from the left on a metal with conductivity σ. Assuming the fields vary only along the z direction, the two Maxwell equations, together with $\boldsymbol{j} = \sigma \boldsymbol{E}$, reduce to

$$\nabla \times \boldsymbol{E} = -\frac{1}{c}\frac{\partial \boldsymbol{B}}{\partial t} \quad \begin{cases} \dfrac{\partial E_x}{\partial z} = -\dfrac{1}{c}\left(\dfrac{\partial H_y}{\partial t} + 4\pi\dfrac{\partial M_y}{\partial t}\right) \\ \dfrac{\partial E_y}{\partial z} = \dfrac{1}{c}\left(\dfrac{\partial H_x}{\partial t} + 4\pi\dfrac{\partial M_x}{\partial t}\right) \end{cases}, \tag{5.147}$$

$$\nabla \times \boldsymbol{H} = \frac{4\pi}{c}\boldsymbol{j} \quad \begin{cases} \dfrac{\partial H_x}{\partial z} = \dfrac{4\pi}{c}j_y = \dfrac{4\pi}{c}\sigma E_y \\ \dfrac{\partial H_y}{\partial z} = -\dfrac{4\pi}{c}j_x = -\dfrac{4\pi}{c}\sigma E_x \end{cases}, \tag{5.148}$$

Introducing the circular components $M_+ = M_y + iM_y$, $H_+ = H_x + iH_y$, and assuming solutions of the form

$$M_+ = m\,e^{i\omega t - \kappa z},$$
$$H_+ = h\,e^{i\omega t - \kappa z}, \tag{5.149}$$

we obtain

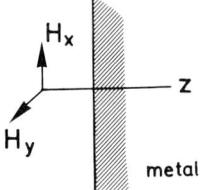
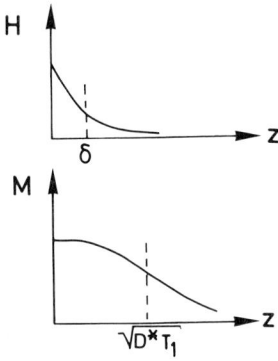

Fig. 5.12. Geometry and field distributions associated with conduction-electron spin resonance

$$\frac{m}{h} = \chi = -\frac{1}{4\pi}(1 + \tfrac{1}{2} i\delta^2\kappa^2), \tag{5.150}$$

where $\delta = c/\sqrt{2\pi\sigma\omega}$ is the skin depth. Self-consistency between the susceptibility (5.142) derived from the equation of motion for the magnetization and that derived above from Maxwell's equations gives a quadratic equation for the parameter κ. This has two roots, one of the order of δ, the other of the order of $\sqrt{D^*T_1}$. The latter characterizes how far a spin diffuses before relaxing. The amplitudes of the fields associated with these solutions are determined from the boundary conditions. The results are sketched in Fig. 5.12. The magnetic field falls off in a distance of the order of the skin depth. The magnetization, however, penetrates a distance $\sqrt{D^*T_1}$ which may be much larger than δ.

In the standard reflection—type resonance experiment the metal is placed at one end of a microwave cavity. The power absorbed by the cavity, which is easily calculated from the above results, is shown in Fig. 5.13.

We immediately notice the asymmetric nature of the absorption. This is associated with the fact that although an electron may only spend a time of the order of $T_D = \delta^2/2D^*$ in the skin depth, it returns to the skin depth many times before its spin "memory" is destroyed. Thus it experiences a set of microware pulses whose intervals are random but whose phases are coherent. This can lead to destructive as well as constructive interference with the result that the power absorbed at some values of the dc field may be *less* than that ordinarily absorbed

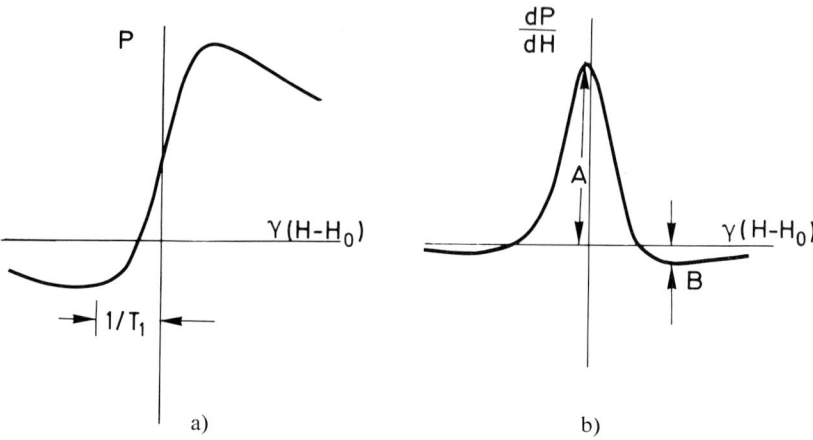

Fig. 5.13a, b. Power absorption (**a**) and its derivative (**b**) due to electron spin resonance in a metal for a ratio of $T_D/T_1 \simeq 0.1$

in zero field. Since *Dyson* [5.15] was the first to derive this characteristic shape, such lines are referred to as *Dysonian*. The asymmetry may be quantified by the ratio A/B which depends upon the ratio of T_D to T_1 as shown in Fig. 5.14. The fact that the width of the line is proportional to $1/T_1$ is also due to the fact that the electron returns to the skin depth many times.

The fact that the magnetization penetrates much farther into the metal than the driving magnetic field is the basis of the transmission technique. In this technique the sample takes the form of a thin film whose thickness is large compared with δ but smaller than $\sqrt{D^*T_1}$. This film is placed between two microwave cavities. Off resonance no power reaches the second cavity. At resonance, however, the precessing spins diffuse across the film and radiate into the second

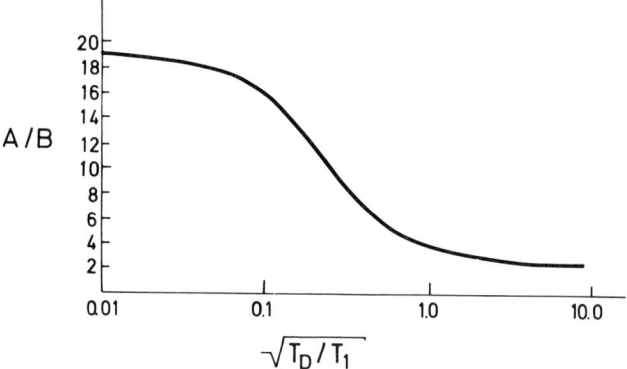

Fig. 5.14. Dependence of the asymmetry ratio A/B on the ratio of the spin diffusion time T_D to the spin relaxation time T_1

cavity. The intensity of the magnetization excited depends upon the susceptibility within the skin depth. Thus it has been found that the sensitivity of the transmission technique can be greatly enhanced by depositing a thin ferromagnetic film on the front surface of the film being studied.

Spin Waves. Let us now consider the effect of the electron–electron interactions. In particular, if

$$\left| \frac{(B_0 - B_1)\omega_s \tau}{1 + B_0} \right| \gg 1 ,$$

then D^* becomes pure imaginary and modes appear at $\omega = \omega_s - D^* k^2$. Thus, the interactions convert the transport of the magnetization from diffusive to wavelike. These modes are referred to as *spin waves*, although they bear only a qualitative resemblence to the spin waves we shall discuss in the next chapter. In a slab of thickness L the wave vector takes on values which are multiples of π/L. The modes then appear as sidebands of the main conduction-spin resonance line. Such *spin–wave* sidebands have been observed in various metals [5.16]. The transmission spectrum for sodium is shown in Fig. 5.15. From careful measurements of the angular dependence of the spin–wave positions the Fermi liquid parameters can be deduced. For sodium $B_0 = -0.18$ and $B_1 = 0.05$.

Notice that D^* may be positive or negative, depending on the direction of q with respect to the dc field. In the next chapter we shall find that the spin-wave spectrum in a ferromagnetic insulator always curves upward. The fact that the spectrum may curve downward in the metal is due to the "repulsion" of these spin–wave modes by the higher-lying orbital modes, whose $q = 0$ frequencies we have derived in (5.141).

Since the Platzmann-Wolff solution (5.142) is valid only for small q near ω_s, it does not include the single-particle excitations illustrated in Fig. 5.10. In the presence of interactions these modes are shifted upward by a self-energy that is proportional to $N_\downarrow - N_\uparrow$. Therefore the spectrum has the form shown in Fig. 5.16. When the spin waves enter the single-particle band, they can decay into single-particle excitations. In this region they become critically damped.

Local Moments in Metals. It is interesting to consider the case when a "potentially" magnetic ion is dissolved in a metal. We say "potentially" because in some cases the magnetic response of the impurity is destroyed by the conduction electrons, a situation referred to as the Kondo effect. We shall discuss this problem in Chap 7. For our present purposes we shall assume that the local moment exists and that its coupling to the conduction electrons may be described by the exchange interaction

$$\mathscr{H} = -J\,\boldsymbol{S}\cdot\boldsymbol{s} . \tag{5.151}$$

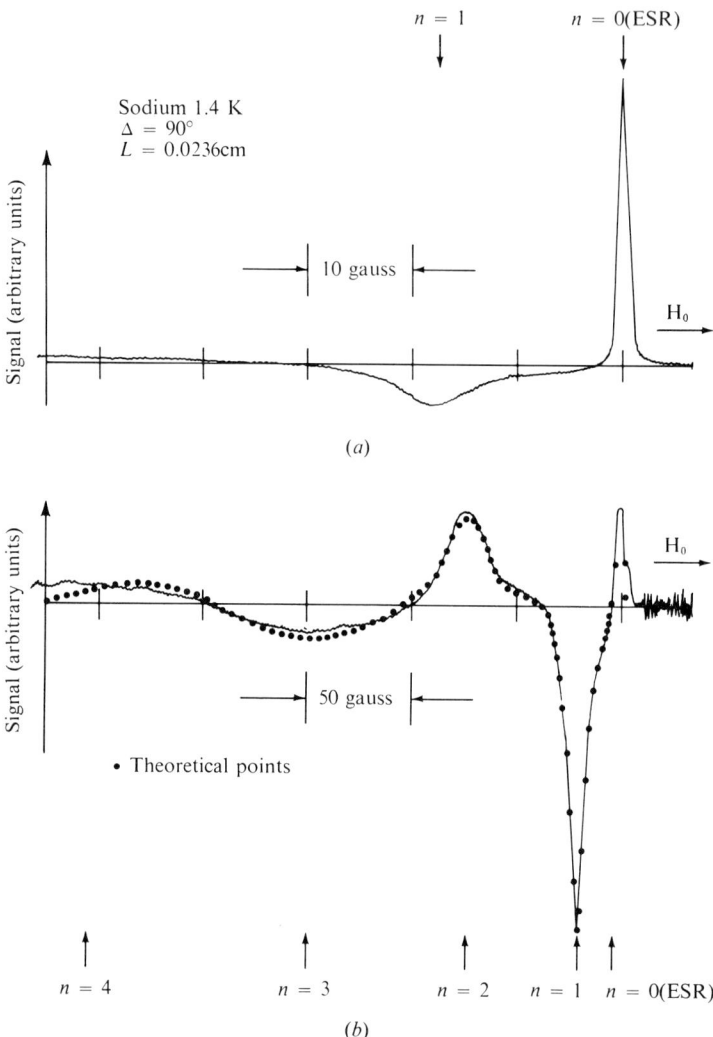

Fig. 5.15a, b. Typical spin wave signals as a function of applied dc magnetic field ($H_0 \approx 3{,}250$ G). (a) The $n = 0$ mode (usual conduction-electron spin resonance) and the $n = 1$ mode are clearly shown. (b) The gain and field sweep have been increased to display the first four spin-wave modes beyond the CESR. The theoretical points have been obtained from the susceptibility given in (5.142) [5.16]

As a result of this interaction the magnetization associated with the local impurity will relax with a rate

$$\frac{1}{T_{ie}} = \frac{\pi}{\hbar} J^2 N(\epsilon_F)^2 k_B T . \tag{5.152}$$

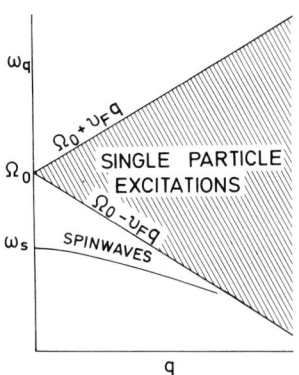

Fig. 5.16. Magnetic excitation spectrum of an interacting electron system

(This result is analogous to the relaxation of nuclear spins in a metal through their hyperfine coupling with conduction electrons. In that case J is replaced by the hyperfine constant A, and we refer to the result as the Korringa relaxation rate). The impurity spins also relax the conduction-electron magnetization with the rate

$$\frac{1}{T_{ei}} = \frac{2\pi}{\hbar} J^2 N(\epsilon_F) \frac{S(S+1)}{3} c \qquad (5.153)$$

where c is the impurity concentration. Notice that

$$\frac{T_{ie}}{T_{ei}} = \frac{\chi_i}{\chi_d} \qquad (5.154)$$

where χ_i is the Curie susceptibility and χ_e the Pauli susceptibility. This relation is actually very general, being a consequence of the principle of detailed balance. Considering that both spin systems can also relax directly to the lattice we have the situation illustrated in Fig. 5.17. The magnetizations are described by a pair of coupled Bloch-like equations [5.17],

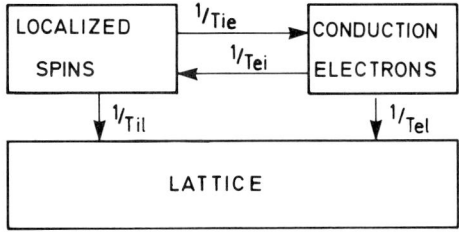

Fig. 5.17. Schematic representation of the energy resevoirs associated with a localized spin impurity in a metal

$$\frac{dM_e}{dt} = \gamma_e M_e \times (H + \lambda M_i) - \frac{M_e}{T_{ei}} + \frac{M_i}{T_{ie}} - \frac{(M_e - M_e^0)}{T_{el}},$$
$$\frac{dM_i}{dt} = \gamma_i M_i \times (H + \lambda M_e) - \frac{M_i}{T_{ie}} + \frac{M_e}{T_{ei}} - \frac{(M_i - M_i^0)}{T_{il}}.$$
(5.155)

Here λ is an effective field parameter which is proportional to the exchange J. For transition-metal impurities, such as Mn in Cu the relaxation rates $1/T_{ei}$ and $1/T_{ie}$ are much larger than the relaxation rate to the lattice. This corresponds to a "bottleneck" condition. If the coupling J is strong, then the two magnetizations remain parallel to one another and the transverse components of (5.155) become

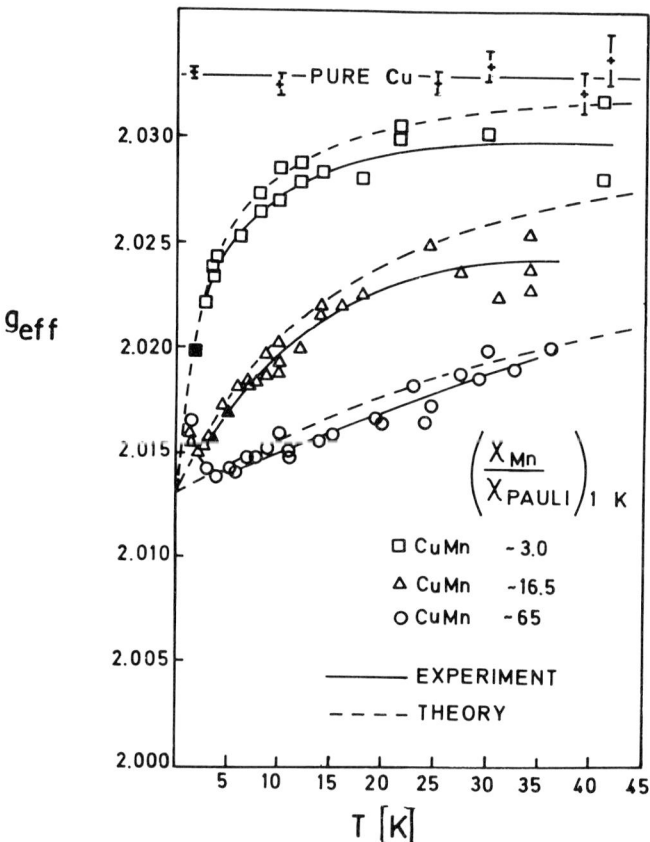

Fig. 5.18. Plot of the g value obtained from transmission resonance versus temperature for three Cu Mn samples. The theoretical curves were obtained from (5.157) using $g_{Cu} = 2.033$ and $g_{Mn} = 2.013$ and the susceptibility ratios shown. A susceptibility ratio of 3 corresponds to 13 ppm of Mn in Cu

$$\frac{d}{dt}\begin{pmatrix} M_e^+ \\ M_i^+ \end{pmatrix} = \begin{pmatrix} -i\gamma_e H - \dfrac{1}{T_{ei}} & \dfrac{1}{T_{ie}} \\ \dfrac{1}{T_{ei}} & -i\gamma_i H - \dfrac{1}{T_{ie}} \end{pmatrix} \begin{pmatrix} M_e^+ \\ M_i^+ \end{pmatrix}. \tag{5.156}$$

Diagonalizing this equation we find that the effective g value for the strongly coupled case is

$$g_{\text{eff}} = \frac{g_i \chi_i + g_e \chi_e}{\chi_i + \chi_e}. \tag{5.157}$$

At low impurity concentrations or high temperatures χ_i is small and the observed g value should be close to that of the pure metal. As one increases the impurity concentration or goes to lower temperatures, g_{eff} should shift towards a value characteristic of the impurity. The transmission electron spin resonance data [5.18] on **Cu Mn** shown in Fig. 5.18 has these features.

5.3 Faraday Effect

The Faraday effect refers to the rotation of the plane of polarization of a linearly polarized electromagnetic wave as it propagates through a magnetic or magnetized material. Assuming for the moment that the propagation through such a medium is adequately characterized by the susceptibility (or permeability) tensor, Maxwell's equations tell us that the rotation per unit length of sample is given by

$$\frac{\Delta \theta}{\Delta z} = \frac{2\pi \omega}{v} \chi''_{xy}.$$

In the case of a Kramers ground state we found (p.150) that when $\omega \gg \omega_0$, χ''_{xy} decreases as $1/\omega$ and $\Delta \theta / \Delta z$ reduces to

$$\frac{\Delta \theta}{\Delta z} = -\frac{\omega_M}{2c/n}$$

where n is the real part of the complex index of refraction in the absence of a static magnetization and $\omega_M = 4\pi \gamma M$. For a magnetic material with a saturation magnetization of $4\pi M = 2000$ G and a dielectric constant of 13, this gives a rotation of approximately $140°$/cm. Such rotation forms the basis for various nonreciprocal microwave devices.

As one goes beyond microwave frequencies, one must also consider the time-varying electric field which accompanies the "probing" magnetic field and brings the dielectic, or conductivity, tensor into play. The Faraday rotation associated with this "polarization" response is

$$\frac{\Delta\theta}{\Delta z} = -\frac{2\pi}{cn}\sigma'_{xy} = \frac{2\pi\omega}{cn}\varepsilon''_{xy}.$$

Symmetry arguments, similar to those used in deriving the Onsager relation (1.89), tell us that ε''_{xy} is a linear function of M_z. Physically, this dependence of the dielectric function on the magnetization arises through the spin–orbit interaction. The important point, however, is that the polarization-induced rotation increases with frequency while the magnetic contribution is independent of frequency. Thus there will be a frequency beyond which the polarization contribution dominates. This typically occurs in the near infrared. For this reason, one generally finds magnetooptical phenomena described entirely in terms of the conductivity. *Pershan* [5.19], in fact, has argued that so long as the electromagnetic wavelength is less than the sample dimensions, one can always describe the wave properties in terms of an effective dielectric function, setting the magnetic susceptibility to zero. However, as *Pershan* has also noted, one must pay careful attention to the boundary conditions when applying such a description. On the other hand, there may be situations in which the microscopic behavior of the Faraday rotation may be more conveniently understood, or calculated, in terms of magnetic dipole matrix elements.

6. The Dynamic Susceptibility of Strongly Interacting Systems

We have just seen that in nonferromagnetic metals the presence of interactions leads to magnetic excitations of a wavelike nature called spin waves. In this chapter we shall investigate these excitations in magnetically ordered systems.

6.1 Broken Symmetry

At one time spin waves were only associated with local moment systems. But they have a far more fundamental basis. In the paramagnetic state the system is rotationally invariant. In the ferromagnetically ordered state this symmetry is broken—there is a uniform magnetization M in a particular direction. In principle, this direction is arbitrary, but in practice it is determined by spin–orbit effects. In Chap. 4 we argued that the energy associated with a gradual variation in the magnetization has the form

$$\frac{\delta E}{V} = \frac{A}{M_0^2} |\nabla M|^2 .$$

This energy implies that the magnetization in any macroscopically small region is acted on by an exchange torque coming from an effective field

$$H_{\text{eff}} = -\frac{2A}{M_0^2} \nabla^2 M$$

Assuming that M responds to this effective field just as it does to a real magnetic field, we have

$$\frac{dM}{dt} = -\gamma M \times H_{\text{eff}} .$$

Thus, a transverse fluctuation in the magnetization with a wave vector q has a frequency

$$\omega_q = \gamma \frac{2A}{M_0} q^2 .$$

Such modes are a direct consequence of having broken the continuous rotational symmetry. A system characterized by an Ising Hamilontian does not possess such a continuous symmetry and, indeed, there are no spin-wave-like excitations in the ordered state. The behavior of these modes at shorter wavelengths and the form of the exchange stiffness parameter depend upon the microscopic details which we shall consider in this chapter.

6.2 Insulators

Let us begin by considering a lattice of spins whose interactions may be described by the Heisenberg exchange interaction (2.89). Suppose we apply a uniform static field which serves to define a z axis. We now wish to determine how this system responds to the time- and space-dependent field $H_1 \cos(q \cdot r) \cos \omega t$. If this field is in the x direction, the total Hamiltonian becomes

$$\mathcal{H} = -\sum_i \sum_{j \neq i} J_{ij} S_i \cdot S_j + g\mu_B H_0 \sum_i S_i^z + g\mu_B H_1 \sum_i S_i^x \cos(q \cdot R_i) \cos \omega t. \tag{6.1}$$

Introducing the Fourier components defined by (4.4), we obtain

$$\mathcal{H} = -\sum_{q'} J(-q') S(q') \cdot S(-q') + g\mu_B H_0 S_z(0) + \tfrac{1}{2} g\mu_B H_1 \\ \times [S_x(q) + S_x(-q)] \cos \omega t. \tag{6.2}$$

Since this Hamiltonian is space dependent as well as time dependent, the appropriate equation of motion, analogous to (5.4), is

$$\frac{dM(q)}{dt} = -\frac{i}{\hbar V} \langle [\mathcal{M}(q), \mathcal{H}] \rangle. \tag{6.3}$$

Since we are expressing the spin in units of $\hbar$, the commutation relations have the form

$$[S_x(q), S_y(q')] = iS_z(q + q'). \tag{6.4}$$

The commutator involving the x component of the spin with the Hamiltonian then becomes

$$[S_x(q), \mathcal{H}] = -i\sum_{q'} J(q')[S_z(q + q') S_y(-q') + S_y(q') S_z(q - q') \\ - S_y(q + q') S_z(-q') - S_z(q') S_y(q - q')] - ig\mu_B H_0 S_y(q). \tag{6.5}$$

In the *random-phase approximation* each Fourier component is independent. This implies that $\langle S_z(q + q') S_y(-q') \rangle = \langle S_z(q + q') \rangle \langle S_y(-q') \rangle$. Furthermore, in the low-temperature region we may linearize (6.3) by making the approximation

$$\langle S_z(\boldsymbol{q}')\rangle = NS\,\varDelta(\boldsymbol{q}')\,. \tag{6.6}$$

With these approximations, (6.3) becomes

$$\frac{dM_x(\boldsymbol{q})}{dt} = -\left\{\gamma H_0 + \frac{2NS}{\hbar}[J(0) - J(\boldsymbol{q})]\right\}M_y(\boldsymbol{q}) \tag{6.7}$$

where $\gamma = g\mu_B/\hbar$ is the gyromagnetic ratio, which we take as positive for electrons. This equation is analogous to (5.6a), and the equation for $M_y(\boldsymbol{q})$ is analogous to (5.6b). Therefore all the arguments leading to the susceptibility (5.18) also apply in this case. Thus we find that the imaginary part of the susceptibility of a ferromagnet at low temperatures has the form

$$\chi''_{xx}(\boldsymbol{q},\omega) = \frac{\pi g^2 \mu_B^2 NS}{2\hbar}\{\delta[\omega - \omega(\boldsymbol{q})] - \delta[\omega + \omega(\boldsymbol{q})]\} \tag{6.8}$$

where

$$\omega(\boldsymbol{q}) = \gamma H_0 + \frac{2NS}{\hbar}[J(0) - J(\boldsymbol{q})]\,. \tag{6.9}$$

These are the spin-wave modes. Classically, these excitations correspond to a precession of the spin system in which the phase of one spin relative to another is determined by the wave vector $\boldsymbol{q}$. An example of these modes is shown in Fig. 6.1. The $\boldsymbol{q} = 0$, or uniform-precession, mode corresponds to ferromagnetic resonance. Quantum mechanically, a spin wave consists of a single spin flip propagating throughout the lattice. Since the spins are quantized, these excitations are also quantized. Consequently, they are often referred to as *magnons*. Because these modes are collective, in the sense that they involve all the spins in the lattice, the excitation energy varies continuously from $\gamma\hbar H_0$ up to the exchange energy. If the only excitations were single-particle excitations, the energy required to produce a spin deviation would be of the order of the exchange energy.

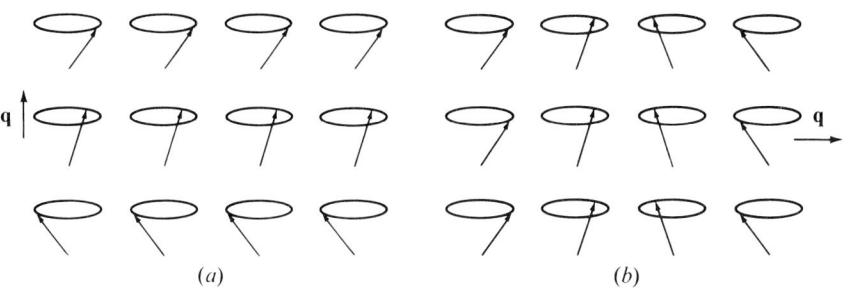

Fig. 6.1a, b. Schematic representation of a spin wave (a) propagating parallel to the applied magnetic field, and (b) propagating perpendicular to this field

186   6. The Dynamic Susceptibility of Strongly Interacting Systems

For a simple cubic lattice with lattice parameter a, and a nearest-neighbor exchange J, the long wavelength limit of $J(0) - J(\mathbf{q})$ is Ja^2q^2/N. Therefore the exchange stiffness is

$$A = S^2 a^2 J/v$$

where v is the atomic volume, V/N.

Combining (4.16; 6.9), we see that the ratio of the maximum spin-wave frequency to $k_B T_C/\hbar$ for a simple cubic lattice is $6/(S+1)$. Therefore the high-temperature approximation used in writing (4.45) is valid only for $T/T_C > 6/(S+1)$.

6.2.1. Spin-Wave Theory

Notice that our low-temperature approximation (6.6) has led to a set of modes that behaves just like a system of independent harmonic oscillators. In 1940 Holstein and Primakoff introduced a very useful technique which exploits this harmonic-oscillator analog. Their approach was based on the expansion of spin operators in terms of the creation and annihilation operators of the harmonic oscillator. For example, consider the effect of S_i^- acting on $|M_S\rangle$,

$$S_i^- |M_S\rangle = \sqrt{(S+M_S)(S-M_S+1)}\,|M_S - 1\rangle. \tag{6.10}$$

If we introduce the *spin-deviation operator* $\hat{n}_i = S - S_i^z$, with the quantum numbers $n_i = S - M_S$, this relation becomes

$$S_i^- |n_i\rangle = \sqrt{2S}\,\sqrt{n_i+1}\,\sqrt{1 - \frac{n_i}{2S}}\,|n_i + 1\rangle. \tag{6.11}$$

The harmonic-oscillator operator $a^\dagger$ has the property (as a consequence of its commutation relations) that

$$a^\dagger |n\rangle = \sqrt{n+1}\,|n+1\rangle. \tag{6.12}$$

Therefore we make the associations

$$\begin{aligned}
S_i^- &= \sqrt{2S}\,a_i^\dagger f_i(S), \\
S_i^+ &= \sqrt{2S}\,f_i(S)a_i, \\
S_i^z &= -S + a_i^\dagger a_i,
\end{aligned} \tag{6.13}$$

where $f_i(S) \equiv \sqrt{1 - (a_i^\dagger a_i/2S)}$. Notice that this factor makes the transformation nonlinear. It is an easy matter to show that these combinations satisfy the required commutation relations.

As we have seen, in dealing with coupled systems it is convenient to introduce Fourier components. This suggests that we expand the spin-deviation operators according to

$$a_i = \frac{1}{\sqrt{N}} \sum_k e^{i\mathbf{k}\cdot\mathbf{R}_i} a_k , \qquad (6.14)$$

$$a_i^\dagger = \frac{1}{\sqrt{N}} \sum_k e^{-i\mathbf{k}\cdot\mathbf{R}_i} a_k^\dagger . \qquad (6.15)$$

We shall refer to a_k and $a_k^\dagger$ as *magnon annihilation* and *magnon creation operators*, respectively. The spin operators may be written in terms of these operators as

$$S_i^+ = \sqrt{\frac{2S}{N}} \sum_k e^{i\mathbf{k}\cdot\mathbf{R}_i} a_k - \frac{1}{N\sqrt{8NS}} \sum_{k,k',k''} e^{-i(\mathbf{k}-\mathbf{k}'-\mathbf{k}'')\cdot\mathbf{R}_i} a_k^\dagger a_{k'} a_{k''} + \cdots , \qquad (6.16)$$

$$S_i^- = \sqrt{\frac{2S}{N}} \sum_k e^{-i\mathbf{k}\cdot\mathbf{R}_i} a_k^\dagger - \frac{1}{N\sqrt{8NS}} \sum_{k,k',k''} e^{-i(\mathbf{k}+\mathbf{k}'-\mathbf{k}'')\cdot\mathbf{R}_i} a_k^\dagger a_{k'}^\dagger a_{k''} + \cdots , \qquad (6.17)$$

$$S_i^z = -S + \frac{1}{N} \sum_{k,k'} e^{-i(\mathbf{k}-\mathbf{k}')\cdot\mathbf{R}_i} a_k^\dagger a_{k'} . \qquad (6.18)$$

The Hamiltonian for a system of independent harmonic oscillators has the form $\sum \hbar\omega_k a_k^\dagger a_k$. Therefore the objective of the *Holstein-Primakoff approach* is to determine the quadratic part of the Hamiltonian resulting from substitutions (6.16–18). Let us consider this procedure for a ferromagnet (other systems are treated in [6.1]).

The Zeeman and exchange interactions were considered above. Let us add to these the dipole–dipole interaction (2.50). The total Hamiltonian is then

$$\mathcal{H} = \mathcal{H}_z + \mathcal{H}_{ex} + \mathcal{H}_{dip} . \qquad (6.19)$$

The low-temperature or spin–wave approximation (6.6) corresponds to taking $f_i(S) = 1$. This is because at very low temperatures the number of spin waves excited is much less than the total number of spins. Therefore the average spin deviation n_i is much less than 1.

In terms of the magnon operators, the Zeeman interaction becomes

$$\mathcal{H}_z = \gamma H_0 \sum_i S_i^z = -\gamma H_0 \sum_i \left(S - \frac{1}{N} \sum_{k,k'} e^{-i(\mathbf{k}-\mathbf{k}')\cdot\mathbf{R}_i} a_k^\dagger a_{k'} \right). \qquad (6.20)$$

If the lattice has translational invariance, then

$$\sum_i e^{-i(\mathbf{k}-\mathbf{k}')\cdot\mathbf{R}_i} = N\Delta(\mathbf{k} - \mathbf{k}') . \qquad (6.21)$$

Therefore (6.20) reduces to

$$\mathcal{H}_z = -\gamma H_0 NS + \gamma H_0 \sum_k a_k^\dagger a_k. \tag{6.22}$$

Similarly, if the exchange coupling between an ion i and those ions at positions $\boldsymbol{\delta}$ relative to i has the value J, then the exchange interaction is

$$\mathcal{H}_{ex} = -J \sum_{i,\delta} \mathbf{S}_i \cdot \mathbf{S}_{i+\delta} = -J \sum_{i,\delta} [\tfrac{1}{2} S_i^+ S_{i+\delta}^- + \tfrac{1}{2} S_i^- S_{i+\delta}^+ + S_i^z S_{i+\delta}^z], \tag{6.23}$$

$$\begin{aligned}\mathcal{H}_{ex} = -J \sum_{i,\delta} \Bigg(& \frac{S}{N} \sum_{k,k'} e^{i\mathbf{k}\cdot\mathbf{R}_i} e^{-i\mathbf{k}'\cdot(\mathbf{R}_i+\boldsymbol{\delta})} a_k a_{k'}^\dagger \\ & + \frac{S}{N} \sum_{k,k'} e^{-i\mathbf{k}\cdot\mathbf{R}_i} e^{i\mathbf{k}'\cdot(\mathbf{R}_i+\boldsymbol{\delta})} a_k^\dagger a_{k'} \\ & + \left[S - \frac{1}{N} \sum_{k,k'} e^{-i(\mathbf{k}-\mathbf{k}')\cdot\mathbf{R}_i} a_k^\dagger a_{k'} \right] \\ & \times \left[S - \frac{1}{N} \sum_{k,k'} e^{-i(\mathbf{k}-\mathbf{k}')\cdot(\mathbf{R}_i+\boldsymbol{\delta})} a_k^\dagger a_{k'} \right] \Bigg). \end{aligned} \tag{6.24}$$

In keeping with our spirit of setting $f_i(S) = 1$, we also neglect the fourth-order terms in this expansion. Thus, if z is the number of interacting neighbors,

$$\mathcal{H}_{ex} = -JS \sum_k \sum_\delta e^{-i\mathbf{k}\cdot\boldsymbol{\delta}} a_k a_k^\dagger - JS \sum_k \sum_\delta e^{i\mathbf{k}\cdot\boldsymbol{\delta}} a_k^\dagger a_k \\ - NzJS^2 + 2zJS \sum_k a_k^\dagger a_k. \tag{6.25}$$

We introduce

$$\gamma_k = \frac{1}{z} \sum_\delta e^{i\mathbf{k}\cdot\boldsymbol{\delta}}.$$

For example, with nearest-neighbor interactions in a simple cubic crystal having a lattice parameter a,

$$\gamma_k = \tfrac{1}{3}(\cos k_x a + \cos k_y a + \cos k_z a).$$

Notice that

$$\sum_k \gamma_k = 0.$$

Thus

$$\mathcal{H}_{ex} = -NzJS^2 + 2zJS \sum_k (1 - \gamma_k) a_k^\dagger a_k. \tag{6.26}$$

Let us now consider the dipole–dipole interaction, whose treatment is more tedious,

$$\mathcal{H}_{\text{dip}} = \tfrac{1}{2} g^2 \mu_B^2 \sum_{i,j} \left[\frac{\mathbf{S}_i \cdot \mathbf{S}_j}{r_{ij}^3} - \frac{3(\mathbf{S}_i \cdot \mathbf{r}_{ij})(\mathbf{S}_j \cdot \mathbf{r}_{ij})}{r_{ij}^5} \right], \tag{6.27}$$

$$\mathcal{H}_{\text{dip}} = \tfrac{1}{2} g^2 \mu_B^2 \sum_{i,j} \left\{ \frac{S_i^+ S_j^-}{2 r_{ij}^3} + \frac{S_i^- S_j^+}{2 r_{ij}^3} + \frac{S_i^z S_j^z}{r_{ij}^3} - \frac{3}{r_{ij}^5} \right.$$
$$\left. \times \left[(\tfrac{1}{2} S_i^+ r_{ij}^- + \tfrac{1}{2} S_i^- r_{ij}^+ + S_i^z z_{ij})(\tfrac{1}{2} S_j^+ r_{ij}^- + \tfrac{1}{2} S_j^- r_{ij}^+ + S_j^z z_{ij}) \right] \right\}, \tag{6.28}$$

$$\mathcal{H}_{\text{dip}} \simeq \tfrac{1}{2} g^2 \mu_B^2 \sum_{i,j} \left[S_i^+ S_j^- \frac{1}{2 r_{ij}^3} \left(1 - \frac{3}{2} \frac{r_{ij}^+ r_{ij}^-}{r_{ij}^2}\right) + S_i^- S_j^+ \frac{1}{2 r_{ij}^3} \left(1 - \frac{3}{2} \frac{r_{ij}^+ r_{ij}^-}{r_{ij}^2}\right) \right.$$
$$\left. + S_i^z S_j^z \frac{1}{r_{ij}^3} \left(1 - 3 \frac{z_{ij}^2}{r_{ij}^2}\right) - \tfrac{3}{4} S_i^+ S_j^+ \frac{(r_{ij}^-)^2}{r_{ij}^5} - \tfrac{3}{4} S_i^- S_j^- \frac{(r_{ij}^+)^2}{r_{ij}^5} \right], \tag{6.29}$$

where $r_{ij}^\pm = x_{ij} \pm i y_{ij}$.

The approximation symbol in (6.29) indicates that we have kept only those terms which will lead to quadratic magnon terms. In addition to neglecting three- and four-magnon terms, we have also neglected linear terms. The presence of such terms is an indication that our assumed ground-state spin configuration is not correct. To find the correct configuration we assume that the ground-state spin orientation at site i has some direction, characterized by the angles θ_i and φ_i with respect to the crystallographic axes. Setting the coefficients of the one-magnon terms to 0 determines these angles. In the dipolar case this is a small effect, and we shall neglect it. However, in certain canted spin systems care must be exercised in choosing the proper configuration.

Introducing the magnon operators, we obtain

$$\mathcal{H}_{\text{dip}} = \tfrac{1}{2} g^2 \mu_B^2 \sum_{k,k'} \sum_{i,j} \left[\frac{S}{N r_{ij}^3} \left(1 - \frac{3}{2} \frac{r_{ij}^+ r_{ij}^-}{r_{ij}^2}\right) a_k a_{k'}^\dagger e^{i \mathbf{k} \cdot \mathbf{r}_{ij}} e^{i(\mathbf{k} - \mathbf{k}') \cdot \mathbf{R}_j} \right.$$
$$+ \frac{S}{N r_{ij}^3} \left(1 - \frac{3}{2} \frac{r_{ij}^+ r_{ij}^-}{r_{ij}^2}\right) a_k^\dagger a_{k'} e^{-i \mathbf{k} \cdot \mathbf{r}_{ij}} e^{-i(\mathbf{k} - \mathbf{k}') \cdot \mathbf{R}_j} + \frac{S^2}{r_{ij}^3} \left(1 - 3 \frac{z_{ij}^2}{r_{ij}^2}\right)$$
$$- \frac{2S}{N r_{ij}^3} \left(1 - 3 \frac{z_{ij}^2}{r_{ij}^2}\right) a_k^\dagger a_{k'} e^{-i(\mathbf{k} - \mathbf{k}') \cdot \mathbf{R}_j}$$
$$- \frac{3S}{2N} \frac{(r_{ij}^-)^2}{r_{ij}^5} a_k a_{k'} e^{i \mathbf{k} \cdot \mathbf{r}_{ij}} e^{i(\mathbf{k} + \mathbf{k}') \cdot \mathbf{R}_j}$$
$$\left. - \frac{3S}{2N} \frac{(r_{ij}^+)^2}{r_{ij}^5} a_k^\dagger a_{k'}^\dagger e^{-i \mathbf{k} \cdot \mathbf{r}_{ij}} e^{-i(\mathbf{k} + \mathbf{k}') \cdot \mathbf{R}_j} \right]. \tag{6.30}$$

To evaluate the spatial sums we neglect boundary effects, so that the sums may be carried out independently; that is,

$$\sum_{R_j} \sum_{R_i} \rightarrow \sum_{R_j} \sum_{r = R_i - R_j}. \tag{6.31}$$

The dipole Hamiltonian then reduces to

$$\mathcal{H}_{\text{dip}} = \tfrac{1}{2}g^2\mu_B^2 \sum_k \left\{ N^2 S^2 \sum_r \frac{1}{r^3}\left(1 - \frac{3z^2}{r^2}\right) + S\sum_r e^{i\mathbf{k}\cdot\mathbf{r}}\frac{1}{r^3}\left(1 - \frac{3}{2}\frac{r^+r^-}{r^2}\right)\right.$$

$$- \left[2S \sum_r \frac{1}{r^3}\left(1 - \frac{3z^2}{r^2}\right) - 2S\sum_r e^{i\mathbf{k}\cdot\mathbf{r}}\frac{1}{r^3}\left(1 - \frac{3}{2}\frac{r^+r^-}{r^2}\right)\right] a_k^\dagger a_k$$

$$\left. - \left[\tfrac{3}{2}S \sum_r e^{i\mathbf{k}\cdot\mathbf{r}} \frac{(r^-)^2}{r^5}\right] a_k a_{-k} - \left[\tfrac{3}{2}S \sum_r e^{-i\mathbf{k}\cdot\mathbf{r}} \frac{(r^+)^2}{r^5}\right] a_k^\dagger a_{-k}^\dagger \right\}. \tag{6.32}$$

We now break up the sum over $\mathbf{r}$ into a *sum* within a sphere of radius a plus an *integral* over the rest of the sample,

$$\sum_r = \sum_r' + \frac{N}{V}\int d\mathbf{r}. \tag{6.33}$$

The reason for doing this is that if $a \ll \lambda$, the magnon wavelength, then the sum for a *cubic* crystal is 0.

Now consider the remaining integral associated with the first term in the Hamiltonian,

$$\frac{N}{V}\int \frac{1}{r^3}\left(1 - \frac{3z^2}{r^2}\right) d\mathbf{r} = \frac{N}{V}\int \boldsymbol{\nabla} \cdot \left(\frac{\mathbf{z}}{r^3}\right) d\mathbf{r}$$

$$= \frac{N}{V}\int_{\text{sphere}} \frac{\mathbf{z}\cdot d\mathbf{S}}{r^3} - \frac{N}{V}\int_{\substack{\text{sample}\\\text{surface}}} \frac{\mathbf{z}\cdot d\mathbf{S}}{r^3}$$

$$= \frac{N}{V}\left(\frac{4\pi}{3} - 4\pi N_z\right), \tag{6.34}$$

where the *demagnetization factor*

$$N_z \equiv \frac{1}{4\pi}\int_{\substack{\text{sample}\\\text{surface}}} \frac{\mathbf{z}\cdot d\mathbf{S}}{r^3}$$

gives rise to the demagnetization field and the term $4\pi/3$ gives rise to the *Lorentz field*.

Now consider the integral

$$\frac{N}{V}\int e^{i\mathbf{k}\cdot\mathbf{r}} \frac{1}{r^3}\left(1 - \frac{3z^2}{r^2}\right) d\mathbf{r} \tag{6.35}$$

which involves a plane-wave factor. Expanding the plane wave in spherical harmonics and recognizing that $1 - 3(z^2/r^2)$ is a spherical harmonic, we have

$$-\frac{N}{V}\int \sum_{l,m} 4\pi(i)^l j_l(kr)\, Y_l^{m*}(\Omega)\, Y_l^m(\Omega') \frac{2}{r^3}\sqrt{\frac{4\pi}{5}}\, Y_2^0(\Omega) r^2\, dr\, d\Omega$$

$$= -\frac{N}{V} 4\pi \left[\frac{j_1(kr)}{kr}\right]_a^{\text{surface}} (3\cos^2\theta_k - 1) \tag{6.36}$$

where θ_k is the angle that k makes with the z axis. If the sample dimension d is large, then $kd \gg 1$ and $j_1(kd)/kd \to 0$. If $ka \ll 1$, then $j_1(ka)/ka \to \tfrac{1}{3}$. Therefore (6.35) becomes

$$\frac{N}{V}\left(\frac{8\pi}{3} - 4\pi \sin^2 \theta_k\right). \tag{6.37}$$

The condition $kd \gg 1$ gives a zero result because the magnetization on the surface of the sample alternates rapidly, averaging to 0. This condition does not hold for small wave vectors. In particular, let us consider the $k = 0$ mode. We shall return later to a discussion of the nonzero, but small, wave vectors. For the case $k = 0$ it is convenient to rewrite those terms involving $r^\pm$ as x^2, y^2, and xy. The xy terms give 0. The x^2 and y^2 terms are similar to those we encountered involving z^2. Therefore we obtain transverse demagnetizing factors N_x and N_y. After we have carried out all the integrals, the dipole–dipole Hamiltonian becomes

$$\begin{aligned}\mathcal{H}_{\text{dip}} = \sum_k \Bigg[& \tfrac{1}{2}MV\left(4\pi N_z M - \frac{4\pi}{3}M\right) + \begin{Bmatrix}\tfrac{1}{2}\omega_M(N_x + N_y) - N_z\omega_M \\ \tfrac{1}{2}\omega_M \sin^2 \theta_k - N_z\omega_M\end{Bmatrix} a_k^\dagger a_k \\ & + \begin{Bmatrix}\tfrac{1}{4}\omega_M(N_x - N_y) \\ \tfrac{1}{4}\omega_M \sin^2 \theta_k \exp(-i\,2\varphi_k)\end{Bmatrix} a_k a_{-k} \\ & + \begin{Bmatrix}\tfrac{1}{4}\omega_M(N_x - N_y) \\ \tfrac{1}{4}\omega_M \sin^2 \theta_k \exp(i\,2\varphi_k)\end{Bmatrix} a_k^\dagger a_{-k}^\dagger \Bigg]\end{aligned} \tag{6.38}$$

where $M = Ng\mu_B S/V$ and $\omega_M = 4\pi\gamma M$. The upper terms in the braces apply to the $k = 0$ mode and the lower terms apply to the $k \neq 0$ modes. The total Hamiltonian may now be written as

$$\mathcal{H} = E_0 + \sum_k (A_k a_k^\dagger a_k + B_k a_k a_{-k} + B_k^* a_k^\dagger a_{-k}^\dagger) \tag{6.39}$$

where A_k and B_k are determined from (6.22, 26, 38).

This is not yet the desired form, since the dipole–dipole interaction has produced a coupling between the $+k$ and $-k$ spin waves. This means that the plane waves are not the correct normal modes. To find the correct modes we must diagonalize $\mathcal{H}$. Since the Hamiltonian (6.39) is the product of *operators*, its diagonalization is slightly different from that of a product of c numbers. In particular, suppose we write the Hamiltonian as

$$\mathcal{H} = x^\dagger H x \tag{6.40}$$

where x is a column vector whose components are the operators entering the Hamiltonian and H is a c-number matrix. For example, if we neglect the zero-point term, we may write the Hamiltonian (6.39) as

$$\mathscr{H} = \tfrac{1}{2} \sum_k x_k^\dagger H_k x_k \tag{6.41}$$

where

$$x_k = \begin{bmatrix} a_k \\ a_{-k}^\dagger \end{bmatrix} \quad \text{and} \quad H_k = \begin{bmatrix} A_k & 2B_k^* \\ 2B_k & A_{-k} \end{bmatrix}. \tag{6.42}$$

The operator nature of x may be specified by the commutator

$$[x, x^\dagger] \equiv x(x^{*T}) - (x^* x^T)^T \equiv g. \tag{6.43}$$

Note that when the elements of matrices are operators, the transpose of a product of such matrices is *not* equal to the product of the transposes in reverse order.

It is generally desired that the new modes that diagonalize the Hamiltonian have the same commutation relations as the original modes; that is, the transformation must be canonical. If the new modes are written as

$$x = Sx' \tag{6.44}$$

then the condition that the transformation S diagonalizes the Hamiltonian and also preserves the commutation relations is [6.2]

$$HS = g^{-1} S g \Omega_H \tag{6.45}$$

where Ω_H is the eigenvalue matrix. Solving this eigenvalue problem for our case leads to the transformation

$$\begin{bmatrix} a_k \\ a_{-k}^\dagger \end{bmatrix} = \begin{bmatrix} \sqrt{\dfrac{A_k + \omega_k}{2\omega_k}} & \sqrt{\dfrac{A_k - \omega_k}{2\omega_k}} \exp(i 2\varphi_k) \\ -\sqrt{\dfrac{A_k - \omega_k}{2\omega_k}} \exp(-i 2\varphi_k) & \sqrt{\dfrac{A_k - \omega_k}{2\omega_k}} \end{bmatrix} \begin{bmatrix} \alpha_k \\ \alpha_{-k}^\dagger \end{bmatrix}. \tag{6.46}$$

In terms of these new modes, the Hamiltonian is

$$\mathscr{H} = \sum_k \hbar \omega_k \alpha_k^\dagger \alpha_k \tag{6.47}$$

where

$$\omega_k = \sqrt{A_k^2 - 4|B_k|^2},$$
$$= \begin{cases} \sqrt{(\gamma H_i + N_x \omega_N)(\gamma H_i + N_y \omega_M)} & k = 0 \\ \sqrt{[\gamma H_i + 2zJS(1 - \gamma_k)][\gamma H_i + 2zJS(1 - \gamma_k) + \omega_M \sin^2 \theta_k]}, & k \neq 0 \end{cases} \tag{6.48}$$

and H_i is the internal field $H_0 - 4\pi N_z M$. These modes are plotted in Fig. 6.2 as a function of wave vector. Notice that as a result of the demagnetization fields

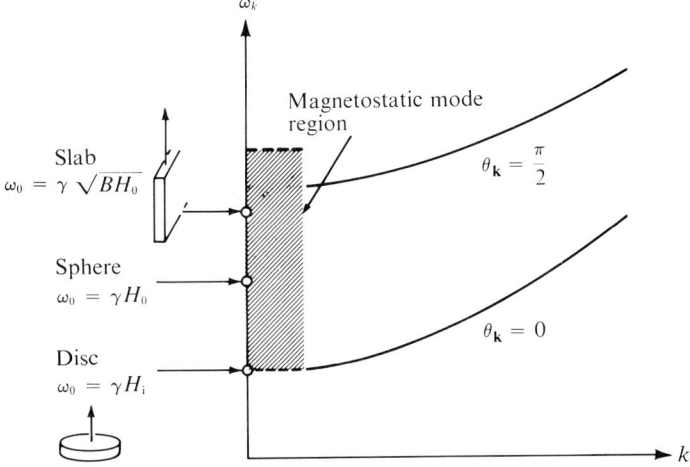

Fig. 6.2. Spin–wave frequencies as a function of their wave vector

the ferromagnetic resonance frequency depends on the shape of the sample. This was first pointed out by *Kittel* [6.3].

The individual spin precessions associated with these modes may be obtained by transforming back to the original components S_i^x and S_i^y. We find that the precession associated with the modes a_k is circular and that associated with the α_k is, in general, elliptical. Owing to the nature of the coefficient B_k, the uniform precession is elliptical when the transverse demagnetization factors are different. We also find that the ellipticity associated with a spin–wave mode depends on the direction of propagation with respect to the dc field. The most elliptical spin waves are those which propagate at right angles to the dc field.

6.2.2 Magnetostatic Modes

Let us now consider those modes which, although not uniform, have a spatial variation comparable to the sample dimensions. We use the fact that at such wavelengths the exchange interaction is negligible. Therefore the problem is a classical one. Furthermore, since $kc \gg \omega$, these modes are essentially static. Thus they are described by the magnetostatic Maxwell equations

$$\boldsymbol{\nabla} \times \boldsymbol{H} = 0, \tag{6.49}$$

$$\boldsymbol{\nabla} \cdot \boldsymbol{H} = -4\pi \boldsymbol{\nabla} \cdot \boldsymbol{M}, \tag{6.50}$$

plus the constitutive torque equation

$$\frac{d\boldsymbol{M}}{dt} = -\gamma \boldsymbol{M} \times \boldsymbol{H}. \tag{6.51}$$

The solution to (6.51) under the assumption of small transverse fields is

$$\boldsymbol{B} = \begin{bmatrix} \mu & -i\kappa & 0 \\ i\kappa & \mu & 0 \\ 0 & 0 & 1 \end{bmatrix} \boldsymbol{H} \tag{6.52}$$

where

$$\mu = 1 + \frac{4\pi\gamma^2 M_0 H_i}{\gamma^2 H_i^2 - \omega^2} \tag{6.53}$$

and

$$\kappa = \frac{4\pi\gamma M_0 \omega}{\gamma^2 H_i^2 - \omega^2} \tag{6.54}$$

If we introduce a magnetic scalar potential ϕ, defined by $\boldsymbol{H} = \boldsymbol{\nabla}\phi$, Maxwell's equations become

$$\mu\left(\frac{\partial^2 \phi}{\partial x^2} + \frac{\partial^2 \phi}{\partial y^2}\right) + \frac{\partial^2 \phi}{\partial z^2} = 0 \quad \text{(inside)},$$
$$\boldsymbol{\nabla}^2 \phi = 0 \quad \text{(outside)}. \tag{6.55}$$

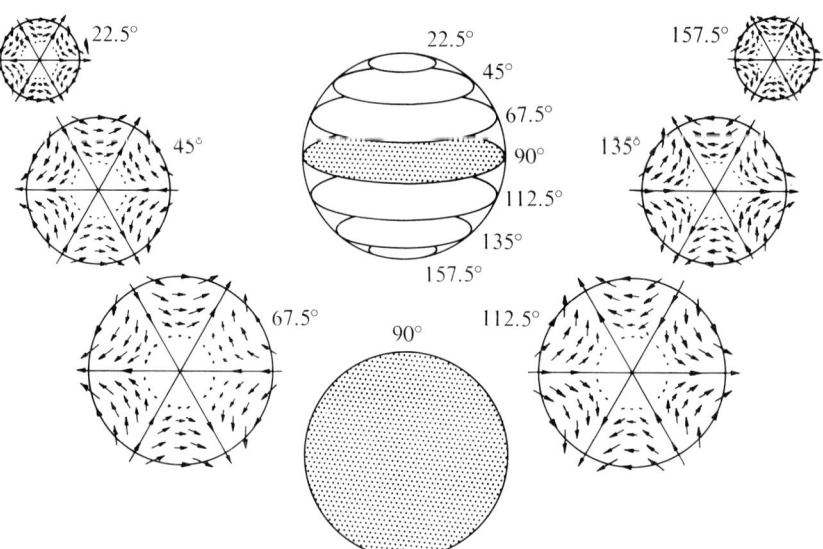

Fig. 6.3. The (4, 3, 0) Walker mode of a sphere showing the instantaneous positions of the radio-frequency magnetization vectors in the planes transverse to the dc magnetic field. This illustrates the indexing scheme and the concept of nonuniform precession [6.4]

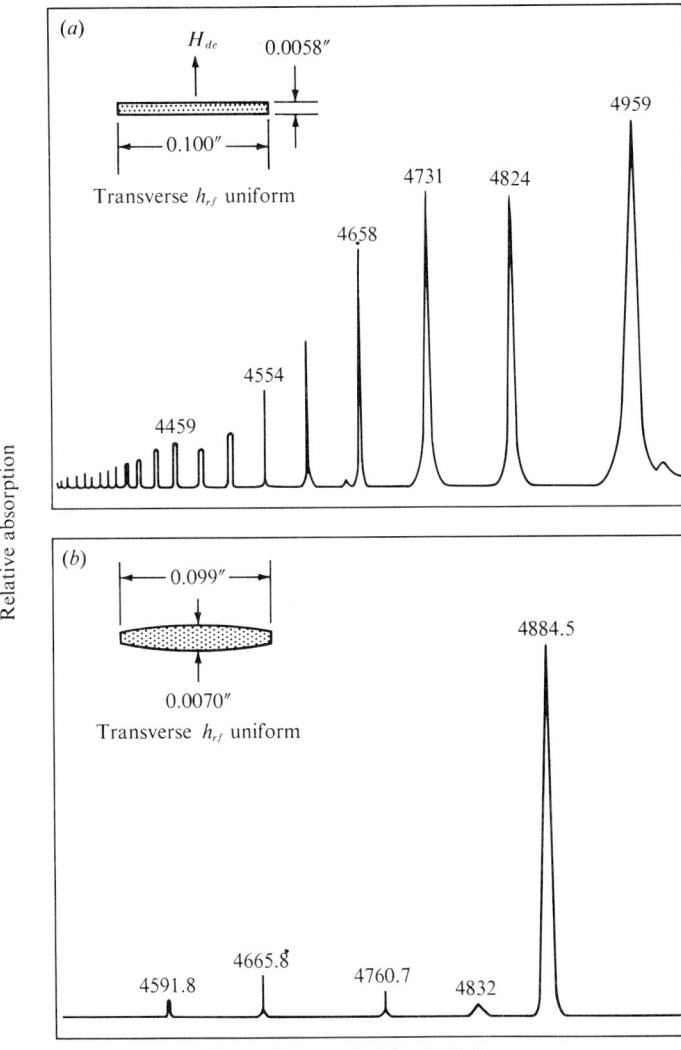

Fig. 6.4a, b. Absorption spectrum of a yttrium-iron garnet disk. The sharp edges in (a) produce nonuniform fields which excite many modes. Rounding these edges (b) suppresses the modes [6.6]

Solving these equations for an ellipsoidal sample gives the so-called *magnetostatic*, or *Walker*, *modes*. These modes are described by a set of indices (n, m, r) instead of (k_x, k_y, k_z). For example, the (4, 3, 0) mode is shown in Fig. 6.3. Since these modes have an inhomogeneous magnetization, they may be excited by an inhomogeneous microwave field. In fact, if the sample is very large or if it is mounted on a dielectric post, the resulting inhomogeneity in the field may be enough to excite magnetostatic modes. These modes were discovered in this way [6.5]. A typical absorption spectrum is shown in Fig. 6.4.

Notice that the magnetostatic mode spectrum in Fig. 6.2 extends above the spin–wave spectrum when it is extrapolated to long wavelengths. The nature of these modes became clear from the work of *Damon* and *Eshbach* [6.7] who calculated the magnetostatic modes of a thin slab magnetized in its plane. The modes which lie outside the spin–wave manifold correspond to *surface* modes, i.e., modes that have their maximum amplitude at the surface and decay exponentially into the slab as illustrated in Fig. 6.5. These modes have the interesting feature that with reversal of the direction of propagation the maximum amplitude shifts to the opposite surface. That is, on one side one only finds surface modes propagating in one direction relative to the direction of the applied field.

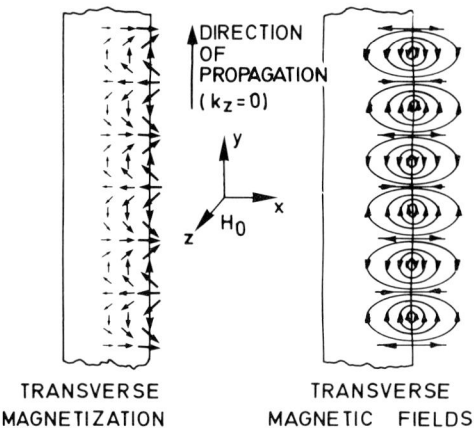

Fig. 6.5. Mode amplitudes for a short-wavelength surface mode with $k_z = 0$

6.2.3 Solitons

The sine-Gordon wave equation, $\varphi_{zz} - \varphi_{tt}/c^2 = m^2 \sin \varphi$, describes a variety of physical phenomena. There are three distinct types of solutions: unbounded (in space) oscillatory solutions, bounded oscillatory solutions of finite amplitude called "breathers" and fixed amplitude "kink" solutions, corresponding to a 2π change in φ, that propagate undistorted and are called solitons. The large amplitude breathers may be thought of as bound states of a soliton and antisoliton. Under certain conditions, the equations of motion for magnetic systems have the sine-Gordon form. The unbounded oscillatory solutions are then spinwaves and the solitons correspond to magnetic domain walls (see Fig. 6.6) In three dimensions, the soliton, or domain wall, energies are much greater than $k_B T$. However, in one dimension, soliton energies are comparable to $k_B T$ and solitons behave as elementary excitations. The presence of solitons in the linear chain system CsNiF₃ has been inferred from quasi-elastic neutron scattering [6.8] and specific heat measurements [6.9].

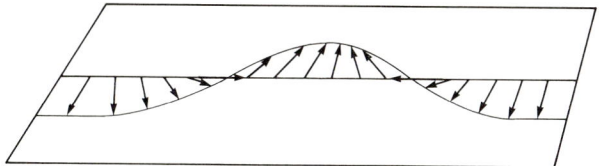

Fig. 6.6. Schematic illustration of a soliton in an easy-plane ferromagnet

6.2.4 Thermal Magnon Effects

Fortunately the density of spin–wave modes is much greater than that of the magnetostatic modes, so that most of the properties of magnetic materials may be understood by considering only the simpler spin–wave modes. The importance of spin waves in determining the low-temperature properties of ferromagnets was first recognized by *Bloch* [6.10]. Let us first consider the saturation magnetization. In the spin–wave region this is given by

$$M(T) = -\frac{g\mu_B}{V}\sum_i \langle S_i^z \rangle = \frac{g\mu_B}{V}\sum_i (S - \langle n_i \rangle) = M(0) - \frac{g\mu_B}{V}\sum_k \langle n_k \rangle \ . \quad (6.56)$$

In principle, there is no limit to the number of magnons that may be in any given mode. Their thermal occupation number is given by the *Bose–Einstein distribution* as

$$\langle n_k \rangle = 1/[\exp(\hbar\omega_k/k_B T) - 1] \ . \quad (6.57)$$

Since most of the modes lie in the exchange region of the spectrum, we shall neglect dipolar effects. Thus a_k is the correct mode, and we do not have to transform to α_k. If we also use the small k expansion of (6.26) and neglect the external field, then

$$\hbar\omega_k = Dk^2 \quad (6.58)$$

where, for example, D is $2JSa^2$ for a simple cubic lattice. The sum over $\mathbf{k}$ is now replaced by an integral, and we have

$$M(0) - M(T) = -\frac{g\mu_B}{V}\frac{4\pi V}{(2\pi)^3}\int_0^\infty \frac{k^2\,dk}{\exp(Dk^2/k_B T) - 1} \ . \quad (6.59)$$

The integral may be written in the dimensionless form

$$\int_0^\infty \frac{k^2\,dk}{\exp(Dk^2/k_B T) - 1} = \frac{1}{2}\left(\frac{k_B T}{D}\right)^{3/2}\int_0^\infty \frac{x^{1/2}\,dx}{e^x - 1} \ . \quad (6.60)$$

The evaluation of this integral is simplified by using the expansion

$$\frac{1}{e^x - 1} = \sum_{r=1}^{\infty} e^{-rx}. \tag{6.61}$$

The resulting magnetization is

$$M(0) - M(T) = \zeta\left(\frac{3}{2}\right) \frac{g\mu_B}{M(0)} \left(\frac{k_B}{4\pi D}\right)^{3/2} T^{3/2} \tag{6.62}$$

where ζ is the Riemann zeta function.

This is referred to as the *Bloch $T^{3/2}$ law*. The fact that it accurately describes the magnetization at low temperatures, as shown in Fig. 6.7, is evidence of the importance of such collective modes. If we keep those terms of order k^4, k^6, etc., in the dispersion relation, they lead to contributions to the magnetization that vary as $T^{5/2}$, $T^{7/2}$, etc. If only single-particle excitations were important, the low-temperature magnetization would vary exponentially, with an activation energy of the order of the exchange energy.

The spin–wave contribution to the specific heat is given by

$$C_V = \frac{\partial}{\partial T} \sum_k \hbar\omega_k \langle n_k \rangle . \tag{6.63}$$

With the same approximations employed in obtaining the magnetization, (6.63) gives

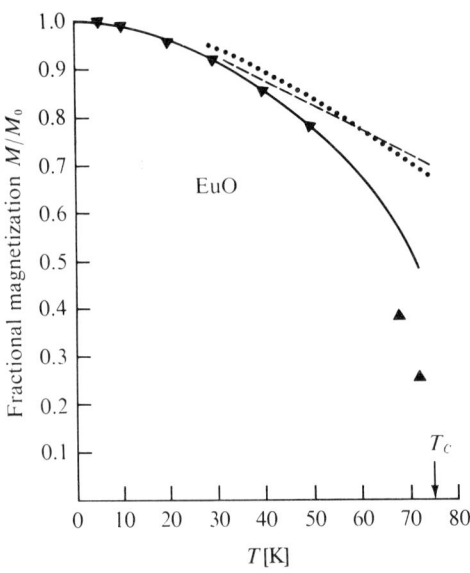

Fig. 6.7. The temperature dependence of the magnetization of EuO. The solid curve, calculated from spin–wave theory, includes the effects of dynamical interactions between pairs of magnons and subsequent renormalization of the spin-wave energies used in evaluating the magnon populations. The broken curve corresponds to noninteracting spin waves and the dotted curve corresponds to the usual series expansion in terms dependent on $T^{3/2}$ and $T^{5/2}$. Experimental points: (▲) observations corrected to infinite applied field; (▼) observations corrected to zero applied field [6.11]

$$C_V = \frac{15}{4}\zeta\left(\frac{5}{2}\right)k_B\left(\frac{k_B}{4\pi D}\right)^{3/2}T^{3/2}. \tag{6.64}$$

So far our treatment has been entirely linear. That is, we have restricted ourselves to those terms in the Hamiltonian that involve quadratic products of magnon operators and may therefore be transformed into a set of non-interacting harmonic oscillators. The presence of higher-order terms corresponds to coupling between these oscillators. This coupling has important consequences for certain properties of ordered systems.

The lowest-order nonlinear terms arising from the isotropic exchange interaction are the four-magnon terms

$$\mathcal{H}_{ex}^{(4)} = \tfrac{1}{2}\sum_{k,k',k''}[J(k) + J(k+k'-k'') - 2J(k-k'')]a_k^\dagger a_{k'}^\dagger a_{k''} a_{k+k'+k''}. \tag{6.65}$$

The effect of such spin–wave interactions on the magnetization was first analyzed by *Dyson* [6.12], who found that, in addition to the terms discussed above, there is a contribution to $M(0) - M(T)$ which is proportional to T^4. Although it is not yet possible experimentally to distinguish this contribution from $T^{3/2}$ and $T^{5/2}$, it has important theoretical consequences in that any alternative approach to computing $M(T)$ must reproduce Dyson's result at low temperatures.

A quantity that is more directly affected by the nonlinear terms of (6.65) is the magnon frequency itself. Let us consider only the exchange interaction. In the linear approximation the magnon frequency as given by (6.26) is independent of temperature. Applying the random-phase approximation and considering only nearest-neighbor interactions, we may write (6.65) as

$$-(2zJS^2 N)^{-1}\sum_k \omega_k \langle n_k \rangle \sum_{k'} \omega_{k'}\langle n_{k'}\rangle . \tag{6.66}$$

If we add this expression to the linear part $\sum \omega_k \langle n_k \rangle$ and minimize the free energy with respect to $\langle n_k \rangle$, we obtain the same form as (6.57), but with the renormalized frequency

$$\omega_k(T) = \omega_k[1 - (zJS^2 N)^{-1}\sum_{k'}\omega_{k'}\langle n_{k'}\rangle]. \tag{6.67}$$

We recognize the sum in this expression as the total energy in the spin–wave system. As the temperature increases this energy also increases, causing a decrease in the spin–wave frequencies. This decrease has been observed by neutron scattering, which we shall discuss in Chap. 8.

Nonlinearity also plays an important role in spin–wave relaxation phenomena. Consider, for example, the lowest-order nonlinear terms arising from the dipolar interaction. These are the three-magnon terms

$$\mathcal{H}_{dip}^{(3)} = \sum_{\substack{k,k' \\ k\neq 0}}(g_k \alpha_k \alpha_{k'}\alpha_{k+k'}^\dagger + g_k^* \alpha_k^\dagger \alpha_{k'}^\dagger \alpha_{k+k'}) \quad \text{where} \tag{6.68}$$

$$g_k = -\frac{4\pi}{V}\sqrt{2NS}\,g^2\,\mu_B^2\cos\theta_k\sin\theta_k\exp(i\phi_k)\,. \tag{6.69}$$

Since $a_k^\dagger$ corresponds to the creation of a magnon with wave vector $\boldsymbol{k}$ and α_k corresponds to the annihilation of a magnon with wave vector $\boldsymbol{k}$, the processes involved in (6.68) may be represented schematically as shown in Fig. 6.8. Figure 6.8a shows the confluence of a magnon $\boldsymbol{k}$ with a magnon $\boldsymbol{k}'$ to produce a third magnon $\boldsymbol{k}+\boldsymbol{k}'$. This type of process constitutes a relaxation channel for the magnon $\boldsymbol{k}$ (or $\boldsymbol{k}'$).

Since we are dealing with a ferromagnet, the exchange-narrowing effect, which was discussed in Chap. 5, will be quite pronounced. The origin of this narrowing has to do with the fact that when the frequency spectrum of the effective field is very broad, only a narrow band of nearly degenerate modes influences the relaxation. Under this condition the relaxation frequency may be calculated by time-dependent perturbation theory. This is a particularly convenient way of computing magnon relaxation frequencies. As an illustration of this approach consider the three-magnon process of Fig. 6.8. The assumption that the magnon occupation number, if disturbed, relaxes exponentially to its equilibrium value $\langle n_k \rangle$ enables us to define a relaxation frequency η_k by the rate equation

$$\frac{dn_k}{dt} = -\eta_k(n_k - \langle n_k \rangle)\,. \tag{6.70}$$

The rate of change of the number of magnons in mode $\boldsymbol{k}$ is also given by

$$\frac{dn_k}{dt} = W_{n_k \to n_k+1} - W_{n_k \to n_k-1} \tag{6.71}$$

where W is the transition probability per unit time. For the three-magnon interaction (6.68), this is given by

$$W_{n_k \to n_k+1} = \frac{2\pi}{\hbar^2}\sum_{k'}|g_k^* + g_{k'}^*|^2(n_k+1)(n_{k'}+1)n_{k+k'}\delta(\omega_k+\omega_{k'}-\omega_{k+k'}) \tag{6.72}$$

(the details of this calculation may be found in [Ref. 6.13, Sect. 5.3]).

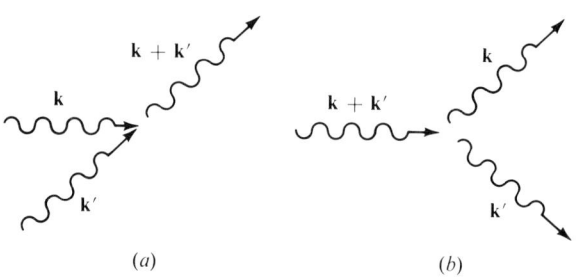

Fig. 6.8a, b. Representation of (a) three-magnon confluence process and (b) three-magnon splitting process

The result is that this three-magnon confluence process gives a relaxation frequency which, for small wave vectors, is proportional to the magnitude of the wave vector and is linear in temperature. These features, as well as the order of magnitude, are found to agree well with experimental measurements of η_k in the ferromagnetic insulator yttrium-iron garnet. This process does not relax the $k = 0$, or ferromagnetic resonance, mode. It appears that surface and volume inhomogeneities provide the dominant relaxation channel for this mode [6.13].

6.2.5 Parametric Excitation

Since the $k \neq 0$ magnons have such short wavelengths, they cannot be directly excited by an electromagnetic field. However, there are a number of mechanisms by which they may be indirectly excited. Some of these mechanisms involve the nonlinearities discussed above. For example, consider the three-magnon terms in (6.68), in which $k' = -k$, that is,

$$\sum_k (g_k \alpha_k \alpha_{-k} \alpha_0^\dagger + g_k^* \alpha_k^\dagger \alpha_{-k}^\dagger \alpha_0) .$$

These terms describe the relaxation of the $k = 0$, or uniform-precession, mode by decay into two magnons with equal and opposite wave vectors. Since energy must be conserved in this process, the resultant magnons will each have a frequency of $\omega_0/2$, as illustrated in Fig. 6.9.

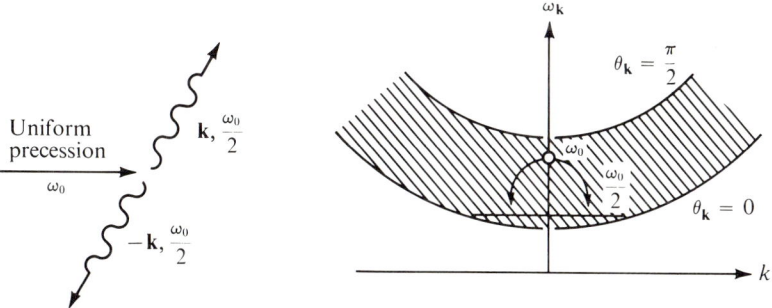

Fig. 6.9. Representation of the first-order Suhl instability in which uniform precession magnons split into $(k, -k)$ pairs

The essential feature of this relaxation process is that it is nonlinear. That is, the rate at which the uniform-precession mode decays in this fashion depends on the occupation number $\langle n_k \rangle$ of the final-state magnons, as seen for example in (6.72). This occupation number, however, depends on the magnon relaxation rate η_k, as well as on the amplitude of the uniform precession $\langle n_0 \rangle$ which is driving it. There is a critical value of $\langle n_0 \rangle$ at which the rate of increase in the number of magnons in mode k exceeds their rate of decrease through relaxation. At this point the number of magnons $\langle n_k \rangle$ increases abruptly, producing a corresponding increase in the relaxation rate of the uniform precession. Since the amplitude of the uniform precession $\langle n_0 \rangle$ is proportional to the amplitude of the driving microwave field H_1, this process appears as a saturation phenomenon. It turns out, however, that the field at which this happens is lower than that associated with the longitudinal relaxation time T_1 (see page 161). The observation of this "premature" saturation in ferrites by *Bloembergen* and *Damon* in 1952 opened an era of exploration of nonlinear phenomena in microwave ferrites [6.14]. The explanation of this saturation was given by *Suhl* and is referred to as the *first-order Suhl instability* [6.15]. There is also a second-order Suhl instability which arises from terms in (6.65) of the form $a_k^\dagger a_{-k}^\dagger a_0 a_0$, in which *two* uniform precession magnons decay into a $(k, -k)$ magnon pair.

Since all these processes involve the magnon relaxation rate η_k, they serve to measure this quantity. However, the onset of these processes is sometimes difficult to determine experimentally. A more accurate determination of η_k is obtained by means of the so-called *parallel-pumping instability*. In parallel pumping, as the name implies, a microwave magnetic field $H_1 \cos \omega t$ is applied parallel to the dc magnetic field. From (6.18) we find that the interaction Hamiltonian is

$$\mathcal{H}_1 = -g\mu_\mathrm{B} H_1 \cos \omega t \sum_k a_k^\dagger a_k . \tag{6.73}$$

We saw that because of the dipolar interaction the a_k are not the normal modes. When (6.46) is used to express the interaction in terms of the normal modes α_k we obtain terms of the form

$$g\mu_\mathrm{B} H_1 \cos \omega t \sum_k \frac{|B_k|}{\omega_k} \exp(\mathrm{i}2\varphi_k) \alpha_k^\dagger \alpha_{-k}^\dagger . \tag{6.74}$$

These terms correspond to the creation of $(k, -k)$ magnon pairs. Again, since this is a nonlinear process, the rate at which energy flows into these magnons depends on their occupation number $\langle n_k \rangle$. This, in turn, depends upon the amplitude of the driving field, as well as on the relaxation rate η_k. When the driving rate exceeds the relaxation rate, the magnon occupation number becomes very large. This results in an increased absorption. The threshold for this process is given by

$$(H_1)_\mathrm{crit} = \min \left(\frac{2\omega \eta_k}{\gamma \omega_M \sin^2 \theta_k} \right) \tag{6.75}$$

where min means that we use that value of **k** giving the minimum.

Notice that those magnons which propagate perpendicular to the dc field, and hence have the greatest ellipticity, are the first to become unstable. Since there is essentially no absorption below this threshold, its onset provides us with an accurate means of determining the spin–wave relaxation frequency η_k.

6.2.6 Optical Processes

Infrared Absorption [6.16]. There are other mechanisms for exciting magnons which do not involve nonlinearities. For example, consider the antiferromagnet MnF_2. The ground state of the Mn^{2+} ion is an orbital singlet. However, since $S = \frac{5}{2}$, there are six spin states, which in the magnetically ordered crystal are split by the exchange interaction. Those ions which occupy, say, the A sublattice have $M_S = -\frac{5}{2}$ as their ground state, while the ground state for those ions on the B sublattice is $M_S = +\frac{5}{2}$. A $(\boldsymbol{k}, -\boldsymbol{k})$ magnon pair may be excited by the process illustrated in Fig. 6.10. First, the applied radiation induces an *electric dipole transition* to a high-lying state of one of the Mn^{2+} ions. Since this is a virtual process, it need not conserve energy. However, it must conserve parity, which means that the high-lying state must be opposite in parity from the ground state. Then, as a result of the Coulomb interaction between this ion and a neighboring ion on the opposite sublattice, the excited ion drops down to its first excited *spin* state ($M_S = -\frac{3}{2}$), while the second ion is simultaneously excited up to its first excited spin state ($M_S = +\frac{3}{2}$). Thus we end up with two spin deviations. Since any pair of ions may be excited, the final state will be a linear combination of such pair excitations. This results in the excitation of two magnons. The absorption coefficient for this process is proportional to the density of final states. Therefore, since the magnon density of states is largest at the Brillouin zone boundary, the absorption is strongest at a frequency approximately twice that of a Brillouin zone magnon, corresponding to an infrared frequency. However, the detailed shape of the absorption is complicated by the interaction between the two output magnons.

Spin–wave sidebands [6.17]. In addition to this infrared two-magnon absorption, there is an optical *exciton-magnon absorption*. In this process the ion on sublattice

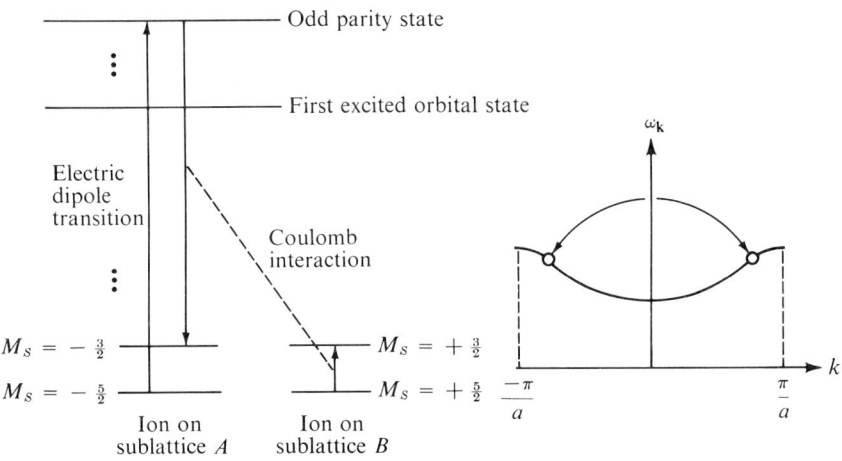

Fig. 6.10. Representation of the mechanism for the excitation to two magnons

A in Fig. 6.10 ends up in its first excited *orbital* state rather than its first excited spin state. This orbital excitation then propagates throughout the lattice, forming what is known as a *Frenkel exciton*. The absorption frequency for this process is approximately that of the exciton plus a magnon. In MnF_2 the first excited orbital state lies in the green part of the optical spectrum. The absorption of a single Mn^{2+} ion into this state appears as a very weak magnetic-dipole transition. The exciton-magnon absorption appears as a somewhat stronger electric-dipole sideband on the high-frequency side of this "no-magnon" line.

Light Scattering [6.18]. Magnons may also be excited by scattered light. One mechanism for such scattering is illustrated in Fig. 6.11. The incident photon is virtually absorbed in an electric dipole process. The spin–orbit interaction then produces a spin flip in the excited state and a second photon is virtually emitted leaving the system with a spin excitation. Macroscopically this gives rise to a dielectric function which depends upon the magnetization or spin density,

$$\epsilon^{\alpha\beta} = \epsilon_0 \delta_{\alpha\beta} + \delta\epsilon^{\alpha\beta}(\mathbf{S}(\mathbf{r}, t)). \tag{6.76}$$

The spin-dependent part may be expanded in powers of the spin density

$$\delta\epsilon^{\alpha\beta} = \sum_{\gamma} K_{\alpha\beta\gamma} S_{\gamma}(\mathbf{r}, t) + \ldots . \tag{6.77}$$

For a cubic material $K_{\alpha\beta\gamma} = \pm K$ if all the indices are different, and zero otherwise. For the mechanism described in Fig. 6.11 K would be proportional to

$$\frac{\lambda \langle 0^* | \beta | 2 \rangle \langle 2 | L_\gamma | 1 \rangle \langle 1 | \alpha | 0 \rangle}{(E_1 - E_0)(E_2 - E_0)} . \tag{6.78}$$

The static contributions to $\delta\epsilon$ lead to magnetooptic effects while the fluctuating contributions are responsible for light scattering. Thus, in a cubic material, at least, magnetooptical effects and light scattering are governed by the same parameter K.

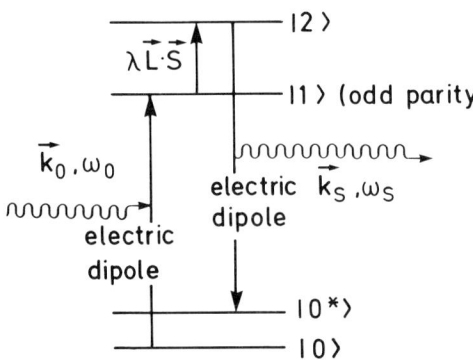

Fig. 6.11. Schematic representation of a perturbation process contributing to magnon scattering. The electric dipole transitions are "virtual" in the sense that they do not conserve energy, i.e., $E_1 - E_0 \neq \hbar\omega_0$

To see how the intensity of the scattered light depends upon the magnetic fluctuations consider an incident wave

$$\varepsilon_{\text{inc}}^{\alpha}(\mathbf{r}, t) = \varepsilon_0^{\alpha} \exp\left[i(\mathbf{k}_0 \cdot \mathbf{r} - \omega_0 t)\right]. \quad (6.79)$$

Inside the medium the total electric field will consist of this incident wave plus the scattered wave,

$$\varepsilon^{\alpha}(\mathbf{r}, t) = \varepsilon_{\text{inc}}^{\alpha}(\mathbf{r}, t) + \varepsilon_{\text{scat}}^{\alpha}(\mathbf{r}, t) \quad (6.80)$$

which is related to the displacement vector by

$$D^{\alpha}(\mathbf{r}, t) = \epsilon^{\alpha\beta} \varepsilon^{\beta}(\mathbf{r}, t). \quad (6.81)$$

Assuming $\delta\epsilon^{\alpha\beta}$ and $\varepsilon_{\text{scat}}^{\alpha}$ are small, a linearization of Maxwell's equations leads to a driven wave equation for $\varepsilon_{\text{scat}}^{\alpha}$ where the driving, or source, term involves the product $\delta\epsilon^{\alpha\beta}(\mathbf{r}, t)\, \varepsilon_{\text{inc}}^{\beta}(\mathbf{r}, t)$. The field at the detector is obtained by multiplying this source term by the so-called Green's function for the wave equation, a function that describes how a disturbance propagates, and integrating over the sample volume and over all time,

$$\varepsilon_{\text{scat}}^{\alpha}(\mathbf{r}, t) \sim \left(\frac{\omega_0}{c}\right)^2 \int d\mathbf{r}' \int dt\, G_{\alpha\beta}(\mathbf{r}, \mathbf{r}'; t, t')\, \delta\epsilon^{\beta\gamma}(\mathbf{r}', t')\, \varepsilon_{\text{inc}}^{\gamma}(\mathbf{r}', t'). \quad (6.82)$$

The scattered intensity is found by squaring this field and then performing an ensemble average $\langle \ldots \rangle$ over the fluctuations in $\delta\epsilon$. This results in the correlation function

$$\langle \delta\epsilon^{\beta'\gamma'}(\mathbf{r}, t)\, \delta\epsilon^{\beta\gamma}(\mathbf{r}', t') \rangle. \quad (6.83)$$

For a bulk geometry this is a function only of $\mathbf{r} - \mathbf{r}'$ and $t - t'$. Introducing the time and spatial dependences in G and ε_{inc} one finds, for example, that the intensity of light scattered with a polarization z, when the incident polarization is x, is

$$\int d\mathbf{r} \int dt\, e^{i(\mathbf{q}\cdot\mathbf{r} - \Omega t)} \langle S_y(\mathbf{r}, t)\, S_y(0, 0) \rangle \quad (6.84)$$

where $\mathbf{q}$ is the momentum transfer, $\mathbf{q} = \mathbf{k}_s - \mathbf{k}_0$, and Ω is the excitation energy, $\Omega = \omega_s - \omega_0$. The maximum momentum transfer occurs for back scattering,

$$q_{\max} = 2|\mathbf{k}_0| = \frac{2n\omega_0}{c} \quad (6.85)$$

which, for typical optical frequencies is of the order of 3×10^5 cm^{-1}. This is to be compared with the Brillouin zone wave vector, $q_{\text{BZ}} \sim 3 \times 10^8$ cm^{-1}. Thus light scattering probes only the very long-wavelength excitations. If the frequencies exceed 1 cm^{-1} we speak of Raman scattering, while scattering at smaller frequencies is referred to as Brillouin scattering, the distinction being largely based on experimental technique.

If we expand the correlation function $\langle S_y(\mathbf{r}, t)S_y(0,0)\rangle$ in terms of the spin–wave amplitudes introduced above, we obtain two terms: $\langle a_q a_q^\dagger\rangle = n_q + 1$ and $\langle a_q^\dagger a_q\rangle = n_q$. The first corresponds to the creation of a spin wave. The scattered light appears at a frequency $\omega_0 - \Omega_q$ with an intensity proportional to $n_q + 1$ and is referred to as the Stokes line. The second term gives an anti-Stokes line at the frequency $\omega_0 + \Omega_q$.

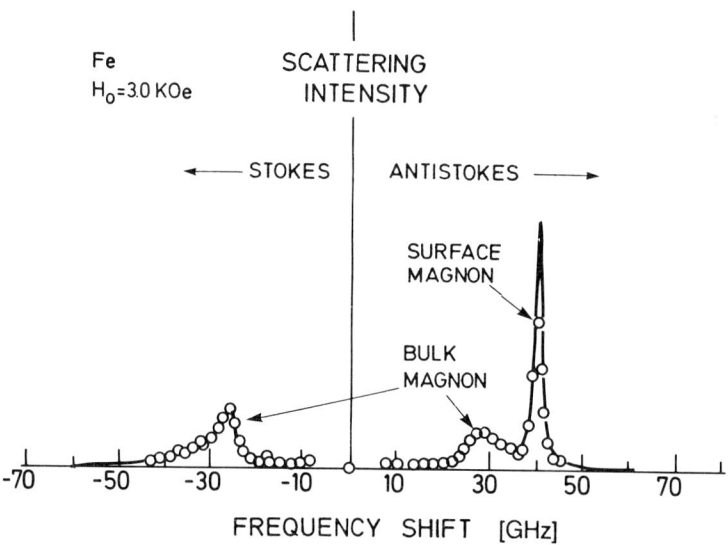

Fig. 6.12. Brillouin spectrum of light scattered from magnons in iron

In Fig. 6.12 we show the result for light scattering from iron. Since the skin depth is relatively small, the intensity of the surface mode is relatively strong. Notice that it only appears on one side (the anti-Stokes side in this case) of the exciting frequency. This is a direct consequence of the nonreciprocal nature of surface magnons which we indicated above.

6.2.7 High Temperatures

We have seen how effectively the spin–wave concept describes the low-temperature properties of magnetically ordered systems. However, near the transition temperature, and particularly above it, the lifetime of the spin waves becomes too short for them to be a useful concept. At very high temperatures the relaxation-function approach developed in Chap. 5 is applicable. There we found that a strong exchange modulation of the magnetic dipolar field led to a long-time spin–spin correlation function which has the form

$$\langle \mathcal{M}_x(t)\mathcal{M}_x\rangle_{T\to\infty} = \tfrac{1}{3}NS(S+1)g^2\mu_B^2 \exp(i\omega_0 t)\exp[-(2M_2/\omega_{ex})t] \qquad (6.86)$$

where M_2 is the Van Vleck second moment and ω_{ex} characterizes the rate at which the exchange interaction modulates the dipolar field. This result was derived under the assumption that the exchange energy was small in comparison with the Zeeman energy. When the exchange energy is very large, as in the case we are now considering, the off-diagonal elements of the dipole–dipole interaction become important, with the result that the exponent $2M_2/\omega_{ex}$ in (6.86) is larger by a factor of about $\tfrac{10}{3}$ (the factor is exactly $\tfrac{10}{3}$ for a polycrystalline sample with cubic symmetry).

The theory of relaxation phenomena at finite temperatures is a subject of active research. Consequently, we shall restrict ourselves to a few qualitative remarks that give some indication of the problems involved (most of this discussion is based on [6.19]). Let us begin by considering the temperature dependence of the second moment. In the finite-temperature region this is

$$M_2 = -\frac{\langle[\mathcal{H}'_{dip}, \mathcal{M}_x]^2\rangle}{\hbar^2 \langle \mathcal{M}_x^2\rangle} \qquad (6.87)$$

where $\mathcal{H}'_{dip}$ is given by (5.50). After computing this commutator and squaring the result, we are left with a numerator involving terms such as $\langle S_i^x S_j^y S_m^z S_n^y\rangle$. Making the random-phase approximation and using the high-temperature form of the fluctuation-dissipation theorem,

$$\langle \mathcal{M}_x^2\rangle = 2\pi k_B T \chi(q=0, \omega=0),$$

we obtain

$$M_2 = \frac{9g^2\mu_B^2 \sum_q |F(q)|^2 \langle S_z(q) S_z(-q)\rangle^2}{2\pi\hbar^2 k_B T \chi} \qquad (6.88)$$

where

$$F(q) = Ng^2\mu_B^2 \sum_r \frac{3\cos^2\theta - 1}{r^3} e^{iq\cdot r}. \qquad (6.89)$$

From (4.45) we have

$$\langle S_z(q)S_z(-q)\rangle = \frac{k_B T V}{g^2\mu_B^2}\chi(q, \omega=0). \qquad (6.90)$$

Converting the sum over q in (6.88) into an integral, we find that for a ferromagnet the second moment varies as $(T - T_C)^{1/2}$. If the linewidth were described by the second moment alone, this would indicate that the paramagnetic resonance line should narrow as the temperature approaches the Curie point. For an antiferromagnet, however, the uniform static susceptibility does not diverge, with the result that the second moment increases as $(T - T_N)^{-1/2}$ where T_N is the Néel temperature.

This is not the complete story, for the exchange frequency is also temperature dependent. The exchange frequency was defined in (5.64) as

$$\langle \Delta\omega(\tau)\Delta\omega \rangle = M_2(1 - \omega_{ex}^2 \tau^2) . \tag{6.91}$$

With the same approximations used in obtaining (6.88), this may also be written as

$$\langle \Delta\omega(\tau)\Delta\omega \rangle = \sum_q |F(q)|^2 \langle S_z(q, \tau)S_z(-q) \rangle^2 . \tag{6.92}$$

Therefore the exchange frequency is given by

$$\omega_{ex}^2 = -\frac{1}{M_2} \sum_q |F(q)|^2 \langle S_z(q) S_z(-q) \rangle \frac{d^2}{d\tau^2} \langle S_z(q, \tau)S_z(-q) \rangle . \tag{6.93}$$

We see that this expression involves the second derivative of the correlation function. Without evaluating this quantity it is possible to say something very important about (6.93), namely, that its numerator does not diverge or go to 0 as the temperature approaches the critical temperature.

To see this let us rewrite the derivative of the correlation function in terms of individual spins as

$$\langle (d^2/d\tau^2)S_i^z(\tau)S_j^z \rangle e^{i q \cdot R_{ij}} .$$

We now apply a thermodynamic result derived by *Kubo* [6.20] which says that this derivative is proportional to $\langle [[\mathcal{H}_0, S_i^z], S_j^z] \rangle$. The consequence of these commutators is that the derivative of the correlation function is nonzero only for spins very close to one another. This, is turn, implies that the correlation function in the numerator of (6.93) involving the second derivative shows no drastic temperature dependence as the critical temperature is approached, since only short-range static correlations are involved, and these do not change suddenly near the critical point. These considerations lead us to the result that the aquare of the exchange frequency also goes to 0 as $(T - T_C)^{1/2}$. Therefore, if we assume that the method of moments provides an adequate description of a ferromagnet in its paramagnetic region, we are led to the conclusion that the resonance linewidth remains exchange narrowed as the temperature decreases toward the Curie point, decreasing as $(T - T_C)^{1/4}$. Evidence of such behavior is indicated for the ferromagnet CrBr$_3$ in Fig. 6.13.

The above arguments assume a zero applied field. In a finite field the divergences are masked, and the simple power-law dependencies are not expected to hold in the immediate vicinity of T_C. Since resonance linewidths are measured in finite fields of appreciable magnitude, we may expect considerable departure from these simple relations. In particular, the ferromagnetic-resonance linewidth does not go to 0 at the critical temperature but in typical situations decreases by only about an order of magnitude from its high-temperature value (explicit calculations of the moments, taking the finite field into account, are reported in [6.21]). In an antiferromagnet the concept of exchange narrowing breaks down. Our simple ideas predict that the linewidth should diverge as $(T - T_N)^{-1/4}$. The experimental data on MnF_2, shown in Fig. 6.14, indicate that the zero-field linewidth does, in fact, increase as $(T - T_N)^{-3/8}$.

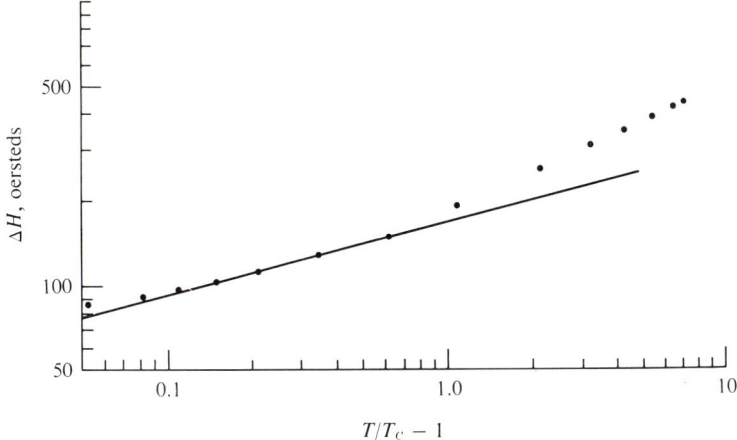

Fig. 6.13. Linewidth as a function of reduced temperature in $CrBr_3$. The solid line is proportional to $[(T/T_C) - 1]^{1/4}$. [6.19]

6.3 Metals

Let us now consider the response of a ferromagnetic metal to a time-dependent field. The Fermi liquid formulation, which we used in the last chapter to study the response of nonmagnetic metals, is not rigorous in this case. This is because Fermi liquid theory rests on the assumption that the quasiparticle excitation spectrum lies close to the Fermi surface. In a ferromagnet this may not be the case. For example, when the spin of a quasiparticle at the Fermi surface is reversed, the resulting quasiparticle state will not, in general, lie close to the Fermi surface and will therefore have a finite decay time. Therefore we shall

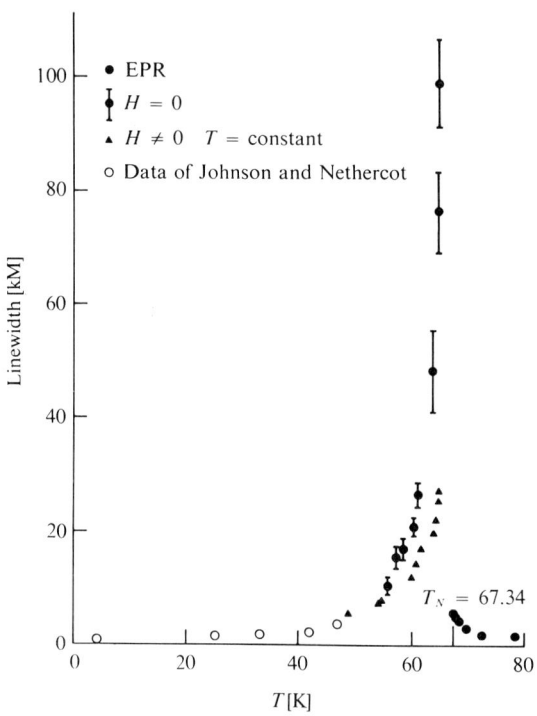

6.14. Linewidth as a function of temperature in MnF_2 [6.22]

consider the microscopic model discussed in Chaps. 4 and 5 which was characterized by the delta-function interaction. In Chap. 5 we obtained the generalized susceptibility,

$$\chi_{-+}(\boldsymbol{q}, \omega) = \frac{2\mu_B^2 \, \Gamma(\boldsymbol{q}, \omega)}{1 - 2I\Gamma(\boldsymbol{q}, \omega)}, \qquad (6.94)$$

where

$$\Gamma(\boldsymbol{q}, \omega) = \frac{1}{V} \sum_k \frac{n_{k\uparrow} - n_{k+q\downarrow}}{\hbar\omega - (\tilde{\epsilon}_{k\uparrow} - \tilde{\epsilon}_{k+q\downarrow})} \qquad (6.95)$$

and $\tilde{\epsilon}_{k\sigma}$ is the Hartree–Fock energy

$$\tilde{\epsilon}_{k\sigma} = \epsilon_k - \frac{I}{N} N_\sigma .$$

Adding the constant I to this energy does not change the difference in $\Gamma(\boldsymbol{q}, \omega)$ and enables us to write

$$\tilde{\epsilon}_{k\sigma} = \epsilon_k + \frac{nI}{2} - \sigma \frac{nI}{2} \zeta \qquad (6.96)$$

where $\zeta = (n_\uparrow - n_\downarrow)/n$ is the relative magnetization and $\sigma = \pm 1$. For parabolic energy bands $\Gamma(q, \omega)$ may be evaluated exactly for $T = 0$:

$$\Gamma(q, \omega) = -\frac{N(\epsilon_F)}{4\tilde{q}} \left\{ A(\tilde{q} - \tilde{\omega}/\tilde{q}) + \tfrac{1}{2}[A^2 + (\tilde{q} - \tilde{\omega}/\tilde{q})^2] \ln \frac{\tilde{q} - \tilde{\omega}/\tilde{q} + A}{\tilde{q} - \tilde{\omega}/\tilde{q} - A} \right. $$
$$\left. + B(\tilde{q} + \tilde{\omega}/\tilde{q}) + \tfrac{1}{2}[B^2 - (\tilde{q} + \tilde{\omega}/\tilde{q})^2] \ln \frac{\tilde{q} + \tilde{\omega}/\tilde{q} + B}{\tilde{q} + \tilde{\omega}/\tilde{q} - B} \right\}, \qquad (6.97)$$

where

$$\tilde{q} = q/2q_F, \qquad \tilde{\omega} = (\omega - nI\zeta)/4\epsilon_F,$$
$$A = (1 + \zeta)^{1/3}, \qquad B = (1 - \zeta)^{1/3}.$$

Since $\ln z = \ln r + i\theta$ where $z = re^{i\theta}$, the appearance of the logarithms in (6.97) gives rise for $\zeta < 1$ to four regions in the $\omega - q$ plane. Three of these are characterized by $\Gamma''(q, \omega) \neq 0$ corresponding to single-particle spin-flip, or "Stoner", excitations. These three regions are indicated by the shaded portions of Fig. 6.15. In the fourth region $\Gamma''(q, \omega) = 0$. Therefore, since

$$\chi''_{-+}(q, \omega) = \frac{2\mu_B^2 \Gamma''(q, \omega)}{[1 - 2I\Gamma'(q, \omega)]^2 + 4I^2\Gamma''(q, \omega)^2} \qquad (6.98)$$

in the limit $\Gamma''(q, \omega) \to 0$,

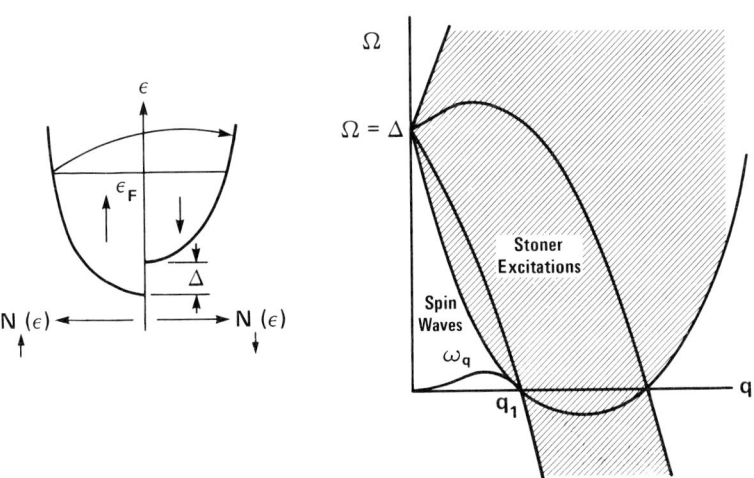

Fig. 6.15. Regions of Ω-q space showing Stoner excitations and spin waves. Here $\Omega = \omega/4\epsilon_F$ and $\Delta = nI\zeta/4\epsilon_F$ is the Stoner gap

$$\chi''(\boldsymbol{q}, \omega) \rightarrow \frac{\mu_B^2}{I} \delta[1 - 2I\Gamma'(\boldsymbol{q}, \omega)]. \tag{6.99}$$

From the property of the delta function, $\delta[f(\omega)] = \delta(\omega - \omega_q)|\partial f/\partial \omega|_{\omega_q}^{-1}$, where ω_q is defined by $f(\omega_q) = 0$, we see that the imaginary part of the susceptibility contains poles at $\omega = \omega_q$. These correspond to collective excitations, or spin waves.

To find the dispersion relation for these spin waves we expand $\Gamma'(\boldsymbol{q}, \omega)$ for small q and ω and set the result equal to $1/2I$. The result is [6.23]

$$\hbar\omega(q) = \frac{1}{N_\uparrow - N_\downarrow} \sum_k \left[\tfrac{1}{2}(n_{k\uparrow} + n_{k\downarrow}) \left(\boldsymbol{q} \cdot \frac{\partial}{\partial \boldsymbol{k}}\right)^2 \epsilon_k - \frac{N(n_{k\uparrow} - n_{k\downarrow})}{I(N_\uparrow - N_\downarrow)} \left(\boldsymbol{q} \cdot \frac{\partial \epsilon_k}{\partial \boldsymbol{k}}\right)^2\right]. \tag{6.100}$$

This has the same quadratic dependence upon q that we found for the spin–wave spectrum of a ferromagnetic insulator. The first term in (6.100) corresponds to the additional kinetic energy that the electrons develop as they change their spin directions to keep up with the change in the macroscopic spin direction of the spin wave. It can be shown [6.24] that this is always positive and vanishes for a filled band. The second term corresponds to a reduction in this kinetic energy

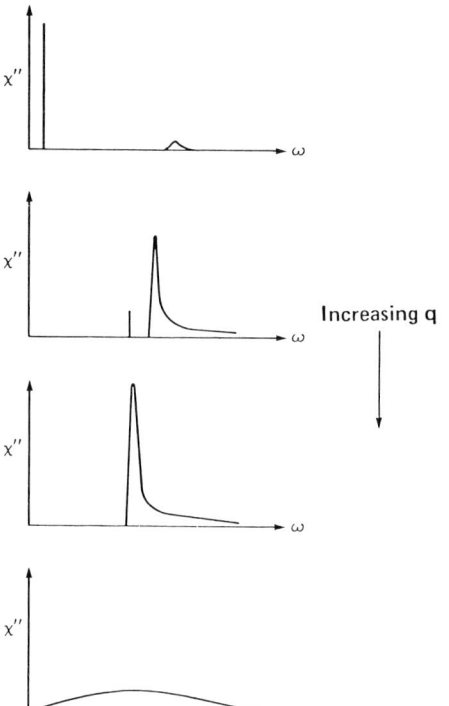

Fig. 6.16. Behavior of the spectral function $\chi''(q, \omega)$ as $\boldsymbol{q}$ increases toward $\boldsymbol{q}_1$, in Fig. 6.15

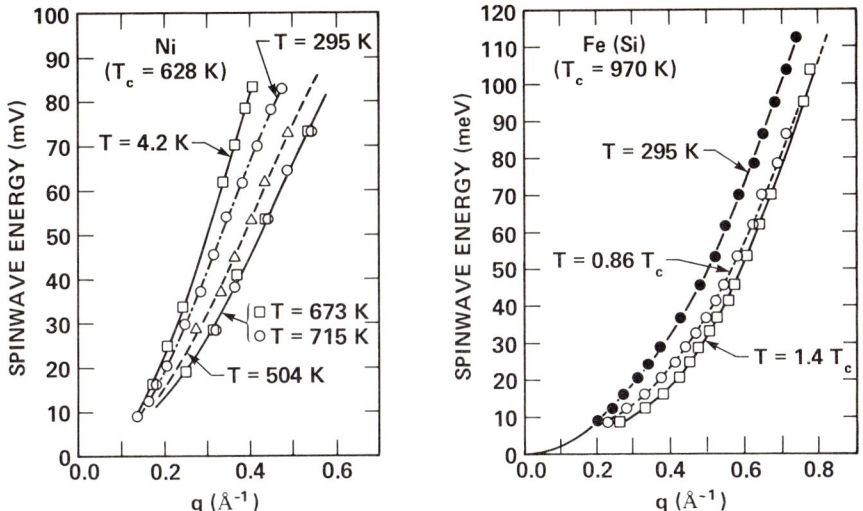

Fig. 6.17. Spin–wave dispersion relations in iron and nickel. The iron sample was alloyed with silicon to bring the Curie temperature down to a convenient value

due to a tilting of the electron spin out of the plane in which the macroscopic magnetization varies.

From Fig. 6.15 we see that for wave vectors for which undamped spin waves exist there will be contributions to $\chi''(\mathbf{q},\omega)$ from the nonzero values of $\Gamma''(\mathbf{q},\omega)$ corresponding to virtual Stoner excitations. If we imagine a vertical cut through the ω-q plane at increasingly larger wave vectors, the susceptibility behaves [6.25] as shown in Fig. 6.16. The spin waves become heavily damped as they approach and merge into the Stoner excitation region. This limits their experimental identification to relatively small wave vectors.

In most ferromagnetic metals the low-temperature magnetization is found to follow a Bloch $T^{3/2}$ law. This is usually taken as evidence of the importance of the spin-wave modes. However, in weak itinerant ferromagnets, such as $ZrZn_2$, the Stoner excitations may dominate, producing a T^2 contribution to the magnetization.

The first direct evidence of spin waves in metals was made by *Lowde*, who observed the inelastic scattering of neutrons from iron [6.26]. Figure 6.17 shows the spin–wave dispersion relations for iron and nickel. In addition to confirming the quadratic dependence on q, these results show an interesting aspect of these spin waves which is not yet completely understood, namely, their existence well *above* the Curie temperature.

Seavey and *Tannenwald* observed the microwave excitation of standing spin waves in thin films of permalloy [6.27], and light scattering has also proven useful in studying spin waves in metals [6.28].

7. Magnetic Impurities

It is probably fair to say that what information we have about the magnetic properties of solids has been obtained by studying magnetic ions dissolved in various hosts. This is certainly true of insulators and has been an underlying motivation for the study of dilute alloys. In our discussion of paramagnetism we considered the problem of a magnetic ion dissolved into a nonmagnetic host. We shall now consider what happens when a magnetic impurity is introduced into a *magnetic* insulating host. This will introduce certain general concepts which bear on our discussion of alloys. Since it is perhaps easier to visualize fixed moments, these concepts may be more readily grasped in this context. We shall then return to the question of magnetic impurities in *nonmagnetic* metals.

7.1 Local Modes

Let us first consider an impurity spin in a ferromagnetic insulator. This problem has a long history of study within the framework of molecular field theory. However, the first exact treatment of local magnon effects at zero temperature was by *Wolfram* and *Callaway* [7.1]. Using Green's function techniques, they considered the conditions for the existence of the localized modes of a ferromagnetically coupled impurity in a ferromagnetic host. *Hone* et al. later explored the thermodynamic properties of such impurity modes [7.2].

To illustrate the approach to this problem let us consider a linear chain with a ferromagnetic ground state. In the absence of an impurity the isotropic exchange Hamiltonian for such a chain of spins with a nearest-neighbor exchange is

$$\mathcal{H}_0 = -J \sum_n \sum_{\delta=\pm 1} \boldsymbol{S}_n \cdot \boldsymbol{S}_{n+\delta} \tag{7.1}$$

We wish to determine the eigenstates of this Hamiltonian when one spin deviation is present in the system. For this purpose let us introduce a complete set of orthonormal single-spin-deviation states

$$|n\rangle \equiv \frac{S_n^-}{\sqrt{2S}}|0\rangle = \begin{bmatrix} 0 \\ 0 \\ \vdots \\ 1 \\ \vdots \\ 0 \\ 0 \end{bmatrix}. \tag{7.2}$$

In this basis the matrix elements of $\mathscr{H}_0$ relative to the fully aligned ground state are

$$\langle n|\mathscr{H}_0|m\rangle = 4JS\Delta(n, m) - 2JS \sum_{\delta=\pm 1} \Delta(n, m + \delta) \tag{7.3}$$

where $\Delta(n, m)$ is the Kronecker delta function. Notice that these spin-deviation states are not eigenfunctions of $\mathscr{H}_0$. Let $|k\rangle$ be such an eigenstate of $\mathscr{H}_0$. Then

$$\mathscr{H}_0|k\rangle = E_0(k)|k\rangle. \tag{7.4}$$

Expanding $|k\rangle$ in terms of the single-spin-deviation states and multiplying by $\langle n|$ gives a system of N homogeneous equations of the form

$$[4JS - E_0(k)]\langle n|k\rangle - 2JS \sum_{\delta=\pm 1} \langle n + \delta|k\rangle = 0. \tag{7.5}$$

If we assume periodic boundary conditions, these equations have the solution

$$\langle n|k\rangle = \frac{1}{\sqrt{N}} e^{ikna} \tag{7.6}$$

where a is the distance between spins and k is equal to some integer times $2\pi/Na$. It is customary to choose the N values from $-(\pi/Na)(N-2)$ up to $+\pi/a$. The corresponding eigenvalue is

$$E_0(k) = 4JS(1 - \cos ka) \tag{7.7}$$

which is the familiar spin–wave dispersion relation.

Now let us replace the spin at l by an impurity spin S' which has an exchange coupling J' with its host neighbors. The Hamiltonian then becomes

$$\mathscr{H} = \mathscr{H}_0 + \mathscr{H}_1 \tag{7.8}$$

where

$$\mathscr{H}_1 = 2(J\mathbf{S}_l - J'\mathbf{S}'_l) \cdot \sum_{\delta=\pm 1} \mathbf{S}_{l+\delta}. \tag{7.9}$$

The matrix elements of $\mathscr{H}_1$ in the basis of single-spin-deviation states are

$$\begin{aligned}\langle n|\mathscr{H}_1|m\rangle = &- 4JS^2\rho\Delta(n, m) + 4JS\epsilon\Delta(l, m)\Delta(n, m) \\ &+ 2JS\rho \sum_{\delta=\pm 1} \Delta(l+\delta, m)\Delta(n, m) \\ &- 2JS\gamma \sum_{\delta=\pm 1} [\Delta(l+\delta, n)\Delta(l, m) \\ &+ \Delta(l, n)\Delta(l+\delta, m)]\end{aligned} \tag{7.10}$$

where

$$\rho = \frac{J'S' - JS}{JS} \quad \epsilon = \frac{J' - J}{J}, \text{ and } \quad \gamma = \frac{J'}{J}\sqrt{\frac{S'}{S}} - 1. \tag{7.11}$$

Let us redefine $\mathcal{H}_0$ and $\mathcal{H}_1$ by incorporating the first term in $\langle n|\mathcal{H}_1|m\rangle$ above into $\langle n|\mathcal{H}_0|m\rangle$. Then $\langle n|\mathcal{H}_1|m\rangle$ is an $N \times N$ matrix, all the elements of which are 0 except for a 3×3 submatrix centered at l. Since all the sites are equivalent, we may choose $l = 2$. Then $\langle n|\mathcal{H}_1|m\rangle$ has the form

$$\langle n|\mathcal{H}_1|m\rangle = 2JS \begin{bmatrix} \rho & -\gamma & 0 & \\ -\gamma & 2\epsilon & -\gamma & 0 \\ 0 & -\gamma & \rho & \\ \hdashline & 0 & & 0 \end{bmatrix} \tag{7.12}$$

Let $|\Psi\rangle$ be the eigenstate of the system in the presence of the impurity. This may be expanded in single-spin-deviation states as

$$|\Psi\rangle = \sum_n \langle n|\Psi\rangle |n\rangle = \begin{bmatrix} \langle 1|\Psi\rangle \\ \langle 2|\Psi\rangle \\ \vdots \\ \langle N|\Psi\rangle \end{bmatrix} \tag{7.13}$$

Thus the Schrödinger equation is the matrix equation

$$(\mathcal{H}_0 + \mathcal{H}_1)|\Psi\rangle = E|\Psi\rangle, \tag{7.14}$$

or

$$[I - (E - \mathcal{H}_0)^{-1}\mathcal{H}_1]|\Psi\rangle = 0 \tag{7.15}$$

where I is the $N \times N$ unit matrix.

We shall find it convenient to introduce the *Green's function operator*

$$G = 2JS(E - \mathcal{H}_0)^{-1}. \tag{7.16}$$

In terms of this Green's function, the eigenvalue equation becomes

$$[I - (2JS)^{-1}G\mathcal{H}_1]|\Psi\rangle = 0. \tag{7.17}$$

Because of the form of $\mathcal{H}_1$, the matrix product $G\mathcal{H}_1$ has elements only in its first three columns. This means that the first three components of $|\Psi\rangle$, corresponding to the impurity and its nearest neighbors, may be solved for independently of the others. The other components are then easily found from the remaining equations. Thus our problem reduces to the 3×3 matrix equation

$$[I - (2JS)^{-1}G\mathcal{H}_1]|\psi\rangle = 0 \tag{7.18}$$

where I, G, and $\mathcal{H}_1$ are now 3×3 matrices and $|\psi\rangle$ is constructed from the first three components of $|\Psi\rangle$.

The matrix elements of G are

$$G_{nm} = 2JS\langle n| \frac{1}{E - \mathcal{H}_0} |m\rangle. \tag{7.19}$$

If the spin–deviation states are expanded in terms of eigenfunctions of $\mathcal{H}_0$, the matrix elements become

$$G_{nm} = \frac{2JS}{N} \sum_k \frac{e^{ik(n-m)a}}{E - E_0(k)}. \tag{7.20}$$

Notice that G_{nm} depends only on $|n - m|$. Thus there are only three different matrix elements of G in this part of our problem. The general matrix element G_n, where $n \equiv |m - n|$, may be evaluated by converting the sum over k into an integral. We introduce the reduced energy

$$\mathcal{E} = \frac{E - 4JS}{4JS}, \tag{7.21}$$

and this integral gives

$$G_n(\mathcal{E}) = \frac{(\sqrt{\mathcal{E}^2 - 1} - \mathcal{E})^n}{2\sqrt{\mathcal{E}^2 - 1}}. \tag{7.22}$$

Notice that for $\mathcal{E} > 1$ the Green's function is real, while it becomes complex when $\mathcal{E} < 1$.

In all problems involving clusters of ions it is convenient to work in a basis which reflects the point-group symmetry. This is accomplished by transforming G and $\mathcal{H}_1$ by a unitary transformation which diagonalizes some symmetry operator of the system. For example, the interchange of particles 1 and 3 is a symmetry operation of the system. The unitary transformation U which diagonalizes this particular symmetry operator leads to the new states

$$U|\psi\rangle = \begin{pmatrix} \langle 2|\Psi\rangle \\ \frac{\sqrt{2}}{2}(\langle 1|\Psi\rangle + \langle 3|\Psi\rangle) \\ \frac{\sqrt{2}}{2}(\langle 1|\Psi\rangle - \langle 3|\Psi\rangle) \end{pmatrix} \tag{7.23}$$

Because of their symmetry the first two modes are referred to as *s-like modes*, while the third is a *p-like mode*. In the S_0 mode the phase of the impurity spin is the same as that of the neighbors, while in the S_1 mode the phase of the impurity spin is opposite that of its neighbors.

Applying this transformation to (7.18) leads to the determinantal equation

$$\begin{vmatrix} 1 - 2\epsilon G_0 + 2\gamma G_1 & \sqrt{2}(\gamma G_0 - \rho G_1) & 0 \\ \sqrt{2}(\gamma G_0 - 2\epsilon G_1 + \gamma G_2) & 1 + 2\gamma G_1 - \rho(G_0 + G_2) & 0 \\ 0 & 0 & 1 - \rho(G_0 - G_2) \end{vmatrix}$$
$$= \begin{vmatrix} D_s & 0 \\ 0 & D_p \end{vmatrix} = 0 \quad (7.24)$$

Notice that the p-like mode is relatively simple, depending only on the parameter ρ. As a result of this simplicity we can obtain analytic expressions for its eigenvalues and eigenfunctions. The eigenvalue is given by

$$D_p = 1 - \rho(G_0 - G_2) = 0 . \quad (7.25)$$

Using the results of G_0 and G_2 obtained from the general relation for G_n, we may write this eigenvalue for $\mathscr{E} > 1$ as

$$\mathscr{E} = \frac{\rho^2 + 1}{2\rho} . \quad (7.26)$$

The behavior of this p-like mode is shown in Fig. 7.1. As ρ becomes large the eigenvalue becomes $\rho/2$. Thus its energy is

$$E = 4JS + 2JS\rho$$
$$= -4S(J'S' - JS) + 4JS + 2(J'S' - JS) . \quad (7.27)$$

The first term is just the correction to the ground-state energy due to the impurity and the remaining expression is the energy that would be obtained from an Ising treatment of this problem. Thus as the local modes move away from the spin–wave band their frequencies approach an Ising limit.

In Fig. 7.1 the spin–wave band, as given by (7.7), extends to $\mathscr{E} = -1$. Had we included a uniaxial anisotropy field the spin–wave spectrum would develop a gap. If the anisotropy field seen by the impurity is less than that of the host, than the S_0 mode appears in this gap.

Notice that the condition for the appearance of a p-like mode outside the band is that $J'S' > 2JS$. In thermodynamic studies of such systems *Hone* et al. found that although such a condition may not be satisfied at absolute zero, if $\langle S_z \rangle$ decreases more rapidly with temperature than $\langle S_z' \rangle$, a temperature may be reached at which the condition is satisfied and a mode "pops out" of the band [7.2].

The eigenfunction of the p-like mode may be obtained exactly. From the first three of our N homogeneous equations we obtain $\langle 2|\Psi\rangle = 0$ and $\langle 3|\Psi\rangle = -\langle 1|\Psi\rangle$. From the nth equation we obtain, for $n > 3$,

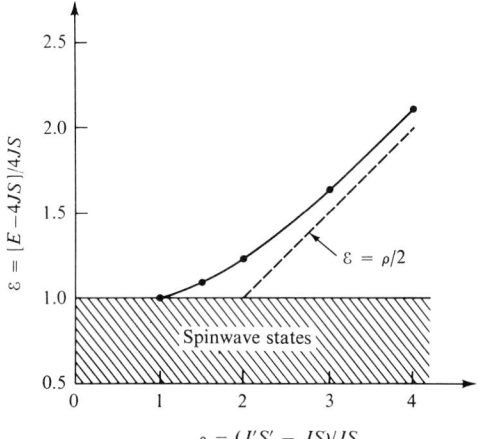

Fig. 7.1. Frequency of the p-like local mode in a linear chain

$$\langle n | \Psi \rangle = \left[\frac{1}{2}(\rho^2 - 1) \right]^{1/2} \left(-\frac{1}{\rho} \right)^{n-2}. \tag{7.28}$$

The resulting wave function is shown in Fig. 7.2. together with the s-like modes. The origin of the terms "s-like" and "p-like" is now evident. The probability of finding the spin deviation at site n is $|\langle n | \Psi \rangle|^2$. As p increases, the p-like mode moves away from the spin–wave band and becomes more localized.

Notice that the eigenvalue (7.26) is defined for all values of ρ. However, for $\rho < 1$ the solution does not correspond to a localized mode. To see this let us consider the eigenvalue condition in its original form involving sums over k. From (7.20, 25) this becomes

$$\sum_{k=2\pi/Na,\,4\pi/Na,\,\ldots}^{\pi/a} \frac{1 - \cos 2ka}{\mathscr{E} + \cos ka} = \frac{N}{\rho}. \tag{7.29}$$

Both sides of (7.29) are plotted schematically in Fig. 7.3. The solutions correspond to the intersections of these two functions. We see that when ρ is large, there is always a solution at point A, with $\mathscr{E} > 1$. As ρ decreases the horizontal line N/ρ moves upward, eventually intersecting at $\mathscr{E} = 1$ (point B). The value of ρ at which this occurs is obtained by evaluating the sum of the left-hand side with $\mathscr{E} = 1$. This may be accomplished by grouping the terms for $k = n2\pi/Na$ with those for $k = (N/2 - n)2\pi/Na$. The sum has the value N. Thus we find that when $\rho = 1$ the solution just begins to enter what was the original unperturbed spin–wave band. At this point the separation of this solution from the next solution at B' becomes comparable to the separation between the spin–wave states themselves. Therefore any formalism employing an integral representation of the Green's functions is not capable of distinguish-

220   7. Magnetic Impurities

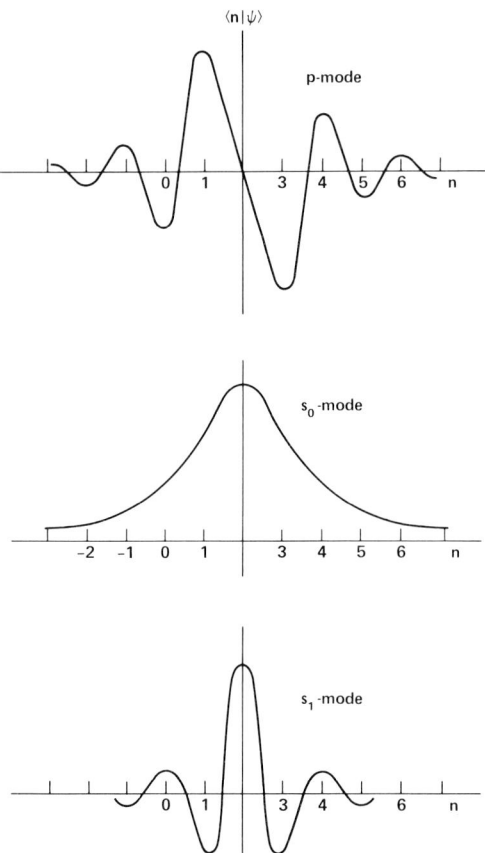

Fig. 7.2. Wave functions of the modes in a linear chain for $p = 2$

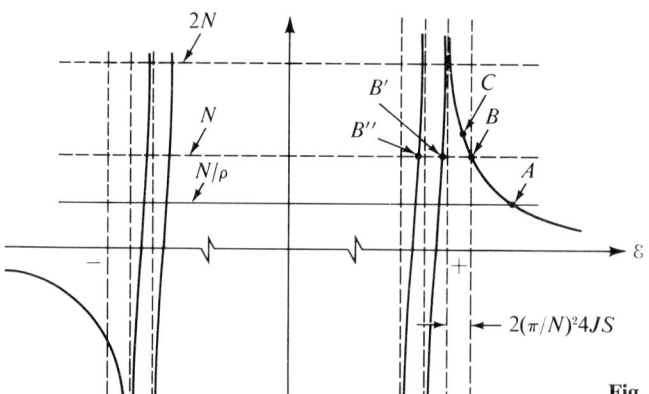

Fig. 7.3. Schematic description of the local-mode solution

ing local modes from band modes beyond this point. In order to describe the situation for $\rho < 1$ we must consider the density of states.

In general, the density of states of a system is the sum of delta functions of the exact eigenvalues,

$$\eta(E) = \sum_n \delta(E - E_n). \tag{7.30}$$

The exact Green's function, to within a constant, is defined as

$$\mathscr{G}(E) = \lim_{s \to 0} \frac{1}{E - \mathscr{H} + is} \tag{7.31}$$

where $\mathscr{H}$ is the total Hamiltonian. The matrix elements of this operator between the exact energy eigenstates are

$$\mathscr{G}_{nn}(E) = \mathscr{P}(E - E_n)^{-1} - i\pi\delta(E - E_n). \tag{7.32}$$

Therefore we have

$$\boxed{\eta(E) = -\frac{1}{\pi} \operatorname{Im}\left\{\operatorname{Tr} \mathscr{G}(E)\right\}.} \tag{7.33}$$

This is a very general result which may be applied to any system.

In our present problem it is convenient to define

$$\mathscr{G}(E) = 2JS \lim_{s \to 0} (E - \mathscr{H} + is)^{-1}.$$

This is related to the "unperturbed" Green's function G by the identity

$$\mathscr{G} = \frac{1}{1 - (2JS)^{-1}G\mathscr{H}_1} G = G + (2JS)^{-1}G\mathscr{H}_1 \frac{1}{1 - (2JS)^{-1}G\mathscr{H}_1} G. \tag{7.34}$$

The trace of $\mathscr{G}$ may be written as

$$\operatorname{Tr}\mathscr{G} = \operatorname{Tr} G + \frac{1}{2D}\frac{d}{d\mathscr{E}} D \tag{7.35}$$

where

$$D = |1 - (2JS)^{-1}G\mathscr{H}_1|. \tag{7.36}$$

Therefore

$$\eta(\mathscr{E}) = \eta_0(\mathscr{E}) - \frac{1}{\pi} \operatorname{Im}\left\{\frac{d}{d\mathscr{E}}\right\} \ln D \tag{7.37}$$

where the unperturbed density of states is given by

$$\eta_0(\mathscr{E}) = -\frac{2}{\pi} \operatorname{Im}\left\{\operatorname{Tr} G(\mathscr{E} + is)\right\} = -\frac{2}{\pi} \operatorname{Im}\left\{NG_0(\mathscr{E} + is)\right\} = \frac{N}{\pi\sqrt{1 - \mathscr{E}^2}}. \tag{7.38}$$

The singularities at $\mathscr{E} = \pm 1$ are due to the one-dimensional nature of our problem. The contribution of the p mode to the density of states may be determined explicitly by recognizing that the determinant D (7.24) is factored into an s part and a p part. Thus

$$\eta_p = -\frac{1}{\pi} \operatorname{Im}\left\{\frac{D'_p}{D_p}\right\}. \tag{7.39}$$

For $\mathscr{E} < 1$, D_p becomes

$$D_p = 1 - p\mathscr{E} + ip\sqrt{1 - \mathscr{E}^2}. \quad \text{Therefore} \tag{7.40}$$

$$\eta_p = \frac{p(\mathscr{E} - p)}{\pi\sqrt{1 - \mathscr{E}^2}(1 - 2p\mathscr{E} + p^2)}.$$

This is plotted in Fig. 7.4 as a function of $\mathscr{E}$ for $p = 4$ and $p = \frac{1}{2}$. The unperturbed density of the states divided by N is indicated by the dashed curve. Notice that

$$\int_{-1}^{+1} \eta_p(\mathscr{E})d\mathscr{E} = \begin{cases} 0 & p < 1 \\ -1 & p > 1 \end{cases}. \tag{7.41}$$

The -1 corresponds to the fact that the localized mode appears outside the band at the expense of one mode inside the band. We see that the construction of the local mode results in a nearly uniform depletion of states within the band. In

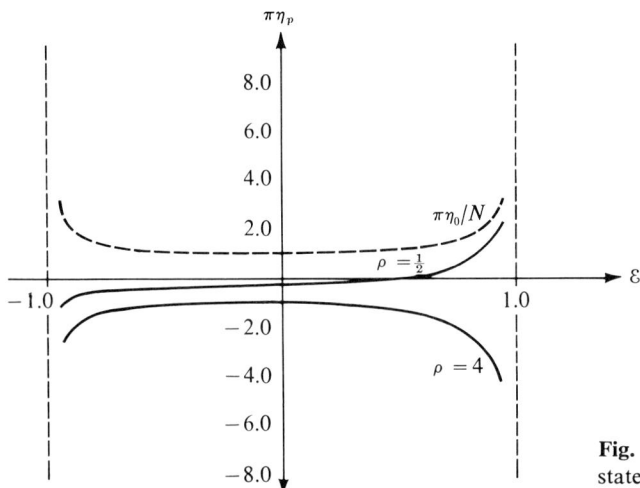

Fig. 7.4. In-band density of states for the p-like mode

three dimensions, when a mode appears inside the band, it manifests itself as a definite "bump" in the density of the states. This bump usually has a Lorentzian shape. Such an in-band mode is called a *resonant mode*.

The relationship between the Green's function and the generalized susceptibility is easily established by means of the fluctuation-dissipation theorem. Since the presence of an impurity destroys the translational invariance, we must use the more general version of (1.89). At $T = 0$ this has the form

$$2\hbar V \chi''_{xx}(\mathbf{q}, \mathbf{k}; \omega) = \int_{-\infty}^{\infty} \langle 0 | \mathscr{M}_x(\mathbf{q}, t) \mathscr{M}_x(-\mathbf{k}) | 0 \rangle e^{i\omega t} dt \qquad (7.42)$$

where $|0\rangle$ is the completely aligned ground state whose energy is defined to be 0; that is, $\mathscr{H}|0\rangle = 0$. Writing

$$\mathscr{M}_x(\mathbf{r}) = -g\mu_B \sum_i S_i^x \delta(\mathbf{r} - \mathbf{R}_i) \qquad (7.43)$$

and recalling that $S_i^x|0\rangle = (S/2)^{1/2}|i\rangle$, we obtain

$$\chi''_{xx}(\mathbf{q}, \mathbf{k}; \omega) = \frac{g^2 \mu_B^2 S}{4\hbar V} \int_{-\infty}^{\infty} \langle \mathbf{q} | \exp[i(\omega - \mathscr{H}/\hbar)t] | \mathbf{k} \rangle dt. \qquad (7.44)$$

In the absence of the impurity, $\mathscr{H} = \mathscr{H}_0$ and the susceptibility is proportional to $\text{Im}\{\langle \mathbf{k}|G|\mathbf{k}\rangle \Delta(\mathbf{q}, \mathbf{k})\}$, which is equivalent to our previous result (5.15). With an impurity present the susceptibility becomes

$$\chi''_{xx}(\mathbf{q}, \mathbf{k}; \omega) = -\frac{g^2 \mu_B^2}{4VJ} \text{Im}\{\langle \mathbf{q} | \mathscr{G}(\hbar\omega) | \mathbf{k} \rangle\}. \qquad (7.45)$$

Since the plane-wave states $|\mathbf{k}\rangle$ are no longer eigenstates, the matrix element in (7.45) is difficult to interpret. If we know the true eigenstates $|\alpha\rangle$, we may write the susceptibility as

$$\chi''_{xx}(\mathbf{q}, \mathbf{k}; \omega) = -\frac{g^2 \mu_B^2}{4VJ} \sum_\alpha \langle \mathbf{q}|\alpha\rangle\langle\alpha|\mathbf{k}\rangle \text{Im}\{\langle \alpha | \mathscr{G}(\hbar\omega) | \alpha \rangle\}$$

$$= \frac{g^2 \mu_B^2}{2\pi V} \sum_\alpha \langle \mathbf{q}|\alpha\rangle\langle\alpha|\mathbf{k}\rangle \delta(\hbar\omega - E_\alpha). \qquad (7.46)$$

Equation (7.46) shows, for example, that a *p*-like mode cannot be excited by a uniform field, since $\langle p|\mathbf{k} = 0\rangle = 0$. From Fig. 7.2 we see that only an S_0 mode can be excited by a long-wavelength probe.

Although our discussion has been based on a hypothetical one-dimensional ferromagnet, the general features carry over to three-dimensional ferro- and antiferromagnets. In MnF_2, for example, Fe^{2+}, Co^{2+}, and Ni^{2+} impurities give rise to an S_0 mode above the spin–wave band which has been excited by neutron scattering, light scattering [7.4], and infrared absorption [7.5].

When Mn^{2+} is dissolved in FeF_2 the S_0 mode lies in the anisotropy gap *below* the spin–wave band. However, the frequency of this mode lies very close to the band. Therefore its wave function is very delocalized. As a result this mode is very efficient in scattering light, much more so than one would expect simply from the impurity concentration [7.6].

Random Fields. It turns out that when a uniform external field is applied to a *dilute* antiferromagnet such as $Co_{1-x}Zn_xF_2$ the magnetic sites experience a random field [7.37]. Recent experiments show that the nature of the long range order in such systems is dramatically affected. This promises to be an active area of future research.

7.2 Local Moments in Metals

When a potentially magnetic atom such as iron is dissolved in a nonmetallic host, the electronic states are localized at the impurity site. The magnetic character of this state is then determined by the intraatomic Coulomb interaction (Hund's rules). If the crystalline fields become large, the magnetic state may be affected. Fe^{2+}, for example, in an octahedral environment can have either a high-spin $^5F_2$ ($t_{2g}^4 e_g^2$) state or a low-spin $^1A_1(t_{2g}^6)$ state. But the low-temperature susceptibility is readily calculated by the formalism described in Chap 3.

In a metallic host, the situation is much more complex because the electronic states associated with the impurity are often not localized. Even when the impurity appears to contain an integral number of electrons, the fact that they are coupled to a Fermi sea, no matter how weak the coupling, can lead to a correlated nonmagnetic ground state.

If we consider the conduction electrons as plane waves, then a convenient formalism for describing the effect of a single impurity is scattering theory. If the impurity is represented by a real spherical potential (i.e., we neglect inelastic scattering), then the problem has symmetry about the axis defined by the incident wave vector k, so the wave function, call it $u_k(r, \theta)$, may be expanded as a sum of products of radial functions and Legendre polynomials,

$$u_k(r, \theta) = \sum_{l=0}^{\infty} (2l + 1) i^l R_l(k, r) P_l(\cos \theta) . \tag{7.47}$$

The coefficient $R_l(k, r)$ is called the partial wave amplitude. This is an important quantity for, when the energy variable k is regarded as a complex number, the poles of $R_l(k, r)$ along the positive imaginary axis correspond to bound states. Far from impurity, this wave function has the form

$$\lim_{r \to \infty} u_k(r, \theta) = e^{ik \cdot r} + f(k, \theta)(e^{ikr}/r) , \tag{7.48}$$

where $f(k, \theta)$ is called the scattering amplitude and is a function of the scattering phase shift δ_l,

$$f(k, \theta) = (1/k) \sum_l [4\pi(2l + 1)]^{1/2} \exp(i\delta_l) \sin \delta_l \, Y_l^0(\theta) . \tag{7.49}$$

Substituting (7.49) into (7.48) shows that $R_l(kr)$ has the asymptotic form

$$R_l(kr) \xrightarrow[r \to \infty]{} \frac{1}{kr} \exp(i\delta_l) \sin(kr - \tfrac{1}{2}l\pi + \delta_l) . \tag{7.50}$$

This phase shift depends upon the energy of the scattering particle and on the details of the impurity potential. If the potential is a delta function of strength V_0, then there is only s-wave scattering and

$$\delta_0(E) = \tan^{-1}\{\pi V_0 N(E)/[1 - V_0 I(E)]\} \tag{7.51}$$

where $N(E)$ is the host density of states and

$$I(E) = \mathscr{P} \int dE' N(E')/(E' - E) , \tag{7.52}$$

where $\mathscr{P}$ means that we take the principal part of the integral. If we define E_0 as the value of E where $V_0 I(E_0) = 1$ and expand $V_0 I(E)$ about $E = E_0$, then

$$\delta_0(E) = \tan^{-1} \frac{\varDelta}{E - E_0} \tag{7.53}$$

where $\varDelta$ is proportional to $N(E_0)$. Notice that when the phase shift goes through $\pi/2$, the scattering amplitude goes through a maximum. This is referred to as a resonance or virtual bound state.

The phase shift tells us how the energies of the scattering states are shifted by the presence of the scattering potential. To see this, we consider the system bounded by a sphere of radius R. If the wave function is to be zero at this surface, then (7.50) tells us that

$$kR - \tfrac{1}{2}l\pi + \delta_l = n\pi \tag{7.54}$$

where n is an integer. Every time n changes by one, another node is introduced into the wave function corresponding to $2(2l + 1)$ additional states (spin degeneracy of two). In the absence of the scattering center, $\delta_l = 0$. When δ_l becomes nonzero, the wave vector corresponding to a particular n and l must change in order to preserve the condition (7.54). This change, $\Delta k = -\delta_l/R$, corresponds to an energy shift $\Delta E = -\hbar^2 k \delta_l/mR$. The unperturbed density of states is $N(E) = mR/\pi\hbar^2 k$. Therefore

$$\Delta E = -\frac{1}{\pi} \frac{\delta_l}{N(E)} . \tag{7.55}$$

Since δ_l/π is the number of extra nodes that have been introduced into the wave function, this corresponds to the number of extra states associated with the impurity. Therefore, there is an additional local density of states given by

$$\eta(E) = \frac{1}{\pi} \sum_l (2l+1) \frac{d\delta_l}{dE}. \tag{7.56}$$

For the phase shift given by (7.53), this virtual bound state is characterized by the Lorentzian form

$$\eta(E) = \frac{1}{\pi} \frac{\Delta}{(E-E_0)^2 + \Delta^2} \tag{7.57}$$

The differential scattering cross section $\sigma(\theta)$ is given by the square of the scattering amplitude, i.e., $\sigma(\theta) = |f(\theta)|^2$. The effective average cross section σ that governs the resistivity is the angular average of $\sigma(\theta)$ weighted by $(1 - \cos\theta)$, which measures the relative change in the component of the electron's velocity along its initial direction of motion:

$$\sigma = 2\pi \int_0^\pi d\theta \sin\theta \, \sigma(\theta)(1 - \cos\theta).$$

The mean free path is then $\Lambda = 1/x\sigma$, where x is the atomic concentration of impurities, and the resistivity associated with the impurities is

$$\Delta\rho = \frac{4\pi x}{ne^2 k_F} \sum_{l=0}^\infty (l+1) \sin^2[\delta_l(E_F) - \delta_{l+1}(E_F)] \tag{7.58}$$

where n is the density of conduction electrons. Thus, again if we only consider s-wave scattering ($l = 0$), the resistivity will show a peak when the Fermi level and the center of the virtual bound state coincide. There is evidence for such behavior in aluminum as shown in Fig. 7.5. This phase shift description has been successfully applied to many alloys, particularly by *Friedel* [7.7] and his collaborators. The problem arises when one begins to consider magnetic properties. Figure 7.6, for example, shows that iron may or may not exhibit a magnetic moment when it is dissolved in 4d transition metals and alloys. This behavior attracted the attention of *Anderson*, who considered how a virtual bound state develops a moment.

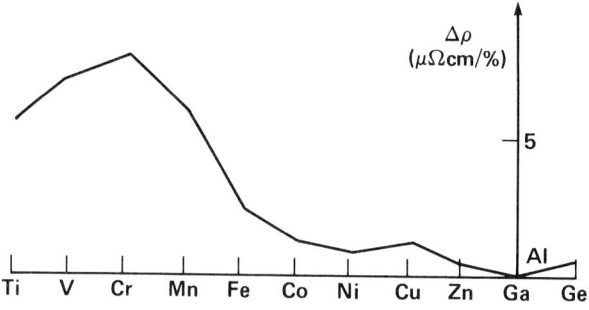

Fig. 7.5. Residual resistivities of transitional impurities in aluminum

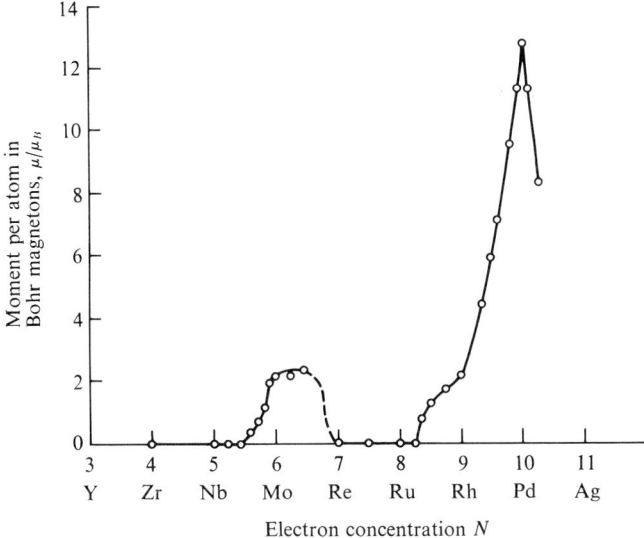

Fig. 7.6. Magnetic moment in Bohr magnetons of an iron atom dissolved in various second-row transition metals and alloys as a function of electron concentration [7.8]

7.2.1 Anderson's Theory of Moment Formation

The presence of an impurity in a metallic host leads to additional terms in the Hamiltonian of the system. Since the size or the charge of the impurity may differ from that of the host, there will be a change $V(\mathbf{r})$ in the crystal potential. Since this potential is screened by the conduction electrons, it is essentially confined to the impurity cell. In addition, the Coulomb interaction between electrons within the impurity cell, v, will be different from that between electrons within host cells. Most of the hosts we shall consider do not exhibit a significant exchange enhancement, and so we shall neglect the Coulomb interactions within the host cells, retaining it only for the impurity cell. Assuming that these are the major effects of the impurity, we write the total Hamiltonian as

$$\mathcal{H} = \mathcal{H}_0 + \sum_i eV(\mathbf{r}_i) + \sum_{\substack{i,j \\ \text{(within} \\ \text{impurity} \\ \text{cell)}}} v(\mathbf{r}_i - \mathbf{r}_j) . \tag{7.59}$$

Here $\mathcal{H}_0$ is the Hamiltonian for the pure host.

The next step is to write this Hamiltonian in its second-quantized form. The actual form will depend on the states we use to expand the field operator. For example, suppose we dissolve a transition metal such as manganese into copper. Since Mn^{2+} has four fewer electrons than Cu^{2+}, it constitutes a strong perturbation. Therefore we might expect the localized state to be pushed out of the d

band of the copper host, as shown in Fig. 7.7. Since the d band is filled, it will not contribute to the properties of the dilute alloy. Therefore we expand our field operator in terms of the s-band states and a localized nondegenerate d-like state. The one-electron terms in (7.59) then become

$$\sum_{k,\sigma} \epsilon_k c_{k\sigma}^\dagger c_{k\sigma} + \sum_\sigma \epsilon_0 c_{0\sigma}^\dagger c_{0\sigma} + \sum_{k,\sigma} (V_{0k} c_{k\sigma}^\dagger c_{0\sigma} + V_{0k}^* c_{0\sigma}^\dagger c_{k\sigma}). \quad (7.60)$$

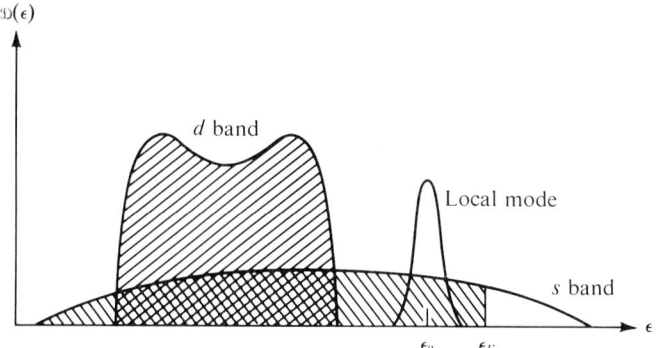

Fig. 7.7. Schematic density of states of copper containing an impurity mode

The first term is the s-band energy, the second term is the contribution from the impurity state, and the last term is the so-called *sd mixing term*. Notice that this mixing is a one-electron effect. It corresponds to the hopping of an electron from the localized d orbital into the conduction band, or vice versa. The interaction part of (7.59) becomes

$$U n_{0\uparrow} n_{0\downarrow} \quad (7.61)$$

where U is the intraatomic Coulomb repulsion between opposite spins in the localized orbital. Equations (7.60, 61) constitute what is known as the *Anderson Hamiltonian* [7.9].

To study the implications of this model, *Anderson* employed the Hartree-Fock approximation in the same form as we encountered in our discussion of the Hubbard model. Namely, the interaction term $Un_{0\uparrow}n_{0\downarrow}$ is replaced by $U(\langle n_{0\uparrow}\rangle n_{0\downarrow} + \langle n_{0\downarrow}\rangle n_{0\uparrow} - \langle n_{0\uparrow}\rangle\langle n_{0\downarrow}\rangle)$. Thus, in the Hartree-Fock approximation the *Anderson Hamiltonian* for a spin σ becomes

$$\mathcal{H}_{\mathrm{HF}}^\sigma = \sum_k \epsilon_k c_{k\sigma}^\dagger c_{k\sigma} + \epsilon_{0\sigma} c_{0\sigma}^\dagger c_{0\sigma} + \sum_k V_{0k}(c_{k\sigma}^\dagger c_{0\sigma} + c_{0\sigma}^\dagger c_{k\sigma}) \quad (7.62)$$

where

$$\epsilon_{0\sigma} = \epsilon_0 + U\langle n_{0,-\sigma}\rangle. \quad (7.63)$$

The Green's function corresponding to the Schrödinger equation for spin σ is

$$G^\sigma(\epsilon) = \lim_{s \to 0} \frac{1}{\epsilon - \mathcal{H}^\sigma + is}. \tag{7.64}$$

Multiplying both sides (7.64) by $(\epsilon - \mathcal{H}^\sigma + is)$ and computing the matrix elements of the resulting equation between the states $|k\rangle$ and $|0\rangle$ gives the following four relations:

$$(\epsilon - \epsilon_{0\sigma} + is)G^\sigma_{00}(\epsilon) - \sum_k V_{0k} G^\sigma_{k0}(\epsilon) = 1, \tag{7.65}$$

$$(\epsilon - \epsilon_k + is)G^\sigma_{kk'}(\epsilon) - V_{k0}G^\sigma_{0k'}(\epsilon) = \delta_{kk'}, \tag{7.66}$$

$$(\epsilon - \epsilon_k + is)G^\sigma_{k0}(\epsilon) - V_{k0}G^\sigma_{00}(\epsilon) = 0, \tag{7.67}$$

$$(\epsilon - \epsilon_{0\sigma} + is)G^\sigma_{0k}(\epsilon) - \sum_{k'} V_{0k'} G^\sigma_{k'k}(\epsilon) = 0. \tag{7.68}$$

By solving these equations we obtain the various matrix elements of the Green's function. We then use (7.33) to compute the local density of states $\eta_{0\sigma}(\epsilon)$. This has the Lorentzian form

$$\eta_{0\sigma}(\epsilon) = \frac{1}{\pi} \frac{\Delta}{(\epsilon - \epsilon_0 - U\langle n_{0-\sigma}\rangle - \delta\epsilon)^2 + \Delta^2}. \tag{7.69}$$

where

$$\Delta = \pi \sum_k |V_{0k}|^2 \delta(\epsilon - \epsilon_k) \tag{7.70}$$

and

$$\delta\epsilon = \mathscr{P} \sum_k \frac{|V_{0k}|^2}{\epsilon - \epsilon_k}. \tag{7.71}$$

From this density of states we can compute the occupation number of the localized state:

$$\langle n_{0\sigma}\rangle = \int_{-\infty}^{\epsilon_F} \eta_{0\sigma}(\epsilon)d\epsilon. \tag{7.72}$$

Since $\eta_{0\sigma}$ is a function of $\langle n_{0-\sigma}\rangle$, we have a set of coupled equations. These may have the single nonmagnetic solution $\langle n_{0\sigma}\rangle = \langle n_{0-\sigma}\rangle$, or the two symmetrical magnetic solutions $\langle n_{0\sigma}\rangle \neq \langle n_{0-\sigma}\rangle$, depending upon the relative values of the parameter ϵ_0, U, and Δ. These regimes are indicated in Fig. 7.8. Notice that integrating (7.56) and comparing it with (7.72) gives $\delta_\sigma(\epsilon_F) = \pi\langle n_{0\sigma}\rangle$. If Z is the total number of localized electrons, then $\delta_\sigma + \delta_{-\sigma} = \pi Z$. This is a special

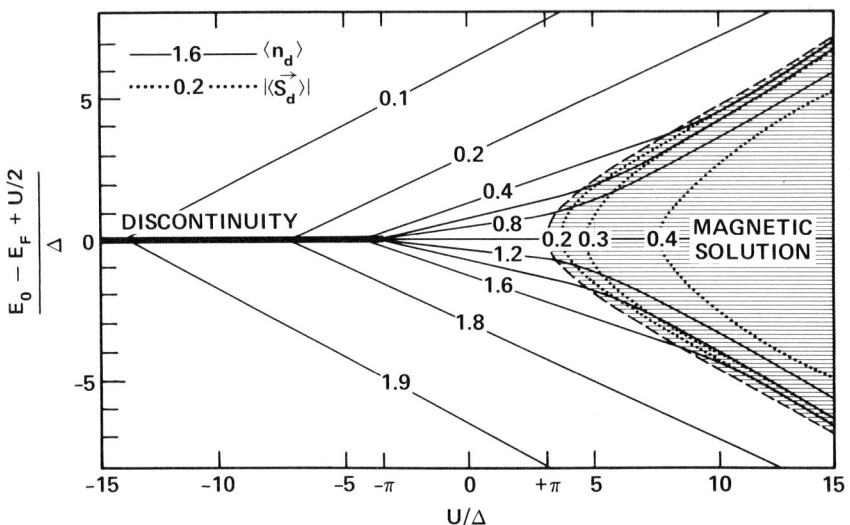

Fig. 7.8. "Phase diagram" corresponding to the Hartree–Fock solution of the Anderson Hamiltonian

case of a general relation known as the Friedel sum rule,

$$\frac{1}{\pi} \sum_{l,\sigma} (2l+1)\delta_{l\sigma}(\epsilon_F) = Z.$$

To determine the boundary between the magnetic and nonmagnetic regions let us introduce the number of localized electrons n_0 and the localized moment m_0 by

$$n_0 = \tfrac{1}{2}(\langle n_{0\uparrow}\rangle + \langle n_{0\downarrow}\rangle), \tag{7.73}$$

$$m_0 = \tfrac{1}{2}(\langle n_{0\uparrow}\rangle - \langle n_{0\downarrow}\rangle). \tag{7.74}$$

Then

$$m_0 = \tfrac{1}{2}f(\epsilon_F; \epsilon_0 + Un_0 - Um_0) - \tfrac{1}{2}f(\epsilon_F; \epsilon_0 + Un_0 + Um_0) \tag{7.75}$$

where f is the $\cot^{-1}$ function obtained from integrating the Lorentzian density of states.

In the limit of large m_0 the right-hand side goes to 0. Therefore, if there is a nonzero solution, the right-hand side must have a slope greater than 1 at $m_0 = 0$. Since

$$\left.\frac{\partial f(\epsilon_F; \epsilon_0 + Un_0 \mp Um_0)}{\partial m_0}\right|_{m_0=0} = \mp U \frac{\partial f(\epsilon_F; \epsilon_0 + Un_0)}{\partial(\epsilon_0 + Un_0)} \tag{7.76}$$

the condition for the appearance of the localized moment is

$$-U\frac{\partial f(\epsilon_F; \epsilon_0 + Un_0)}{\partial(\epsilon_0 + Un_0)} > 1 \tag{7.77}$$

If the localized state lies close to the Fermi surface, then Δ and $\delta\epsilon$ may be approximated by their values at the Fermi surface. Then

$$\frac{\partial f(\epsilon_F; \epsilon_0 + Un_0)}{\partial(\epsilon_0 + Un_0)} = \int_{-\infty}^{\epsilon_F} d\epsilon \frac{\partial \eta_{0\sigma}(\epsilon)}{\partial(\epsilon_0 + Un_0)} = -\eta_{0\sigma}(\epsilon_F). \tag{7.78}$$

The condition (7.77) then becomes

$$\boxed{U\eta_{0\sigma}(\epsilon_F) > 1 .} \tag{7.79}$$

Note the similarity of this result to the Stoner criterion. From (7.68) we see that a localized spin moment appears when U is sufficiently large in comparison with the width Δ of the resonant level. Since $\Delta = \pi V^2 n(\epsilon_F)$ [see (7.70)] a small density of states for the conduction band and a small covalent admixture favor the formation of a moment. Thus we expect to find moments when 3d atoms are dissolved in noble metals but not in aluminum, as is the case.

One of the consequences of Anderson's Hartree–Fock solution is that in the magnetic regime the density of states is characterized by two Lorentzians separated by an energy $U(\langle n_{0\uparrow} \rangle - \langle n_{0\downarrow} \rangle)$. If we again look to the residual resistivity for evidence of such behavior, we do indeed find two peaks for transition-metal impurities dissolved in Cu as shown in Fig. 7.9. More microscopic probes, such as nuclear magnetic resonance [7.10], however, indicate that these solutes behave more atomically than the itinerant Anderson model would suggest. Much more dramatic is the behavior associated with the breakdown of the Hartree–Fock approximation itself, as manifest in the Kondo effect.

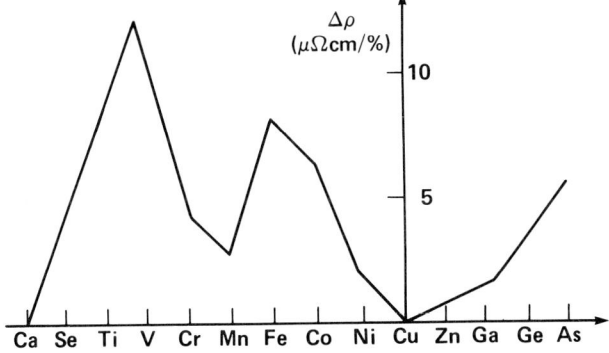

Fig. 7.9. Residual resistivities of transitional impurities in copper

7.3 The Kondo Effect

Figure 7.10 shows the general behavior observed in "local-moment" systems such as those illustrated in Fig. 7.9 with decreasing temperature. The resistivity increases logarithmically and eventually saturates. When this increasing impurity contribution is combined with the decreasing phonon contribution, one obtains a minimum in the resistivity, a phenomenon that was known but not understood for many years. At the same time, the magnetic moment, as measured by the susceptibility, decreases, and the specific heat shows a peak corresponding to an entropy change of approximately $k_B \ln 2$. Although this transition is gradual, it is possible to assign a temperature, called the Kondo temperature, below which these anomalous properties appear. Figure 7.11 shows how this Kondo temperature varies for $3d$ transition metals dissolved in copper.

It is now known that these properties, which are collectively referred to as the "Kondo effect," are associated with the behavior of a single localized electron spin interacting with a degenerate electron system. This seemingly simple problem has required an enormous effort to bring us to the point where we think that we understand what is going on. We shall not cover all this development (see for example [7.11] or the reviews [7.12]), but merely indicate a few aspects that will give the reader a "feel" for the Kondo effect.

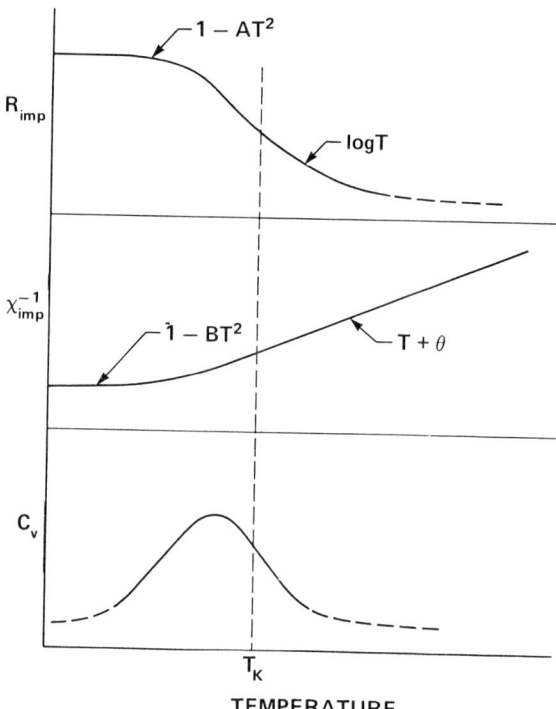

Fig. 7.10. Schematic behavior characteristic of a typical Kondo alloy

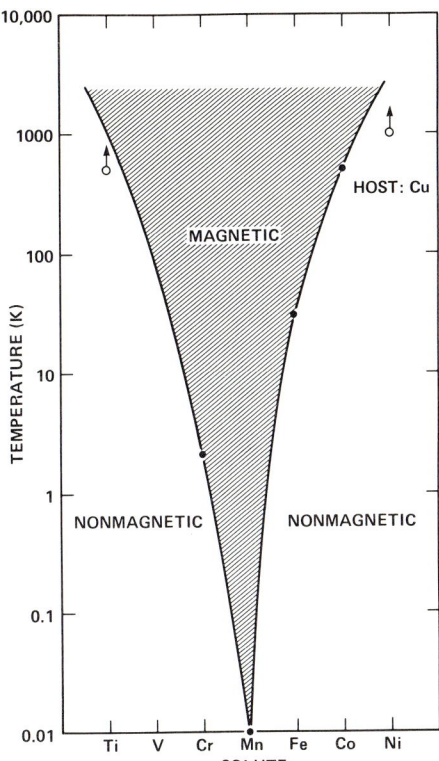

Fig. 7.11. Variation of the Kondo temperature for the $3d$ transition metals in copper

Let us begin by assuming we are in the magnetic regime of the Anderson model, that is $U \gg \Delta$. In this case *Schrieffer* and *Wolff* showed that one could perform a canonical transformation which eliminates V_{0k} to first order [7.13]. They found that the resulting Hamiltonian contains a contribution of the form

$$\mathcal{H}_{sd} = - \sum_{k,k'} J_{k'k}\, \psi_{k'}^\dagger\, s\psi_k \cdot \psi_0^\dagger\, S\psi_0 \tag{7.80}$$

where

$$\psi_k = \begin{bmatrix} c_{k\uparrow} \\ c_{k\downarrow} \end{bmatrix} \quad \text{and} \quad \psi_0 = \begin{bmatrix} c_{0\uparrow} \\ c_{0\downarrow} \end{bmatrix} \tag{7.81}$$

This has the form of an sd exchange interaction. For wave vectors near the Fermi surface the exchange parameter becomes

$$J_{k_F k_F} = 2|V_{0k_F}|^2 \frac{U}{\epsilon_0(\epsilon_0 + U)}. \tag{7.82}$$

In this expression ϵ_0 is measured in relation to the Fermi surface, which makes $J_{k_F k_F}$ negative, corresponding to antiferromagnetic coupling.

As we mentioned above, those alloys possessing local moments are also the ones which show a minimum in their resistivity as a function of temperature. This motivated *Kondo* [7.14] to calculate the spin-flip scattering amplitude of a conduction electron using the *s*–*d* interaction (7.80). Let us outline this calculation. We shall characterize the impurity spin by its spin-projection quantum number M_S and add an extra electron in the state $(\boldsymbol{k}, \sigma)$ to the Fermi sea. The state of the system at time $t = -\infty$ may then be written $c_{k\sigma}^\dagger |FS; M_S'\rangle$. In the absence of an external field the unperturbed Hamiltonian is just

$$\mathcal{H}_0 = \sum \epsilon_k c_{k\sigma}^\dagger c_{k\sigma}.$$

We now adiabatically turn on the interaction $\mathcal{H}_1$ and examine the amplitude for scattering into the state $c_{k'\sigma'}^\dagger |FS; M_S\rangle$. In the notation (5.92), this is

$$\langle k'\sigma'; M_S' | \phi(\infty) \rangle = \delta(\boldsymbol{k} - \boldsymbol{k}')\delta_{\sigma,\sigma'}\delta_{M_S, M_{S'}}$$
$$- 2\pi i \langle k'\sigma'; M_S' | \mathcal{H}_1 + \mathcal{H}_1 \frac{1}{\epsilon_k - \mathcal{H}_0} \mathcal{H}_1$$
$$+ \cdots |\boldsymbol{k}, \sigma; M_S\rangle \delta(\epsilon_k - \epsilon_{k'}). \tag{7.83}$$

Suppose $\mathcal{H}_1$ corresponded to ordinary potential scattering; then this would have the form $\sum c_{l'\sigma}^\dagger V_{l'l} c_{l\sigma}$.

The interesting term in (7.83) is the one involving $\mathcal{H}_1$ twice. There are two distinct ways in which the scattering $\boldsymbol{k} \to \boldsymbol{k}'$ can occur through this term. These are illustrated in Fig. 7.12. The matrix element for process (a) is

$$\langle FS | c_{k'} c_{k'}^\dagger V_{k'l} c_l \frac{1}{\epsilon_k - \mathcal{H}_0} c_l^\dagger V_{lk} c_k c_k^\dagger | FS \rangle \tag{7.84}$$

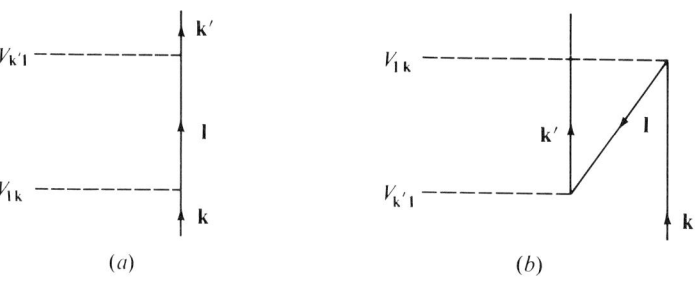

Fig. 7.12a, b. Contributions to the second-order scattering amplitude associated with a spin-independent potential

which reduces to

$$V_{k'l} \frac{1-f(\epsilon_l)}{\epsilon_k - \epsilon_l} V_{lk} . \tag{7.85}$$

Process (b) becomes

$$\langle FS| c_{k'} c_l^\dagger V_{lk} c_k \frac{1}{\epsilon_k - \mathcal{H}_0} c_{k'}^\dagger V_{k'l} c_l c_k^\dagger |FS\rangle \tag{7.86}$$

which contracts to

$$-V_{lk} \frac{f(\epsilon_l)}{\epsilon_k - (\epsilon_k - \epsilon_l + \epsilon_{k'})} V_{k'l} = V_{lk} \frac{f(\epsilon_l)}{\epsilon_k - \epsilon_l} V_{k'l} . \tag{7.87}$$

When these two processes are added together, the Fermi function cancels out. Thus the exclusion principle does not enter into ordinary potential scattering.

This is not the case, however, for a spin-dependent potential. The Hamiltonian (7.80) becomes

$$\mathcal{H}_1 = -J\mathbf{S} \cdot \sum_{l,l'} \sum_{\sigma,\sigma'} c_{l\sigma}^\dagger \mathbf{s}_{\sigma\sigma'} c_{l'\sigma'} . \tag{7.88}$$

The two second-order processes arising from this interaction are illustrated in Fig. 7.13 Process (a) gives

$$J^2 \langle FS; M_S' | S_\nu S_\mu c_{k'\sigma'} c_{k'\sigma'}^\dagger s_{\sigma'\sigma''}^\nu c_{l\sigma''} \frac{1}{\epsilon_k - \mathcal{H}_0} c_{l\sigma''}^\dagger s_{\sigma''\sigma}^\mu c_{k\sigma} c_{k\sigma}^\dagger |FS; M_S\rangle \tag{7.89}$$

which reduces to

$$J^2 \frac{1-f(\epsilon_l)}{\epsilon_k - \epsilon_l} \langle M_S' | S_\nu S_\mu s_{\sigma'\sigma''}^\nu s_{\sigma''\sigma}^\mu | M_S\rangle . \tag{7.90}$$

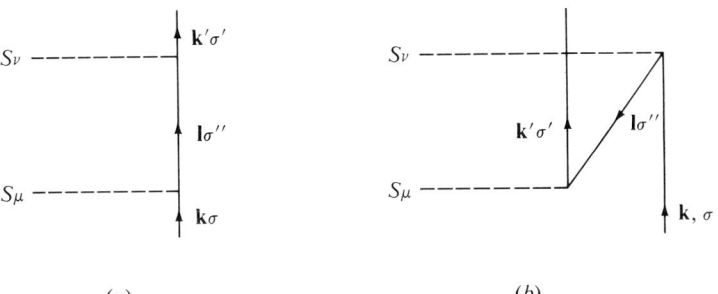

Fig. 7.13a, b. Contributions to the second-order scattering amplitude associated with the s–d interaction

Process (b) gives

$$J^2 \frac{f(\epsilon_l)}{\epsilon_k - \epsilon_l} \langle M'_S | S_\nu \, S_\mu \, s^\nu_{\sigma''\sigma} \, s^\mu_{\sigma'\sigma''} | M_S \rangle . \tag{7.91}$$

When these are added together, we obtain

$$\frac{J^2}{\epsilon_k - \epsilon_l} [\tfrac{1}{4} S(S+1) - \langle M'_S, \sigma' | \mathbf{S}\cdot\mathbf{s} | M_S, \sigma \rangle] + \frac{J^2 f(\epsilon_l)}{\epsilon_k - \epsilon_l} \langle M'_S, \sigma' | \mathbf{S}\cdot\mathbf{s} | M_S, \sigma \rangle . \tag{7.92}$$

The second term in (7.92) arises from the fact that the spin operators may not be interchanged. Therefore we obtain a contribution to the scattering which *does* depend on the exclusion principle. Summing over the intermediate states introduces the factor

$$\sum_l \frac{f(\epsilon_l)}{\epsilon_k - \epsilon_l} .$$

Introducing the conduction-electron density of states $N(\epsilon)$, this sum may be written

$$\int N(\epsilon') \frac{f(\epsilon')}{\epsilon - \epsilon'} d\epsilon' . \tag{7.93}$$

Assuming a simple rectangular band,

$$N(\epsilon) = \begin{cases} N(0) & -D < \epsilon < D \\ 0 & \text{otherwise,} \end{cases} \tag{7.94}$$

the integral for $\epsilon = 0$ gives

$$-N(0) \ln \frac{k_B T}{D} + \text{const} . \tag{7.95}$$

Since the resistivity involves the square of the scattering amplitude, the cross term between the first-order term and this second-order term gives a contribution proportional to $J^3 \ln T$.

For antiferromagnetic exchange ($J < 0$) the resistivity increases with decreasing temperature as shown in Fig. 7.10. The divergence as $T \to 0$, however, indicates that the perturbation theory is not valid in this regime. This remarkable result initiated more than a decade of intense activity. Experimental studies established the systematics of this phenomenon while theoretical work explored the properties of the s–d Hamiltonian as well as the Anderson Hamiltonian.

The first attempts to improve on Kondo's calculation led to a scattering amplitude that diverged at a finite temperature

$$T_K = D \exp[1/2JN(E_F)], \tag{7.96}$$

7.3 The Kondo Effect

which provides a quantitative expression for the Kondo temperature. Although the divergence indicated that the problem was still not being treated properly, it did indicate the formation of a resonance in the scattering amplitude at the Fermi level. That is, the Kondo effect is basically an impurity-induced Fermi surface instability. A great deal of effort has gone into identifying the nature of the ground state. From the very beginning it was suspected that it must be a singlet. However, it has only been with the recent exact diagonalizations of the *s–d* Hamiltonian by *Andrei* [7.15] and *Wiegmann* [7.16] that this singlet nature has been confirmed. *Yosida* and collaborators [7.17] have shown that if one assumes a singlet wave function of the form

$$\psi_{\text{Kondo}} = \frac{1}{\sqrt{2}} (\chi_\uparrow \psi_\downarrow - \chi_\downarrow \psi_\uparrow) \tag{7.97}$$

where χ_σ is the wave function of the impurity spin, then the conduction-electron

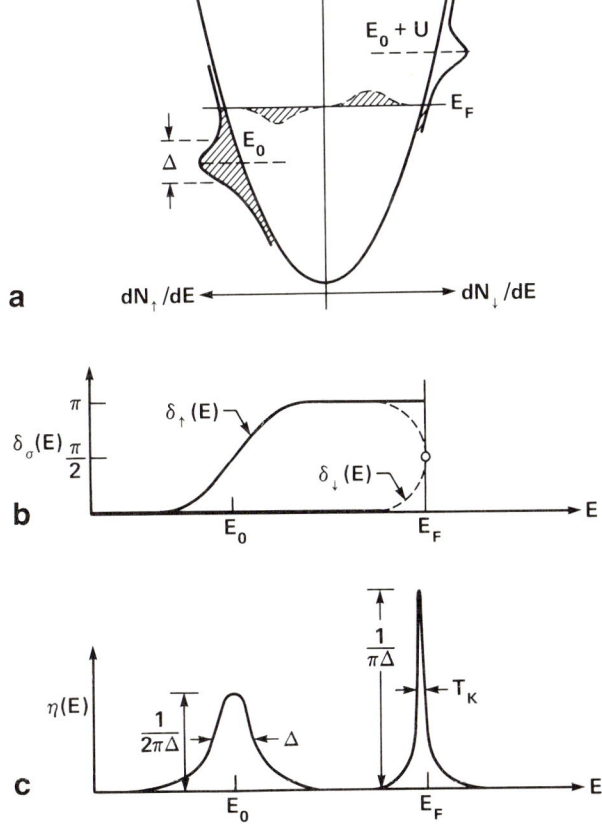

Fig. 7.14. (a) Schematic illustration of how the Hartree–Fock solution is augmented by the appearenace of a resonance at the Fermi level consisting of half a spin-down electron and half a spin-up hole. (b) Variation of the phase shift with energy. (c) Variation in the local density of states

component $\psi_\uparrow$ consists of half an electron with down spin and half a hole with up spin bound to the impurity. This is illustrated schematically in Fig. 7.14.

Anderson [7.18] also considered the implications of a singlet ground state on the conduction-electron phase shifts. In the Hartree–Fock approximation discussed above, in the case of a well-localized moment $\delta_\uparrow = \pi$ and $\delta_\downarrow = 0$. However, if we do have a singlet ground state there is rotational invariance and the phase shift must be independent of σ. *Anderson* postulated that in the vicinity of the Fermi level the phase shifts readjust as indicated in Fig. 7.14. The appearance of $\pi/2$ phase shifts at the Fermi level implies a resonance at the Fermi level. This "Abrikosov–Suhl" resonance, as it is sometimes called, is also shown in Fig. 7.14. Such a peak in the density of states has also been derived [7.19] directly from the *Anderson Hamiltonian*. This result shows that the peak has a width $k_B T_K$ and disappears at the Kondo temperature. If only electrons within a characteristic energy $k_B T_K$ of the Fermi level are involved, this implies a spatial extent of

$$\xi_K = \frac{v_F}{k_B T_K}. \tag{7.98}$$

Nozières [7.20] also noted that if the system is in a singlet ground state, then Fermi liquid theory could be used to obtain relations among various physical properties. That is, in the Kondo state we consider the electrons as having been renormalized into quasiparticles. As we argued in Chap. 4, the physical properties will be functionals of the quasiparticle distribution function n_k. For this impurity configuration it is appropriate to characterize each quasiparticle by its phase shifts. Assuming only s-wave scattering, in analogy with (4.65), we have

$$\begin{aligned}\delta_\sigma(\varepsilon) &= \delta_0(\varepsilon) + \sum_{\varepsilon',\sigma'} \phi_{\sigma\sigma'}(\varepsilon,\varepsilon')\,\delta n_{\sigma'}(\varepsilon') + \ldots \\ &= \delta_0 + \alpha\varepsilon + \phi_{\sigma\sigma}\sum_{\varepsilon'} \delta n_\sigma(\varepsilon') + \phi_{\sigma,-\sigma}\sum_{\varepsilon} \delta n_{-\sigma}(\varepsilon') + \ldots. \end{aligned} \tag{7.99}$$

Defining $\phi_{\sigma,\sigma} \pm \phi_{\sigma,-\sigma} = 2\phi_{s,a}$ where s refers to the symmetric and a to the antisymmetric combinations, $\delta n = \delta n_\uparrow + \delta n_\downarrow$, and $m = \delta n_\uparrow - \delta n_\downarrow$, we obtain

$$\delta_\sigma(\varepsilon) = \delta_0 + \alpha\varepsilon + \phi_s \delta n + \phi_a m \sigma. \tag{7.100}$$

Notice that if the phase shift is close to $\pi/2$, then from an s-wave form similar to that given by (7.53) we see that α is the reciprocal width of the resonance, which in this case is just $1/k_B T_K$.

From (7.56) the change in the density of states is given by

$$\eta_0 = \frac{1}{\pi}\frac{d\delta_\sigma}{d\varepsilon} = \frac{\alpha}{\pi}, \tag{7.101}$$

which corresponds to a change in the electronic specific heat of

$$\frac{C_{\text{imp}}}{C} = \frac{\alpha}{\pi N_0(\epsilon_F)}. \tag{7.102}$$

To find the magnetic susceptibility we note that when a field is applied the difference in phase shifts becomes

$$\delta_\uparrow - \delta_\downarrow = 2\phi_a m + 2g\mu_B H \pi N_0(\epsilon_F). \tag{7.103}$$

But since $\delta_\sigma = \pi \langle n_\sigma \rangle$, and $\delta_\uparrow - \delta_\downarrow = \pi m$

$$\chi = \frac{g\mu_B m}{H} = 2N_0(\epsilon_F)g^2\mu_B^2 \frac{1 + \alpha/\pi N_0(\epsilon_F)}{1 - 2\phi_a/\pi}. \tag{7.104}$$

Notice that this has the same form as our result (4.80) for a homogeneous system. *Nozières* now made two observations that enabled him to relate the change in specific heat to the change in susceptibility. First, because the Abrikosov–Suhl resonance is tied to the Fermi surface, if both ε and the chemical potential μ are shifted by the same amount the phase shift is unchanged. From (7.100) and $\delta n = 2N_0(\epsilon_F)\Delta\varepsilon$ this means

$$\alpha + 2N_0(\epsilon_F)\phi^s = 0. \tag{7.105}$$

Secondly, *Nozières* assumed that in the Kondo state only antiparallel spins are coupled. Thus $\phi_{\sigma\sigma} = 0$ and $\phi^s = \phi^a$. Combining this with (7.105) gives $2N_0(\epsilon_F)\phi^a/\alpha = 1$. Therefore

$$\frac{\chi_{\text{imp}}/\chi}{C_{\text{imp}}/C} = 2, \tag{7.106}$$

which says that in the Kondo state the susceptibility is enhanced *twice* as much as the specific heat. This result was first obtained numerically by *Wilson* [7.21]. For a noninteracting system this ratio is 1.

7.4 Random Exchange

The Kondo effect refers to the behavior of a single impurity. As the impurity concentration increases, the magnetic interactions between impurities become important. Since this interaction is mediated by the conduction electrons, it extends beyond nearest neighbors. As a result, interaction effects begin to appear for concentrations typically of the order of 50 parts per million.

7.4.1. The RKKY Interaction

As we saw in Sect 2.2, there is a contact hyperfine interaction between *s*-state electrons and nuclear moments. *Fröhlich* and *Nabarro* were the first to suggest that this interaction could lead to a polarization of the nuclear moments [7.22]. The actual form of this interaction, however, was obtained by *Ruderman* and *Kittel* [7.23]. In analogy with this nuclear coupling, the exchange interaction between conduction electrons and localized electrons can also lead to indirect coupling between localized electronic moments. *Zener* proposed that this was the origin of ferromagnetism in transition metals *Kasuya* investigated this interaction in more detail, particularly with respect to its effect on spin waves and electrical resistivity [7.25]. *Yosida* also employed this interaction to explain the magnetic properties of **Cu-Mn** alloys [7.26]. As a result of these developments, the indirect coupling of magnetic moments by conduction electrons is referred to as the *Ruderman-Kittel-Kasuya-Yosida* (RKKY) *interaction*.

The form of the RKKY interaction is easily obtained within the framework of our generalized susceptibility. Let us assume that the interaction between a localized spin S_α located at $r = 0$ and the conduction spins s_i has the form

$$- J \sum_i S_\alpha \cdot s_i \delta(r_i) \,.$$

Each conduction spin therefore experiences an effective field given by

$$H_{\text{eff}}(r) = - \frac{J}{g\mu_B} S_\alpha \delta(r) \,. \tag{7.107}$$

The response of an electron gas to such a field is determined by its susceptibility $\chi(q)$. Since the Fourier transform of this field is

$$H_{\text{eff}}(q) = - \frac{J}{g\mu_B} S_\alpha \,, \tag{7.108}$$

the spin density at r is

$$s(r) = \frac{J}{g^2 \mu_B^2 V} \sum_q \chi(q) e^{iq \cdot r} S_\alpha \,. \tag{7.109}$$

For a free-electron gas $\chi(q)$ is given by (3.85). The sum over q is evaluated by converting it into an integral. This integral is easily evaluated by using the integral representation [7.27]

$$\ln \left| \frac{2k_F + q}{2k_F - q} \right| = 2 \int_0^\infty \frac{dx \, \sin(2k_F x) \sin(qx)}{x} \,. \tag{7.110}$$

The result is

$$\frac{1}{V} \sum_q \chi(q) e^{iq \cdot r} = \frac{3g^2 \mu_B^2 (N/V)}{8\epsilon_F} \frac{1}{2\pi^2 r} \int dq \, q F\left(\frac{q}{2k_F}\right) \sin qr$$

$$= \frac{3g^2 \mu_B^2 (N/V)}{8\epsilon_F} \frac{k_F^3}{16\pi} \left[\frac{\sin 2k_F r - 2k_F r \cos 2k_F r}{(k_F r)^4} \right] \,. \tag{7.111}$$

The part of this expression within the brackets is plotted in Fig. 7.15.

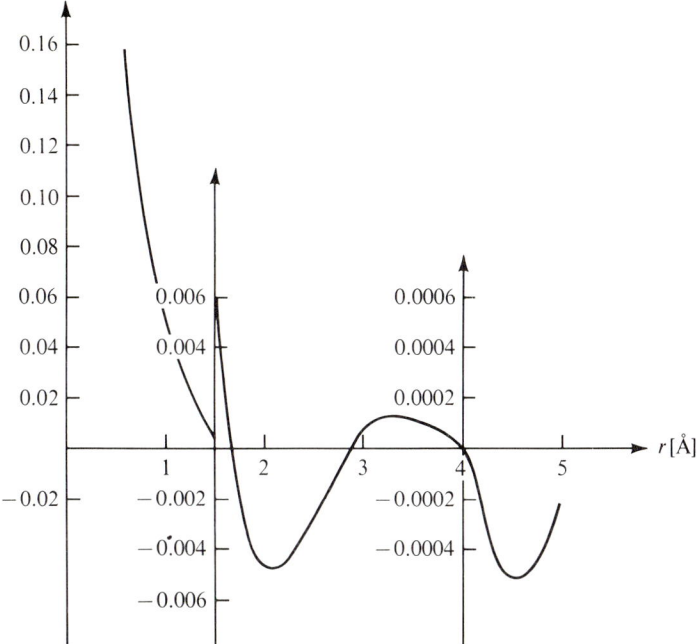

Fig. 7.15. Plot of the conduction-electron spin density produced by an impurity spin in a metallic host

Thus we see that when a localized moment is introduced into a metal, the conduction spins develop an oscillating polarization in the vicinity of this moment. These spin-density oscillations have the same form as the Friedel charge-density oscillations that result when an electron gas screens out a charge impurity.

The most direct evidence for the oscillatory nature of this interaction comes from NMR studies. In **CuFe**, for example, the nuclear resonance spectrum consists of the strong Knight-shifted Cu^{63} line found in pure Cu plus various satellites corresponding to Cu nuclei which lie within the RKKY oscillations around an Fe impurity. These Cu nuclei feel a contact hyperfine field $\delta H(r,T)$ $= -\frac{8}{3}\pi g\mu_B \sigma(r,T)$ due to the conduction-electron spin polarization

$$\sigma(r,T) = \chi_{imp}(T) H f(r),$$

where $\chi_{imp}(T)$ is the susceptibility of a single Fe impurity and $f(r)$ characterizes the RKKY spatial dependence. An external field polarizes the impurity which in turn increases $\sigma(r,T)$. Thus the satellites should move away from the main line with increasing field with a slope that depends upon $f(r)$. Figure 7.16 shows the field dependence of five such satellites [7.28]. The appearance of both positive and negative slopes results from the oscillatory nature of $f(r)$.

242   7. Magnetic Impurities

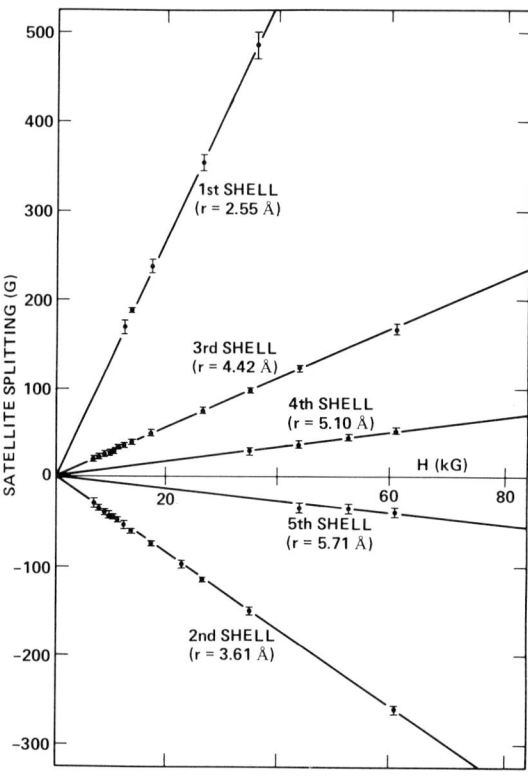

Fig. 7.16. Magnetic field dependence of satellite separations from the main Cu^{63} resonance at 300 K

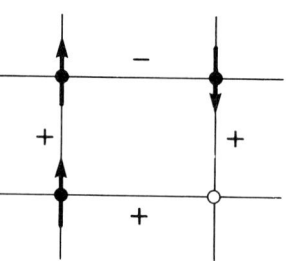

Fig. 7.17. Simple example of frustration. The signs refer to the exchanga. Thus, if the three spins are aligned as indicated, the fourth will be frustrated

If there is another localized spin S_β at r, it interacts with this induced spin density, leading to an effective interaction between the localized spins of the form

$$\mathscr{H}_{\text{RKKY}} = -\frac{J^2}{g^2\mu_B^2 V}\sum_q \chi(q)e^{iq\cdot r}S_\alpha \cdot S_\beta \tag{7.112}$$

This interaction manifests itself in many ways. *Ruderman* and *Kittel* [7.23] showed that it leads to a broadening of the nuclear magnetic resonance absorption. This interaction is also the origin of the exchange coupling between the localized moments in the rare-earth metals. As we saw in Chap. 4, its oscillatory nature leads to helimagnetism in these materials.

7.4.2 Spin Glasses

The fact that the exchange between local moments in a metal can be positive (ferromagnetic) or negative (antiferromagnetic) also means that some spins may find themselves in configurations where it is not possible to satisfy all the exchange interactions. Such a situation is illustrated in Fig. 7.17 and is referred to as "frustration." The first model of a spin system incorporating frustration was suggested by *Edwards* and *Anderson* [7.29]. Their model consists of spins on every site of a regular lattice interacting with z neighbors through an exchange J_{ij}, which is random and governed by a probability distribution $P(J_{ij})$. To see the qualitative implications of this model, consider the mean-field equation (4.38) for the thermal average of a spin $\frac{1}{2}$,

$$\langle S_i \rangle = \tanh \left(\sum_j J_{ij} \langle S_j \rangle / k_B T \right). \tag{7.113}$$

Linearizing gives

$$\langle S_i \rangle = \sum_j J_{ij} \langle S_j \rangle / k_B T. \tag{7.114}$$

Averaging over J_{ij} with the assumption that $\bar{J}_{ij} = 0$ leads to the result that $\overline{\langle S_i \rangle} = 0$. However, if we square (7.114) and again average over the exchange values, we obtain

$$\overline{\langle S_i \rangle^2} = \frac{z}{(k_B T)^2} \overline{J_{ij}^2} \, \overline{\langle S_j \rangle^2}. \tag{7.115}$$

This means that $\overline{\langle S_i \rangle^2}$ becomes nonzero at a critical temperature $T_f = z \overline{J_{ij}^2}/k_B$ and is identified as the order parameter q. The fact that $\overline{\langle S_i \rangle^2} \neq 0$, while $\overline{\langle S_i \rangle} = 0$, implies a spin configuration in which each spin is frozen in some specific direction, but this direction varies randomly throughout the alloy. This is generally what is meant by a "spin glass." The mean-field solution to the Edwards–Anderson model as well as to *Sherrington* and *Kirkpatrick*'s infinite-range, random-exchange, Ising model [7.30] gives a cusp in the low-field ac susceptibility as observed experimentally (Fig. 7.18). However, both theories also predict a cusp in the specific heat at T_f which is *not* observed, as Fig. 7.19 shows. This has raised the question of whether a spin glass has a real phase transition or just a gradual freezing.

Part of the difficulty lies in the technique used to compute thermodynamical quantities for these models. The free energy associated with a particular exchange configuration $\{J_{ij}\}$ is $-k_B T \ln Z\{J_{ij}\}$. The average with respect to $P(J_{ij})$ is then

$$F = -k_B T \int \prod_{ij} P(J_{ij}) \ln Z\{J_{ij}\} \, dJ_{ij}.$$

To simplify averaging over the logarithm, *Edwards* and *Anderson* introduced a technique based on the identity

244   7. Magnetic Impurities

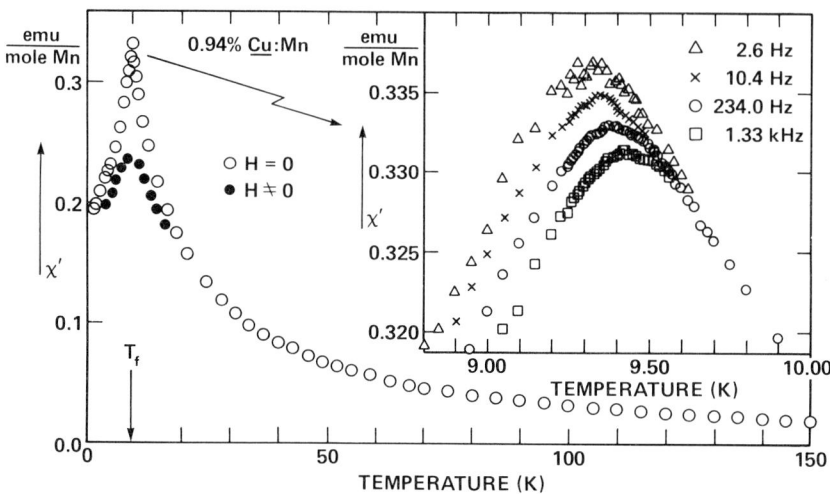

Fig. 7.18. ac susceptibility as a function of temperature for **Cu**: Mn

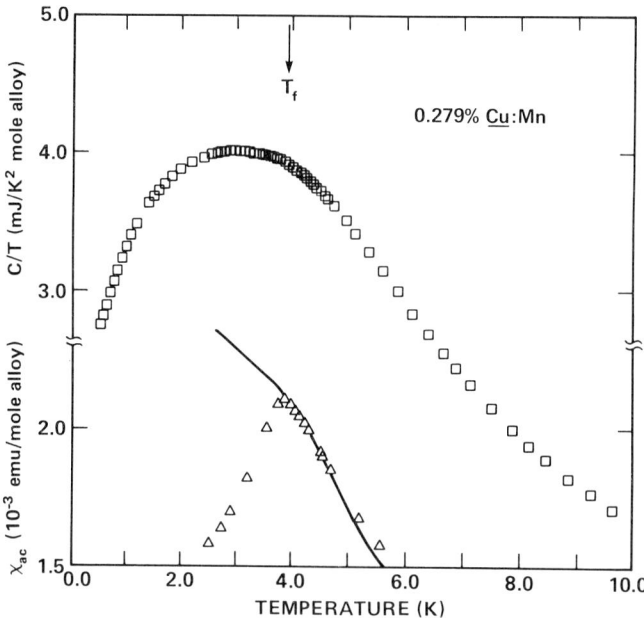

Fig. 7.19. Specific heat (top) and ac susceptibility as a function of temperature for a **Cu**:Mn alloy

$$\ln Z = \lim_{n \to 0} \frac{1}{n}(Z^n - 1),$$

where Z^n corresponds to the partition function for n identical "replicas" of the original system. The average over the exchange is now performed *before* the thermal average over the spins. In doing so, one encounters correlation functions such as $\langle S_i^\alpha S_i^\beta \rangle_n = \langle Q_{\alpha\beta} \rangle_n$ involving spins in *different* replicas. Since the replicas are merely artifacts introduced to facilitate the calculation, one presumes that these correlation functions should all be equal to the Edwards–Anderson order parameter q in the limit $n \to 0$. *De Almeida* and *Thouless* [7.31], however, have shown that this assumption is too restrictive and suggested that the true solution involves "replica symmetry breaking." But there are infinitely many ways of breaking this symmetry, and the replica theory provides neither a clear criterion of which way is correct nor a physical insight to the meaning of the various order parameters and the broken symmetry. The dynamic properties of the Edwards-Anderson model are currently being explored in the hope that they will lead to a better understanding of magnetic systems with random exchange [7.32].

Another difficulty in comparing experimental results with random exchange models is that the data seem to suggest that the actual systems may be much more complex. The ac susceptibility, for example is slightly frequency dependent, as shown in the insert in Fig. 7.18. The dc susceptibility is also marked by time-dependent and irreversible behavior. Figure 7.20 shows the dc susceptibility of Mn in Ag. If the sample is cooled in zero magnetic field (which involves cancellation of the earth's field) and then the susceptibility measured with increasing temperature, one obtains the solid curve. On the other hand, if one cools the sample in the presence of the field which is used to measure the susceptibility, then one obtains the larger values indicated by the dashed lines. These field-cooled susceptibilities are reversible whereas the zero-field-cooled curve is not. Furthermore, if one sits at a given temperature $T < T_f$ on the zero-field-cooled curve, the magnetization increases logarithmically with time, eventually approaching the field-cooled value. This behavior is believed to be associated with the existence of a large number of low-lying states connected by barriers. These barriers restrict the phase space available to the system and the low temperature magnetic properties are governed by the shape of these wells.

7.4.3 Mictomagnetism

Time-dependent and irreversible behavior is also observed in more concentrated alloys. Since the susceptibility resembles that of an antiferromagnet, while the remanence after field cooling is characteristic of a ferromagnet, *Beck* [7.33] termed such alloys "mictomagnets" from the Greek "miktos," meaning mixed. Such mictomagnetic behavior is thought to arise from ferromagnetic clusters. Consider a region of the sample consisting of strongly correlated spins having

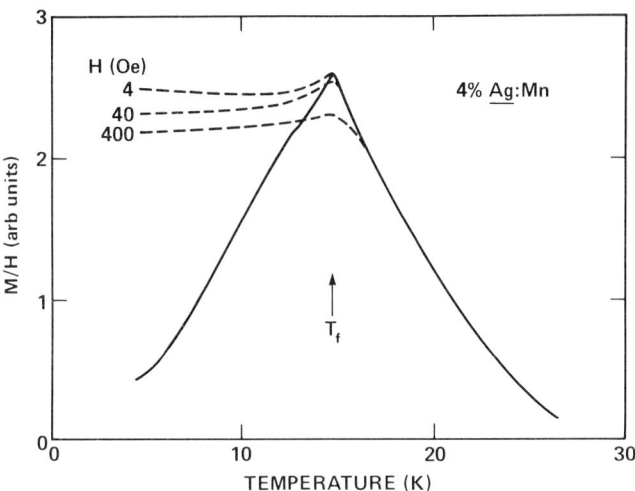

Fig. 7.20. Zero-field-cooled magnetization as a function of temperature for Ag:Mn

a net magnetic moment μ. Let us assume that the orientation of this moment is governed by a uniaxial anisotropy energy $-\mu H_A \cos^2\alpha$, where α is the angle between the moment and the easy axis. The moment is stable for $\alpha = 0$ or π, these two directions being separated by an energy barrier of height μH_A. If the cluster rotates coherently, that is, if all the spins within the cluster maintain their relative orientations during the rotation, the moment will fluctuate with a thermally activated, or Arrhenius, rate

$$\frac{1}{\tau} \sim \exp(-\mu H_A/k_B T) .$$

When $\mu H_A \sim k_B T$, the cluster will behave paramagnetically, a situation called "superparamagnetism" [7.34]. *Schwink* et al. [7.35] have developed a model for concentrated alloys, which consists of superparamagnetic clusters all having the same moment and the same anisotropy field, but whose easy axes are randomly distributed. This model describes much of the experimentally observed $M(t, H, T)$ behavior in systems such as **CuMn** and **AuFe**

As the concentration increases, clustering becomes more pronounced and we eventually reach a point, called the percolation threshold, where each magnetic site has at least one magnetic nearest neighbor, so that long-range order can extend throughout the material. At any given concentration, the degree of atomic clustering will also depend upon sample preparation. *Crane* and *Claus* [7.36], for example, have investigated the effect of annealing on **Au**:Fe. It is known that the reciprocal of the annealing temperature, $1/T_A$, is proportional to the degree of atomic clustering. Figure 7.21 shows the magnetic phase diagram for a 14% alloy. What is particularly interesting is the transition from ferromagnetism to a state with no spontaneous magnetization at low temperatures. This suggests that the ferromagnetic state itself must possess a great deal of randomness.

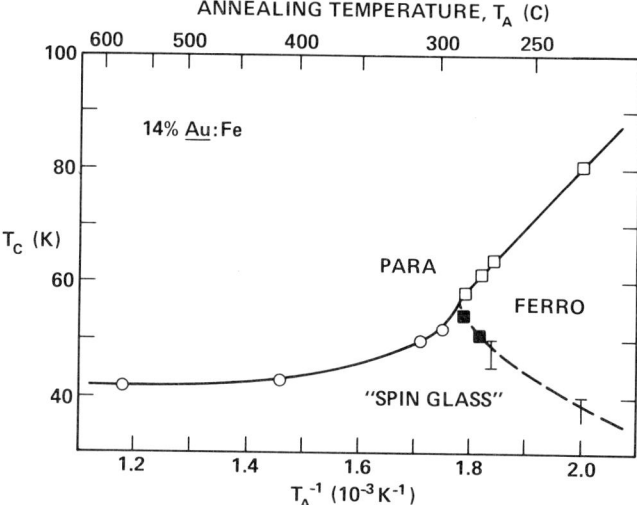

Fig. 7.21. Magnetic phase diagram: T_C vs the reciprocal of the annealing temperature T_A (in degrees K). Open squares paramagnetic to ferromagnetic transition; open circles paramagnetic to spin glass; solid squares ferromagnetic to spin glass; data points with error bars, estimated from ac susceptibility

8. Neutron Scattering

The scattering of thermal neutrons provides us with an exceedingly powerful probe for studying condensed matter. This usefulness arises from the fact that thermal (300 K) neutrons have a wavelength $\lambda \simeq 1.6$ Å. This means that their energy and wavelength are comparable to those of the excitations in solids. In fact, neutron scattering has become almost the last word in determining spin orderings, spin-density distributions, and spin–wave dispersion relations. A detailed discussion of the experimental techniques employed in neutron scattering, with reference to the original papers, may be found in [8.1]

8.1 Neutron Scattering Cross Section

Basically, the neutron experiments consist of sending a collimated beam of monochromatic neutrons into a sample and measuring the energy spectrum of the neutrons scattered in some direction. Thus it is the differential scattering cross section that is measured. For the rest of this section we shall be concerned with how the scattering cross section depends on the state of the crystal [8.2,3].

If the neutrons interact with the scatterer through some interaction $V(r)$, then the probability that the neutron will be scattered from an initial state (k, m_s) into a final state (k', m'_s) while the system goes from state α to α' is, in the Born approximation,

$$\frac{2\pi}{\hbar} \left| \langle \alpha', m'_s | \frac{1}{L^3} \int d\boldsymbol{r}\, e^{i\boldsymbol{\kappa}\cdot\boldsymbol{r}} V(\boldsymbol{r}) | \alpha, m_s \rangle \right|^2 \delta\left(\frac{\hbar^2 k'^2}{2m_0} + E_{\alpha'} - \frac{\hbar^2 k^2}{2m_0} - E_\alpha\right), \qquad (8.1)$$

where $\boldsymbol{\kappa} \equiv \boldsymbol{k} - \boldsymbol{k}'$, L^3 is the volume of the sample, and m_0 is the neutron mass. The number of neutrons scattered into the solid angle $d\Omega'$ in the direction $\boldsymbol{k}'$ with momentum between $\hbar k'$ and $\hbar(k' + dk')$ is obtained by multiplying (8.1) by the number of states in this incremental volume of phase space, which is

$$\frac{k'^2\, dk'\, d\Omega'}{(2\pi/L)^3}. \qquad (8.2)$$

Since the neutron energy and momentum are related by $E' = \hbar^2 k'^2/2m_0$, this becomes

$$\frac{m_0 L^3 k'}{8\pi^3 \hbar^2} d\Omega' dE' . \tag{8.3}$$

The *differential scattering cross section* is defined as the scattering intensity per unit solid angle per unit incident flux. The incident flux associated with a beam of neutrons with velocity v is

$$\frac{v}{L^3} = \frac{\hbar k}{m_0 L^3} . \tag{8.4}$$

Therefore the differential scattering cross section becomes

$$\frac{d^2 \sigma}{d\Omega' dE'}\bigg|_{\alpha, m_s \to \alpha', m_s'} = \frac{k'}{k}\left(\frac{m_0}{2\pi\hbar^2}\right)^2 |\langle \alpha' m_s'| V(-\boldsymbol{\kappa})|\alpha, m_s\rangle|^2$$
$$\times \delta\left(\frac{\hbar^2 k'^2}{2m_0} + E_{\alpha'} - \frac{\hbar^2 k^2}{2m_0} - E_\alpha\right). \tag{8.5}$$

We must now sum over the final states α' and m_s' as well as average over the initial states α, and m_s, using a probability factor P_α. This factor might have the Boltzmann form

$$P_\alpha = \frac{\exp(-E_\alpha/k_\text{B}T)}{\sum_\alpha \exp(-E_\alpha/k_\text{B}T)} \tag{8.6}$$

or some other form such as the Bose–Einstein form in the case of phonons. Thus

$$\frac{d^2 \sigma}{d\Omega' dE'} = \sum_{\alpha, m_s} P_\alpha P_{m_s} \sum_{\alpha', m_s'} \frac{k'}{k}\left(\frac{m_0}{2\pi\hbar^2}\right)^2 |\langle \alpha', m_s'| V(-\boldsymbol{\kappa})|\alpha, m_s\rangle|^2$$
$$\times \delta\left(\frac{\hbar^2 k'^2}{2m_0} + E_{\alpha'} - \frac{\hbar^2 k^2}{2m_0} - E_\alpha\right). \tag{8.7}$$

This is the general result which forms the basis of our discussion of neutron scattering.

The neutron interacts with the crystal in two ways: through the nuclear interaction and through the magnetic dipole–dipole interaction.

8.2 Nuclear Scattering

Although our main interest will be in magnetic scattering, we shall briefly consider the nuclear contribution to see how it complicates the experimental results. Let us begin by considering the nuclear scattering from a *single* fixed zero-spin nucleus located at R. Since the interaction has such a short range, the scattering will be isotropic at these neutron energies. This is because at short

250   8. Neutron Scattering

distances from the nucleus the neutron's angular momentum is very small. Hence we have only *s*-wave scattering. The only form of $V(r)$ which gives isotropic scattering in the Born approximation is a delta function.

Thus we set

$$V(r) = \frac{2\pi\hbar^2}{m_0} a\delta(\mathbf{r} - \mathbf{R}) \tag{8.8}$$

where a is called the *scattering length*. It represents a phenomenological description of the nuclear scattering process which is very difficult to compute directly. Equation (8.8) is called the *Fermi pseudopotential* and gives a total cross section

$$\sigma = 4\pi |a|^2 . \tag{8.9}$$

The scattering length is taken to be a constant, independent of the neutron energy. This is valid at the low neutron energies used.

Now consider scattering from *many* fixed nuclei situated at lattice sites $\mathbf{n}$. If we have a compound or if there are isotopes present, the scattering amplitudes of the different nuclei may be different. In this case the total pseudopotential is

$$V(r) = \frac{2\pi\hbar^2}{m_0} \sum_n a_n \delta(\mathbf{r} - \mathbf{n}) . \tag{8.10}$$

This leads to the differential cross section

$$\frac{d^2\sigma}{d\Omega' dE'} = \sum_{\alpha, m_s} P_\alpha P_{m_s} \sum_{\alpha', m_{s'}} \frac{k'}{k} \left| \langle \alpha', m_s' | \sum_n a_n e^{i\boldsymbol{\kappa}\cdot\mathbf{n}} | \alpha, m_s \rangle \right|^2 \delta(\text{energy}) . \tag{8.11}$$

Here, α refers to all the quantum numbers necessary to describe the crystal. In particular, α and α' might refer to the isotope distribution and the nuclear-spin orientations. To a good approximation the energy is independent of these quantities. Thus,

$$\begin{aligned}\frac{d\sigma}{d\Omega'} &= \sum_{\alpha, m_s} P_\alpha P_{m_s} \sum_{n,m} \exp[i\boldsymbol{\kappa}\cdot(\mathbf{n}-\mathbf{m})] \langle \alpha, m_s | a_m^* a_n | \alpha, m_s \rangle \\ &= \sum_{n,m} \exp[i\boldsymbol{\kappa}\cdot(\mathbf{n}-\mathbf{m})] \langle a_m^* a_n \rangle .\end{aligned} \tag{8.12}$$

Since there is no correlation between a_n and a_m^* for $\mathbf{n} \neq \mathbf{m}$,

$$\langle a_m^* a_n \rangle = \langle a_m^* \rangle \langle a_n \rangle = |\langle a \rangle|^2 .$$

Therefore we write

$$\langle a_m^* a_n \rangle = |\langle a \rangle|^2 + (\langle |a|^2 \rangle - |\langle a \rangle|^2)\Delta(\mathbf{n} - \mathbf{m}) . \tag{8.13}$$

This enables us to separate the cross section into two parts,

$$\frac{d\sigma}{d\Omega'} = \left(\frac{d\sigma}{d\Omega'}\right)_{coh} + \left(\frac{d\sigma}{d\Omega'}\right)_{incoh}, \tag{8.14}$$

where the *coherent cross section* is

$$\left(\frac{d\sigma}{d\Omega'}\right)_{coh} = |\langle a \rangle|^2 |\sum_n e^{i\kappa \cdot n}|^2 \tag{8.15}$$

and the *incoherent cross section* is

$$\left(\frac{d\sigma}{d\Omega'}\right)_{incoh} = N(\langle |a|^2 \rangle - |\langle a \rangle|^2). \tag{8.16}$$

We see that only the mean scattering potential $\langle a \rangle$ gives rise to interference effects and coherent scattering, while the incoherent scattering is proportional to the mean-square deviation $\langle |a - \langle a \rangle|^2 \rangle$. If the nucleus possesses a nonzero spin, then the nuclear scattering amplitude will depend upon the relative orientation of this spin with the neutron spin. Since these nuclear spins are disordered, they lead to a large incoherent scattering. This is beautifully illustrated in Fig. 8.1, where the cross section for thorium hydride is compared with that of thorium deuteride. Since the deuterium has no nuclear spin, the spin-disorder scattering does not contribute to the incoherent scattering.

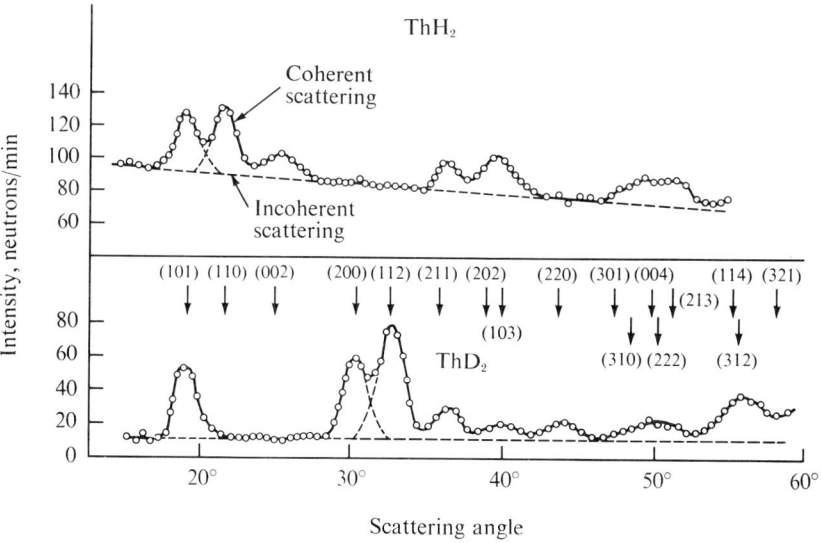

Fig. 8.1. Neutron diffraction patterns for polycrystalline ThH$_2$ and ThD$_2$ [8.4]

8.2.1 Bragg Scattering

The coherent elastic scattering from a rigid lattice is also referred to as *Bragg scattering*. This involves the quantity

$$\sum_n e^{i\boldsymbol{\kappa}\cdot\boldsymbol{n}}$$

which contains N terms. It can be shown that

$$\left|\sum_n e^{i\boldsymbol{\kappa}\cdot\boldsymbol{n}}\right|^2 = \frac{(2\pi)^2 N}{V_0}\sum_G \delta(\boldsymbol{\kappa}-\boldsymbol{G}), \tag{8.17}$$

where V_0 is the volume of the unit cell and $\boldsymbol{G}$ is a reciprocal lattice vector. This is an important result which we shall encounter many times. The coherent, or Bragg, cross section then becomes

$$\left(\frac{d\sigma}{d\Omega'}\right)_{\text{coh}} = |\langle a\rangle|^2 \frac{(2\pi)^3 N}{V_0}\sum_G \delta(\boldsymbol{\kappa}-\boldsymbol{G}). \tag{8.18}$$

If the lattice has more than one atom per unit cell, say r atoms at $\boldsymbol{\rho}_1, \boldsymbol{\rho}_2, \ldots$, relative to some reference point in the unit cell, then

$$\left(\frac{d\sigma}{d\Omega'}\right)_{\text{coh}} = \frac{(2\pi)^3 N}{V_0}\sum_G |F(\boldsymbol{G})|^2 \delta(\boldsymbol{\kappa}-\boldsymbol{G}), \tag{8.19}$$

where

$$F(\boldsymbol{G}) = \sum_{i=1}^{r} \langle a_i\rangle \exp(i\boldsymbol{G}\cdot\boldsymbol{\rho}_i)$$

is the unit cell *structure factor*.

The important point is that there is no coherent scattering unless

$$\boldsymbol{\kappa} = \boldsymbol{k} - \boldsymbol{k}' = \boldsymbol{G}. \tag{8.20}$$

By squaring this relation and making use of the fact that the magnitude of the reciprocal lattice vector in some direction hkl is equal to some multiple of 2π times the inverse spacing of the planes in that direction, $d(\text{hkl})$, we obtain the familiar Bragg law

$$\sin\theta = \frac{n\lambda}{2d(\text{hkl})} \tag{8.21}$$

for the angle 2θ between the incident beam and the diffracted beam.

8.2.2 Scattering by Phonons

Let us now consider the cross section when we allow the ions to deviate from their rigid lattice position n by an amount u_n. The general position is then

$$R_n = n + u_n, \qquad (8.22)$$

and the cross section becomes

$$\frac{d^2\sigma}{d\Omega' dE'} = \sum_{\alpha, m_s} P_\alpha P_{m_s} \sum_{\alpha', m_s'} \frac{k'}{k}$$
$$\times |\langle \alpha', m_s'| \sum_n a_n \exp(i\boldsymbol{\kappa} \cdot \boldsymbol{n}) \exp(i\boldsymbol{\kappa} \cdot \boldsymbol{u}_n)|\alpha, m_s\rangle|^2 \delta(\text{energy}). \qquad (8.23)$$

The system coordinates α and α' now refer to phonon states as well as to the nuclear spin projections m_I and m_I'. Writing out the latter explicitly, we have for the matrix element above

$$\sum_n \langle m_s', m_I'|a_n|m_s, m_I\rangle \exp(i\boldsymbol{\kappa} \cdot \boldsymbol{n})\langle \alpha'|\exp(i\boldsymbol{\kappa} \cdot \boldsymbol{u}_n)|\alpha\rangle. \qquad (8.24)$$

The ionic displacement u_n may be expanded in boson operators corresponding to the creation and annihilation of phonon as [8.5].

$$\boldsymbol{u}_n = \sum_{q,s} \sqrt{\frac{\hbar}{2NM\omega_{qs}}} [a_{qs} e^{i\boldsymbol{q} \cdot \boldsymbol{n}} \hat{\boldsymbol{\epsilon}}_{qs} + a_{qs}^\dagger e^{-i\boldsymbol{q} \cdot \boldsymbol{n}} \hat{\boldsymbol{\epsilon}}_{qs}], \qquad (8.25)$$

where $\hat{\boldsymbol{\epsilon}}_{qs}$ is the unit polarization vector associated with the phonon wave vector $\boldsymbol{q}$ and polarization s. The phonon coordinates α in the matrix element above may now be explicitly written as the number of phonons in each mode. Thus the factor involving $\boldsymbol{u}_n$ becomes

$$\langle n'_{q_1 s_1}, n'_{q_2 s_2}, \ldots|$$
$$\exp\left[i \sum_{q,s} \sqrt{\frac{\hbar}{2NM\omega_{qs}}} (a_{qs} e^{i\boldsymbol{q} \cdot \boldsymbol{n}} \hat{\boldsymbol{\epsilon}}_{qs} \cdot \boldsymbol{\kappa} + a_{qs}^\dagger e^{-i\boldsymbol{q} \cdot \boldsymbol{n}} \hat{\boldsymbol{\epsilon}}_{qs} \cdot \boldsymbol{\kappa})\right]|n_{q_1 s_1}, n_{q_2 s_2}, \ldots\rangle. \qquad (8.26)$$

Using the relation

$$e^{A+B} = e^A e^B e^{-(1/2)[A,B]} \qquad (8.27)$$

where A and B are noncommuting operators, and expanding the exponentials involving the phonon amplitudes, the cross section may be written as the sum of multiphonon processes,

$$\frac{d^2\sigma}{d\Omega' dE'} = \sum_{n=-\infty}^{\infty} \frac{d^2\sigma_n}{d\Omega' dE'} \qquad (8.28)$$

where n is the net number of phonons created.

254   8. Neutron Scattering

Elastic Scattering. Let us consider the case where the net number of phonons created is 0. This corresponds to elastic scattering, but at a *finite* temperature. with $n'_{qs} = n_{qs}$, the matrix element for $n = 0$ in (8.28) becomes, to order $1/N$,

$$\prod_{q,s}\left[1 - \frac{\hbar}{2NM\omega_{qs}}(n_{qs} + \tfrac{1}{2})|\boldsymbol{\kappa}\cdot\hat{\boldsymbol{\epsilon}}_{qs}|^2\right]. \tag{8.29}$$

If this is substituted into the expression for $d^2\sigma_0/d\Omega'\,dE'$, the sum over α, which is the number of phonons in each mode, converts n_{qs} into its thermal equilibrium value $\bar{n}_{qs}$. In the limit $N \to \infty$ this may be written as

$$\exp\left[-\sum_{q,s}\frac{\hbar}{2NM\omega_{qs}}(\bar{n}_{qs} + \tfrac{1}{2})|\boldsymbol{\kappa}\cdot\hat{\boldsymbol{\epsilon}}_{qs}|^2\right]. \tag{8.30}$$

Using this along with (8.23, 24) results in a total elastic cross section,

$$\begin{aligned}\frac{d^2\sigma_0}{d\Omega'\,dE'} &= \sum_{m_I,m_s} P_{m_s} P_{m_I} \sum_{m_I',m_s'} \frac{k'}{k} \sum_{n,m} \langle m_s, m_I | a_m^* | m_s', m_I' \rangle \\ &\quad \langle m_s', m_I' | a_n | m_s, m_I \rangle \exp[i\boldsymbol{\kappa}\cdot(\boldsymbol{n}-\boldsymbol{m})] \\ &\quad \exp\left[-\sum_{q,s}\frac{\hbar}{NM\omega_{qs}}(n_{qs}+\tfrac{1}{2})|\boldsymbol{\kappa}\cdot\hat{\boldsymbol{\epsilon}}_{qs}|^2\right]\delta(\text{energy}).\end{aligned} \tag{8.31}$$

Removing the sum over (m_s', m_I') by closure and integrating over energy, we obtain

$$\frac{d\sigma_0}{d\Omega'} = \sum_{n,m} \exp[i\boldsymbol{\kappa}\cdot(\boldsymbol{n}-\boldsymbol{m})]\langle a_m^* a_n\rangle \exp(-2W), \tag{8.32}$$

where

$$W = \sum_{q,s}\frac{\hbar}{2NM\omega_{qs}}(\bar{n}_{qs}+\tfrac{1}{2})|\boldsymbol{\kappa}\cdot\hat{\boldsymbol{\epsilon}}_{qs}|^2 \tag{8.33}$$

is called the *Debye-Waller factor*. Comparison of (8.32) and (8.12) shows that the coherent and incoherent cross sections are the same as those we obtained for the rigid lattice case except for the factor $\exp(-2W)$. Notice that this is nonzero even at $T = 0$ owing to zero-point vibrations.

One-Phonon Scattering. We now turn to *inelastic* processes involving one phonon. In particular, let us consider those processes in which the number of phonons *annihilated* exceeds by 1 the number of phonons created. The contribution to the matrix element given by (8.26) for these processes is

$$\sum_n \langle m'_s, m'_I | a_n | m_s, m_I \rangle \, e^{i\boldsymbol{\kappa}\cdot\boldsymbol{n}} \, i \sqrt{\frac{\hbar}{2NM\omega_{qs}}} \sqrt{n_{qs}} \, e^{i\boldsymbol{q}\cdot\boldsymbol{n}} \, \boldsymbol{\kappa}\cdot\hat{\boldsymbol{\epsilon}}_{qs}$$

$$\prod_{q',s'}\left[1 - \frac{\hbar}{2NM\omega_{q's'}}(n_{q's'} + \tfrac{1}{2})|\boldsymbol{\kappa}\cdot\hat{\boldsymbol{\epsilon}}_{qs}|^2\right], \tag{8.34}$$

In the computation of the corresponding cross section the sum over the distribution of occupation numbers again introduces the thermal averages of the n_{qs}. Thus we obtain

$$\frac{d^2\sigma_{-1}}{d\Omega' dE'} = \sum_{q,s}\frac{k'}{k}\sum_{n,m}\langle a_m^* a_n\rangle \exp\left[i(\boldsymbol{\kappa} + \boldsymbol{q})\cdot(\boldsymbol{n} - \boldsymbol{m})\right]$$

$$\times \frac{\hbar}{2NM\omega_{qs}}\bar{n}_{qs}|\boldsymbol{\kappa}\cdot\hat{\boldsymbol{\epsilon}}_{qs}|^2 \, e^{-2W} \delta\left[\frac{\hbar^2}{2m_0}(k^2 - k'^2) - \hbar\omega_q\right]. \tag{8.35}$$

From (8.13, 15, 16) we may write

$$\langle a_m^* a_n\rangle = \frac{\sigma_{\text{coh}}}{4\pi} + \frac{\sigma_{\text{incoh}}}{4\pi}\Delta(\boldsymbol{m} - \boldsymbol{n}) \tag{8.36}$$

where $\sigma_{\text{coh}} = 4\pi|\langle a\rangle|^2$ and $\sigma_{\text{incoh}} = 4\pi\{\langle |a|^2\rangle - |\langle a\rangle|^2\}$, so that the coherent part of the cross section becomes

$$\left(\frac{d^2\sigma_{-1}}{d\Omega' dE'}\right)_{\text{coh}} = \frac{(2\pi)^3 N}{V_0}\frac{\sigma_{\text{coh}}}{4\pi}\sum_{q,s}\frac{k'}{k}\sum_G \delta(\boldsymbol{\kappa} + \boldsymbol{q} - \boldsymbol{G})$$

$$\times \frac{\hbar}{2NM\omega_{qs}}\bar{n}_{qs}|\boldsymbol{\kappa}\cdot\hat{\boldsymbol{\epsilon}}_{qs}|^2 \, e^{-2W} \, \delta(\text{energy}). \tag{8.37}$$

Similarly, the incoherent part becomes

$$\left(\frac{d^2\sigma_{-1}}{d\Omega' dE'}\right)_{\text{incoh}} = \frac{N\sigma_{\text{incoh}}}{4\pi}\sum_{q,s}\frac{k'}{k}\frac{\hbar}{2NM\omega_{qs}}\bar{n}_{qs}|\boldsymbol{\kappa}\cdot\hat{\boldsymbol{\epsilon}}_{qs}|^2 \, e^{-2W} \, \delta(\text{energy}). \tag{8.38}$$

We can obtain similar expressions for one-phonon *creation* processes. For more details we refer the reader to [8.2].

In summary, a typical neutron diffraction pattern will consist of Bragg peaks plus a *diffuse* background. The Bragg peaks enable us to determine the crystal structure. The diffuse background arises from three sources: incoherent elastic scattering, coherent inelastic scattering, and incoherent inelastic scattering. By using certain experimental techniques we may separate the coherent inelastic scattering contribution and thereby determine the phonon dispersion relation. If we can also separate the incoherent inelastic scattering, this leads directly to the phonon density of states. In the next section we shall see how magnetic scattering contributes to this spectrum.

The measurement of the energy and momentum transfer associated with neutron scattering is typically accomplished by means of a "triple-axis" spectrometer. The schematic plan of such a spectrometer is shown in Fig. 8.2. The source

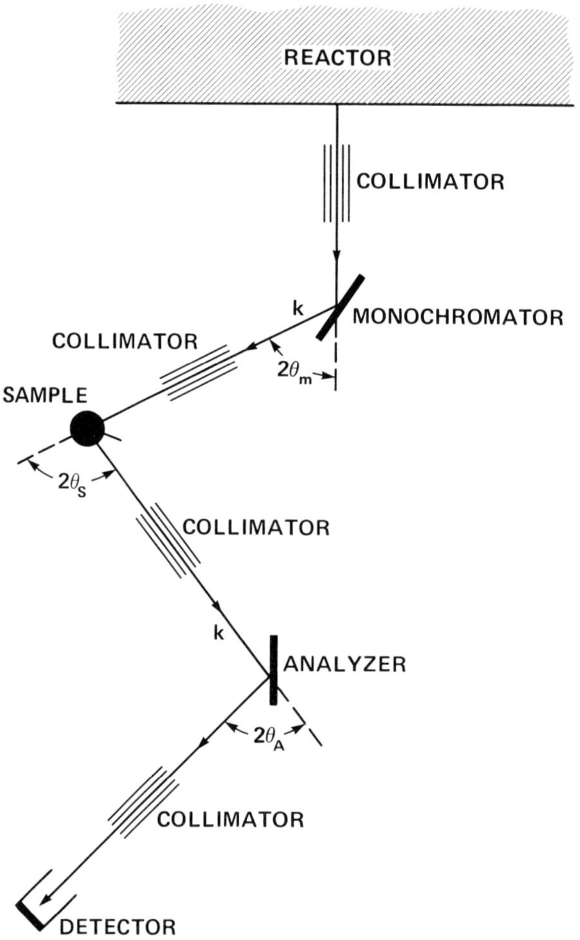

Fig. 8.2. Schematic diagram of a triple-axis spectrometer

of neutrons is usually a nuclear reactor, although particle accelerator sources are being developed. The high-energy neutrons are moderated down to thermal energies and emerge from the reactor with a Maxwellian distribution of energies. A single energy is selected by Bragg reflection from the first monochromator crystal in Fig. 8.2. This monoenergetic beam is incident on the sample with the wave vector k. The scattered beam k' is Bragg reflected by the analyzer crystal into the detector.

8.3 Magnetic Scattering

We now turn our attention to the interaction between the magnetic moment of the neutron and those of the crystal. The vector potential at r_e produced by a neutron's magnetic moment μ at r is

8.3 Magnetic Scattering

$$A = \frac{\mu \times (r_e - r)}{|r_e - r|^3}. \tag{8.39}$$

As we saw in Sect. (2.2), the magnetic interaction between a vector potential A and an electron with momentum p is given by

$$V(x) = -\frac{e}{2mc}[p \cdot A(x) + A(x) \cdot p] - \frac{e\hbar}{2mc}\sigma \cdot \nabla_x \times A(x) \tag{8.40}$$

where $x \equiv r_e - r$ and p operates on r_e. In the Born approximation we consider the matrix element of this interaction between the initial and final states. For the neutron these are just $\exp(i k \cdot r)$ and $\exp(i k' \cdot r)$, respectively. Therefore, if we integrate over the *neutron* coordinates, we have

$$\langle k'|V(x)|k\rangle = -\frac{e}{2mc}\int dr \, e^{-ik'\cdot r}[q \cdot A(x) + A(x)\cdot p]\, e^{ik\cdot r}$$

$$-\frac{e\hbar}{2mc}\sigma \cdot \int dr \, e^{-ik'\cdot r}\nabla_x \times A(x)\, e^{ik\cdot r}. \tag{8.41}$$

The Born approximation is valid in this case because the average interaction energy $\mu_e \mu_N / x^3 \simeq 10^{-3}$ K, which is very small compared with the kinetic energy of the neutron. If we write $r = r_e - x$ and $\kappa = k - k'$, the partial matrix element (8.41) becomes

$$-\frac{e}{2mc}\int dx \, \exp(-i\kappa\cdot x)\exp(i\kappa\cdot r_e)[p\cdot A(x) + A(x)\cdot p]$$

$$-\frac{e\hbar}{2mc}\sigma \cdot \int dx \, \exp(-i\kappa\cdot x)\exp(i\kappa\cdot r_e)\nabla_x \times A(x). \tag{8.42}$$

From the explicit form of $A(x)$ we have that $\nabla_x \cdot A(x) = 0$, as well as

$$\int dx \, e^{-i\kappa\cdot x} A(x) = -\frac{4\pi i}{\kappa^2}(\mu \times \kappa) \tag{8.43}$$

and

$$\int dx \, e^{i\kappa\cdot x}[\nabla_x \times A(x)] = \frac{4\pi}{\kappa^2}\kappa \times (\mu \times \kappa) = 4\pi[\mu - (\hat{\kappa}\cdot\mu)\hat{\kappa}]. \tag{8.44}$$

Therefore (8.41) reduces to

$$i\frac{4\pi e}{mc}\frac{1}{\kappa^2}(\mu \times \kappa)\cdot\exp(i\kappa\cdot r_e)p - \frac{4\pi e\hbar}{2mc}[\sigma\cdot\mu - (\sigma\cdot\hat{\kappa})(\mu\cdot\hat{\kappa})]\exp(i\kappa\cdot r_e) \tag{8.45}$$

The first term represents the interaction between the neutron's moment and the orbital current, and the second term is the spin–spin interaction.

It is interesting to consider what happens when $\kappa \to 0$, that is, the case of forward scattering. We rewrite the first term as

$$i\frac{4\pi e}{mc}\boldsymbol{\mu}\cdot\left[\frac{\exp(i\boldsymbol{\kappa}\cdot\boldsymbol{r}_e)}{\kappa^2}\boldsymbol{\kappa}\times\boldsymbol{p}\right]. \tag{8.46}$$

Expanding the exponential, we have

$$i\frac{4\pi e}{mc}\boldsymbol{\mu}\cdot\frac{\boldsymbol{\kappa}\times\boldsymbol{p}}{\kappa^2} - \frac{4\pi e}{mc}\boldsymbol{\mu}\cdot\left(\frac{\boldsymbol{\kappa}\cdot\boldsymbol{r}_e}{\kappa^2}\boldsymbol{\kappa}\times\boldsymbol{p}\right) + \cdots. \tag{8.47}$$

We take the mean value of the limit as $\kappa \to 0$. In this limit the first term of (8.47) vanishes and the second term becomes

$$-\frac{4\pi e}{mc}\boldsymbol{\mu}\cdot\tfrac{1}{3}(\boldsymbol{r}_e\times\boldsymbol{p}) = \frac{8\pi}{3}\mu_B\boldsymbol{\mu}\cdot\boldsymbol{l}. \tag{8.48}$$

The spin term of (8.45) becomes

$$4\pi\mu_B[\boldsymbol{\sigma}\cdot\boldsymbol{\mu} - \tfrac{1}{3}\boldsymbol{\sigma}\cdot\boldsymbol{\mu}] = \frac{8\pi}{3}\mu_B\boldsymbol{\mu}\cdot\boldsymbol{\sigma}. \tag{8.49}$$

Thus the partial matrix element is

$$\frac{8\pi}{3}\mu_B\boldsymbol{\mu}\cdot(\boldsymbol{l}+2\boldsymbol{s}). \tag{8.50}$$

This shows that the *forward-scattering cross section* is proportional to the *total magnetic moment* of the electron.

Let us return now to the case $\kappa \neq 0$. In many materials the orbital momentum is quenched. Therefore let us consider only the spin term. The matrix element entering (8.7), which is essentially the scattering amplitude, is

$$4\pi\mu_B\langle\alpha', m_s'|\sum_i[\boldsymbol{\sigma}_i\cdot\boldsymbol{\mu} - (\boldsymbol{\sigma}_i\cdot\hat{\boldsymbol{\kappa}})(\boldsymbol{\mu}\cdot\hat{\boldsymbol{\kappa}})]\exp(i\boldsymbol{\kappa}\cdot\boldsymbol{r}_i)|\alpha, m_s\rangle, \tag{8.51}$$

where the sum over i includes all the electrons in the sample. The factor $\boldsymbol{\sigma}_i\exp(i\boldsymbol{\kappa}\cdot\boldsymbol{r}_i)$ may be written as $\int d\boldsymbol{r}\,\boldsymbol{\sigma}_i\,\delta(\boldsymbol{r}-\boldsymbol{r}_i)\exp(i\boldsymbol{\kappa}\cdot\boldsymbol{r})$. By analogy with (1.48),

$$\sum_i\boldsymbol{\sigma}_i\,\delta(\boldsymbol{r}-\boldsymbol{r}_i)$$

is the *spin density* at $\boldsymbol{r}$, which we shall write as $2\boldsymbol{S}(\boldsymbol{r})$. Therefore the total scattering amplitude is proportional to the Fourier transform of the spin density. The neutron's magnetic moment is related to its spin by

$$\boldsymbol{\mu} = -1.91 \frac{e\hbar}{m_0 c} \mathbf{s}$$

and the scattering amplitude becomes

$$-1.91 \frac{4\pi e^2 \hbar^2}{mm_0 c^2} \langle \alpha', m_s' | \mathbf{S}(-\boldsymbol{\kappa}) \cdot \mathbf{s} - [\mathbf{s} \cdot \hat{\boldsymbol{\kappa}}][\mathbf{S}(-\boldsymbol{\kappa}) \cdot \hat{\boldsymbol{\kappa}}] | \alpha, m_s \rangle . \tag{8.52}$$

The cross section is given by

$$\frac{d^2\sigma}{d\Omega' dE'} = 4 \left(\frac{1.91 e^2}{mc^2} \right)^2 \sum_{\alpha, m_s} \sum_{\alpha', m_s'} P_\alpha P_{m_s} \frac{k'}{k}$$
$$\times |\langle \alpha', m_s' | \mathbf{s} \cdot \mathbf{S}(\boldsymbol{\kappa}) - [\mathbf{s} \cdot \hat{\boldsymbol{\kappa}}][\mathbf{S}(\boldsymbol{\kappa}) \cdot \hat{\boldsymbol{\kappa}}] | \alpha, m_s \rangle|^2 \delta(\text{energy}) . \tag{8.53}$$

Since the energy of the system does not depend on the *neutron* polarization, we may sum over m_s' to obtain

$$\frac{d^2\sigma}{d\Omega' dE'} = 4 \left(\frac{1.91 e^2}{mc^2} \right)^2 \sum_{\alpha, \alpha'} P_\alpha \frac{k'}{k}$$
$$\times \sum_{\mu,\nu} \langle \alpha | S_\nu(\boldsymbol{\kappa}) - \hat{\kappa}_\nu [\mathbf{S}(\boldsymbol{\kappa}) \cdot \hat{\boldsymbol{\kappa}}] | \alpha' \rangle \langle \alpha' | S_\mu(-\boldsymbol{\kappa}) - \hat{\kappa}_\mu [\mathbf{S}(-\boldsymbol{\kappa}) \cdot \hat{\boldsymbol{\kappa}}] | \alpha \rangle$$
$$\times \sum_{m_s} P_{m_s} \langle m_s | s_\nu s_\mu | m_s \rangle \, \delta(\text{energy}) . \tag{8.54}$$

Since

$$\sum_{m_s} P_{m_s} \langle m_s | s_\nu s_\mu | m_s \rangle = \tfrac{1}{4} \delta_{\nu\mu} , \quad \text{this becomes}$$

$$\frac{d^2\sigma}{d\Omega' dE'} = \left(\frac{1.91 e^2}{mc^2} \right)^2 \sum_{\mu,\nu} (\delta_{\nu\mu} - \hat{\kappa}_\nu \hat{\kappa}_\mu)$$
$$\times \sum_{\alpha,\alpha'} P_\alpha \frac{k'}{k} \langle \alpha | S_\nu(\boldsymbol{\kappa}) | \alpha' \rangle \langle \alpha' | S_\mu(-\boldsymbol{\kappa}) | \alpha \rangle \, \delta(\text{energy}) . \tag{8.55}$$

Finally, we may incorporate the energy conservation into the matrix elements themselves by writing the delta function as

$$\delta(\text{energy}) = \frac{1}{2\pi} \int_{-\infty}^{\infty} dt \, \exp[i(\omega - E_{\alpha'} + E_\alpha)t] \tag{8.56}$$

where $\omega \equiv \hbar^2(k^2 - k'^2)/2m_0$. Then

$$\frac{d^2\sigma}{d\Omega' dE'} = \left(\frac{1.91 e^2}{mc^2} \right)^2 \sum_{\mu,\nu} (\delta_{\nu\mu} - \hat{\kappa}_\nu \hat{\kappa}_\mu) \frac{k'}{k} \frac{1}{2\pi} \int dt \, e^{i\omega t}$$
$$\times \sum_{\alpha,\alpha'} P_\alpha \langle \alpha | \exp(i\mathcal{H}t) S_\nu(\boldsymbol{\kappa}) \exp(-i\mathcal{H}t) | \alpha' \rangle \langle \alpha' | S_\mu(-\boldsymbol{\kappa}) | \alpha \rangle . \tag{8.57}$$

260   8. Neutron Scattering

The sum over α' may now be removed by closure. Furthermore, the sum over α is just what we mean by the *thermal average*. Therefore we have

$$\frac{d^2\sigma}{d\Omega'dE'} = \left(\frac{1.91e^2}{mc^2}\right)^2 \frac{k'}{k} \sum_{\mu,\nu} (\delta_{\nu\mu} - \hat{\kappa}_\nu\hat{\kappa}_\mu) \frac{1}{2\pi} \int dt\, e^{i\omega t} \langle S_\nu(\boldsymbol{\kappa},t) S_\mu(-\boldsymbol{\kappa}) \rangle. \tag{8.58}$$

It is convenient to separate the cross section into a Bragg part and a diffuse part:

$$\left(\frac{d^2\sigma}{d\Omega'dE'}\right)_{\text{Bragg}} = \left(\frac{1.91e^2}{mc^2}\right)^2 \sum_{\mu,\nu}(\delta_{\nu\mu} - \hat{\kappa}_\nu\hat{\kappa}_\mu)\,\delta(\omega)\,\langle S_\nu(\boldsymbol{\kappa})\rangle\langle S_\mu(-\boldsymbol{\kappa})\rangle, \tag{8.59}$$

$$\left(\frac{d^2\sigma}{d\Omega'dE'}\right)_{\text{diffuse}} = \left(\frac{1.91e^2}{mc^2}\right)^2 \frac{k'}{k}\sum_{\mu,\nu}(\delta_{\nu\mu} - \hat{\kappa}_\nu\hat{\kappa}_\mu)\frac{1}{2\pi}\int dt\, e^{i\omega t}$$
$$\times [\langle S_\nu(\boldsymbol{\kappa},t)S_\mu(-\boldsymbol{\kappa})\rangle - \langle S_\nu(\boldsymbol{\kappa})\rangle\langle S_\mu(-\boldsymbol{\kappa})\rangle]. \tag{8.60}$$

The operators $S_\nu(\boldsymbol{\kappa})$ entering the cross section are the Fourier transform of the spin density of the entire sample:

$$\boldsymbol{S}(\boldsymbol{\kappa}) = \sum_i \exp(i\boldsymbol{\kappa}\cdot\boldsymbol{R}_i)\boldsymbol{s}_i. \tag{8.61}$$

The unpaired electrons on a given site will form a total spin $\boldsymbol{S}_n$ according to Hund's rule. The Wigner-Eckart theorem tells us that a matrix element of the form $\langle\alpha'|\boldsymbol{s}_i|\alpha\rangle$ is proportional to $\langle\alpha'|\boldsymbol{S}_n|\alpha\rangle$. Thus

$$\langle\alpha'|\boldsymbol{S}(\boldsymbol{\kappa})|\alpha\rangle = \sum_{\text{site }n} e^{i\boldsymbol{\kappa}\cdot\boldsymbol{n}} f(\boldsymbol{\kappa})\langle\alpha'|\boldsymbol{S}_n|\alpha\rangle. \tag{8.62}$$

The coefficient of proportionality $f(\boldsymbol{\kappa})$ is the Fourier transform of the normalized spin density associated with the nth site and is referred to as the *magnetic form factor*.

In the *paramagnetic* state the magnetic Bragg contribution vanishes, leaving only the diffuse scattering:

$$\left(\frac{d^2\sigma}{d\Omega'dE'}\right)_{\text{diffuse}} = \left(\frac{1.91e^2}{mc^2}\right)^2 |f(\boldsymbol{\kappa})|^2 \frac{k}{k'}\sum_{\mu,\nu}(\delta_{\mu\nu} - \hat{\kappa}_\mu\hat{\kappa}_\nu)$$
$$\sum_{n,m}\exp[i\boldsymbol{\kappa}\cdot(\boldsymbol{n}-\boldsymbol{m})]\frac{1}{2\pi}\int dt\,\exp(i\omega t)\langle S_{n,\nu}(t)S_{m,\mu}\rangle. \tag{8.63}$$

DeGennes [8.17] analyzed the correlation function appearing in (8.63) by the method of moments including exchange interactions. He found that the second moment is proportional to the square of the exchange. In a perfect paramagnet without exchange, the diffuse peak becomes an elastic peak with a cross section

$$\left(\frac{d\sigma}{d\Omega'}\right)_{\text{diffuse}} = N \left(\frac{1.91 e^2}{mc^2}\right)^2 |f(\boldsymbol{\kappa})|^2 \tfrac{2}{3} S(S+1) . \tag{8.64}$$

The angular dependence of this cross section enables us to determine the magnetic form factor.

Let us now consider the cross sections in ordered magnetic systems.

8.3.1 Bragg Scattering

Let us assume that in the ordered state, the moments are aligned along the z-axis,

$$\langle S_{m,\nu} \rangle = \langle S_z \rangle \delta_{\nu z} .$$

The cross section then becomes

$$\frac{d\sigma}{d\Omega'} = \left(\frac{1.91 e^2}{mc^2}\right)^2 |f(\boldsymbol{\kappa})|^2 (1 - \hat{\kappa}_z^2) \langle S^z \rangle^2 \sum_{n,m} \exp[i\boldsymbol{\kappa} \cdot (\boldsymbol{n} - \boldsymbol{m})] . \tag{8.65}$$

Or, from (8.17),

$$\boxed{\left(\frac{d\sigma}{d\Omega'}\right)_{\text{Bragg}} = \frac{(2\pi)^3 N}{V_0} \left(\frac{1.91 e^2}{mc^2}\right)^2 \langle S^z \rangle^2 \sum_G |f(\boldsymbol{G})|^2 (1 - G_z^2) \delta(\boldsymbol{\kappa} - \boldsymbol{G})} \tag{8.66}$$

where $\boldsymbol{G}$ is a vector in the *magnetic* reciprocal lattice. Thus we see that this cross section consists of peaks at positions determined by the *magnetic structure*. Furthermore, the amplitudes of these peaks are proportional to the magnetic form factor.

As an example of Bragg scattering let us consider the spin arrangement in MnF_2. Chemically, this crystal has the body-centered tetragonal lattice shown in Fig. 2.2. The unit cell has six ions at the following locations:

Mn:

$(0, 0, 0)$ $\left(\frac{a}{2}, \frac{a}{2}, \frac{c}{2}\right)$

F:

$(au, au, 0)$ $(a - au, a - au, 0)$

$\left(\frac{a}{2} + au, \frac{a}{2} - au, \frac{c}{2}\right)$ $\left(\frac{a}{2} - au, \frac{a}{2} + au, \frac{c}{2}\right) .$

The reciprocal lattice vectors are given by

$$\boldsymbol{G} = 2\pi \left(\frac{l}{a}, \frac{m}{a}, \frac{n}{c}\right) \tag{8.67}$$

where l, m, and n are integers, and $u = 0.31$.

The unit cell structure factor for *nuclear* scattering is

$$F(\mathbf{G}) = \sum_{i=1}^{r} \langle a_i \rangle \exp(i\mathbf{G} \cdot \boldsymbol{\rho}_i),$$

$$= \begin{cases} 2\langle a \rangle_{\text{Mn}} + 4\langle a \rangle_{\text{F}} \cos(2\pi u l) \cos(2\pi u m) & l + m + n = \text{even} \\ -4\langle a \rangle_{\text{F}} \sin(2\pi u l) \sin(2\pi u m) & l + m + n = \text{odd} \end{cases}. \quad (8.68)$$

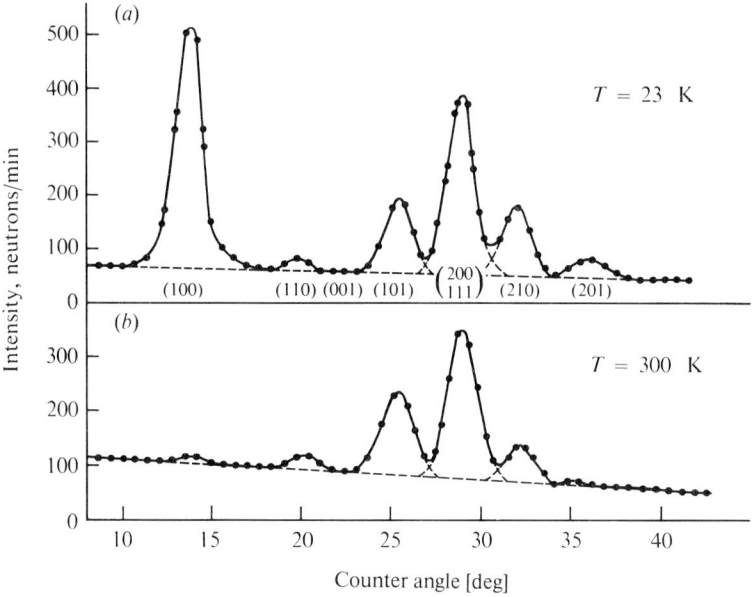

Fig. 8.3a, b. Neutron diffraction pattern for MnF_2 in (a) the paramagnetic state (300 K) and (b) the antiferromagnetic state (23 K) [8.7]

The experimental scattering intensity at a temperature *above* the magnetic ordering temperature is shown in Fig. 8.3b. Notice that there are no peaks corresponding to [100] and [001], in accordance with (8.68). Figure 8.3a shows the scattering intensity in the magnetically ordered state. We immediately notice the presence of additional peaks. These are magnetic peaks. Since they can be indexed in the same way as the nuclear peaks the magnetic unit cell is the same as the chemical unit cell. By their mere existence these peaks tell us that the spin density is different from the electronic charge density; that is, the spin arrangement is not ferromagnetic. Since there are two magnetic ions in the unit cell, we try an antiferromagnetic arrangement. This introduces a sign change into the magnetic form factor. If the up spins are on the corners of the unit cell and the down spins are at the centers, the magnetic structure factor is

$$F_\mathrm{M}(\boldsymbol{G}) = \sum_i (\pm)_i \exp(\mathrm{i}\boldsymbol{G}\cdot\boldsymbol{\rho}_i) = 1 - \mathrm{e}^{\mathrm{i}\pi(l+m+n)} = \begin{cases} 0 & l+m+n = \text{even} \\ 2 & l+m+n = \text{odd}. \end{cases}$$
(8.69)

We see that a peak is expected at [100], as observed in Fig. 8.3a. However, we also expect one at [001]. The fact that this is not observed indicates that the spins are oriented in this direction since the factor $1 - \hat{\kappa}_z^2$ then vanishes.

Izuyama et al. [8.8] have shown that in a metal the Bragg scattering cross section is essentially the same as that given by (8.66), except that $\langle S^z \rangle^2$ is replaced by $(N_\uparrow - N_\downarrow)^2$, where N_σ is the total number of electrons with spin δ.

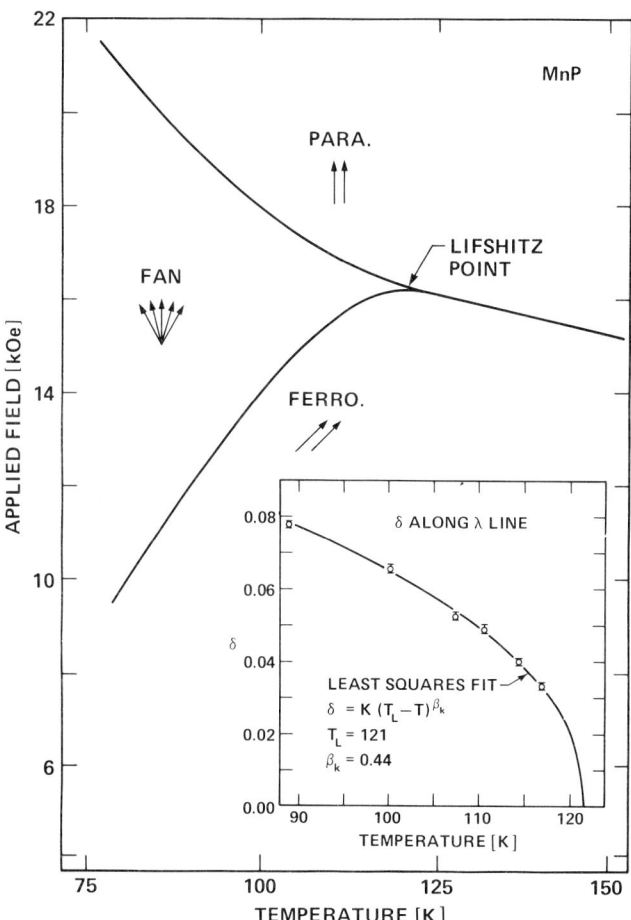

Fig. 8.4. Phase diagram of MnP near the Lifshitz point. The magnetic field is parallel to the b axis. The insert shows the temperature dependence of the fan wave vector along the fan–para phase boundary

Neutron scattering is also particularly useful in studying structures which are not commensurate with the crystal lattice. The rare-earth metals listed in Fig. 4.3, for example, exhibit a variety of incommensurate structures. These were all confirmed by neutron scattering.

An interesting study of incommensurate structures is found in MnP. This orthorhombic material is ferromagnetic below 291 K, with a magnetic phase transition at 50 K. The lower temperature phase is characterized by a spiral spin structure with a wave vector along the a axis. Application of a magnetic field along the b axis produces a fan structure with a periodicity governed by $q = 2\pi(\delta/a)\boldsymbol{a}$. Figure 8.4 shows the phase diagram of MnP [8.9].

The point at which the three phases meet was recognized as a possible Lifshitz point. This particular multicritical point is characterized by three features: 1) it is the meeting point of three phases: a paramagnetic phase, an ordered phase whose order parameter has a fixed wave vector (zero in the ferromagnetic case), and an ordered phase (the fan phase) whose wave vector varies continuously with the thermodynamic variables; 2) the transitions from the paramagnetic phase to both ordered phases must be second order; and 3) the wave vector in the fan phase must approach zero continuously as the Lifshitz point is approached.

Neutron scattering was essential in verifying the last condition. Figure 8.5 shows a triple-axis scan through the (200) position of MnP, indicating the magnetic satellites due to the fan structure. The smooth decrease of δ to zero as shown in the insert in Fig. 8.4 confirms that this is indeed a Lifshitz point [8.10].

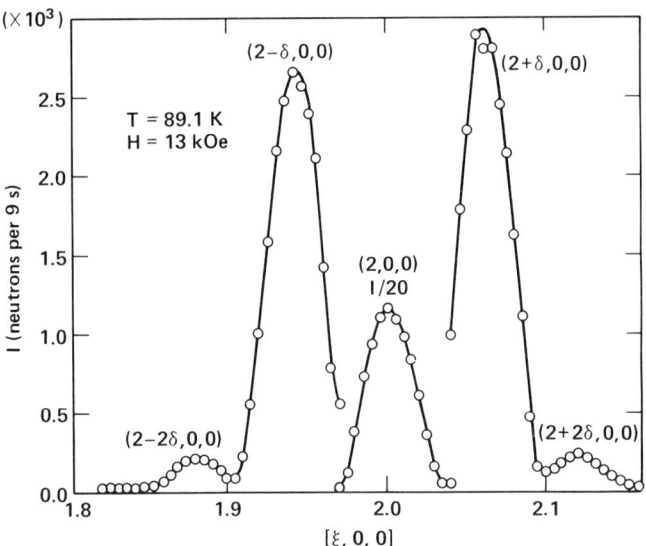

Fig. 8.5. An elastic triple-axis scan through the (200), position of MnP showing first- and second-order magnetic satellites. Note that the central (200) peak is shown on a reduced scale

8.3 Magnetic Scattering   265

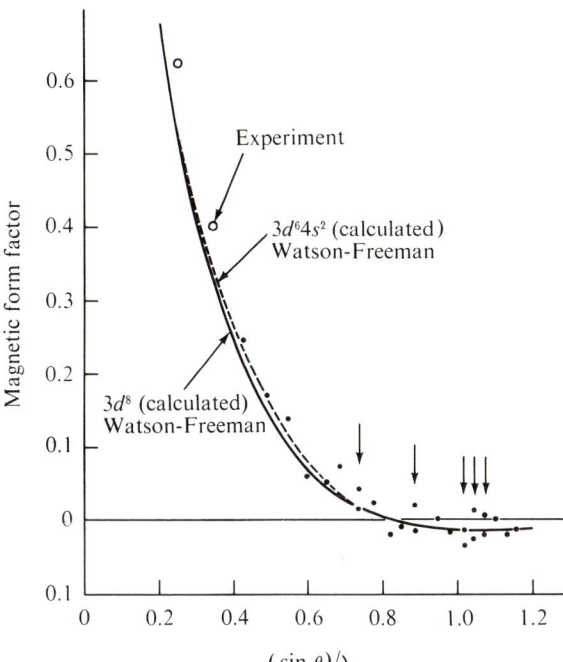

Fig. 8.6. Comparison between experimental and calculated values of the magnetic form factor for iron. The solid curve is that calculated for the spherically symmetric free atom. [8.11]

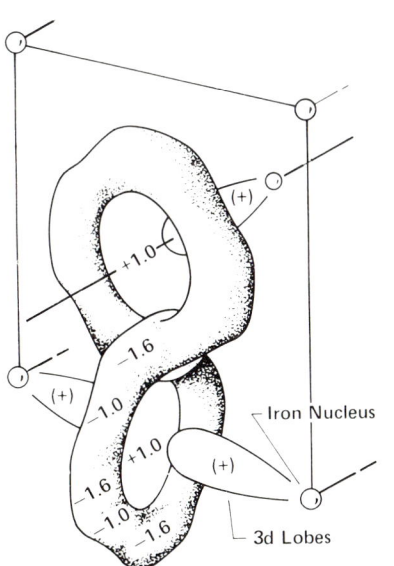

Fig. 8.7. Model of the magnetization distribution in the body-centered-cubic unit cell of iron. The concentration of the very large positive 3d-shell magnetization along the cube edges is overemphasized to show its relationship to the negative magnetization ring [8.12]

From the amplitudes of the Bragg peaks we obtain the magnetic form factor $f(\kappa)$. The experimental values for $f(\kappa)$ for metallic iron are shown in Fig. 8.6. The arrows indicate pairs of reflections for which the scattering angle is the same for the pair. Since the experimental errors are indicated by the size of the circles, these pairs tell us that the form factor, and hence the spin density, is not spherically symmetric. The curves are the theoretical results of a Hartree–Fock calculation which includes exchange effects by Watson and Freeman. The deviations from these spherical calculations are also apparent. By Fourier transforming such data it is possible to construct a spin-density "contour map" of the unit cell. This is illustrated for iron in Fig. 8.7.

Thus we find that from the Bragg scattering we can obtain information about the *static* magnetic state of the crystal—the spin configuration and the spin-density distribution. Let us now turn to the diffuse-scattering contribution. We shall find that this contains information about the *dynamic* properties of the system—spin waves and critical fluctuations.

8.3.2 Diffuse Scattering

Let us consider the diffuse-scattering cross section (8.60). The sum over ν and μ enables us to replace $\langle S_\nu(\boldsymbol{\kappa},t) S_\mu(-\boldsymbol{\kappa}) \rangle$ by the correlation function $\langle \{S_\nu(\boldsymbol{\kappa},t) S_\mu(-\boldsymbol{\kappa})\} \rangle$. In deriving the fluctuation-dissipation theorem, (1.87), we neglected any long-rang order. When this is included, the correlation function in (1.87) is identical to the quantity in brackets in (8.60). Therefore,

$$\left(\frac{d^2\sigma}{d\Omega' dE'}\right)_{\text{diffuse}} = \left(\frac{1.91 e^2}{mc^2}\right)^2 \frac{k'}{k} \sum_{\mu,\nu} (\delta_{\nu\mu} - \hat{\kappa}_\nu \hat{\kappa}_\mu)$$
$$(1/\pi g^2 \mu_B^2) \hbar V \coth(\beta\hbar\omega/2) \chi''_{\nu\mu}(\boldsymbol{\kappa},\omega)_s \,.$$

(8.70)

The importance of the generalized susceptibility has been emphasized throughout this text. Now we see that the diffuse scattering of neutrons provides us with a direct measure of this quantity. This is why neutron scattering plays such an important role in magnetism.

The susceptibility appearing in (8.70) is the susceptibility associated with the *total* spin density of the system. In deriving explicit expressions from this quantity we have assumed throughout that the spin density was either localized or completely itinerant. This gave us what we might call a *reduced susceptibility*. In order to apply these results, we must relate this reduced susceptibility to the total susceptibility. From (8.62) it readily follows that for localized spins

$$\chi(\boldsymbol{q},\omega)_{\text{total}} = |f(\boldsymbol{q})|^2 \, \chi(\boldsymbol{q},\omega)_{\text{reduced}} \,, \tag{8.71}$$

where $f(\boldsymbol{q})$ is the *atomic magnetic form factor*.

The corresponding relation for an itinerant system is essentially the same. To see this, let us second quantize the magnetic-moment operator in terms of Bloch states. The field operator (1.115) then becomes

$$\psi(\boldsymbol{r}) = \frac{1}{\sqrt{V}} \sum_{k,\sigma} e^{i\boldsymbol{k}\cdot\boldsymbol{r}} u_k(\boldsymbol{r}) \eta_\sigma \, c_{k\sigma} \,. \tag{8.72}$$

The Bloch function may be expanded in terms of the Wannier functions introduced in (2.78) as

$$e^{i\mathbf{k}\cdot\mathbf{r}} u_k(\mathbf{r}) = \sum_\alpha e^{i\mathbf{k}\cdot\mathbf{R}_\alpha} \phi(\mathbf{r} - \mathbf{R}_\alpha) \,. \tag{8.73}$$

In terms of these functions the field operator may be written

$$\psi(\mathbf{r}) = \frac{1}{\sqrt{V}} \sum_{k,\sigma} \sum_\alpha e^{i\mathbf{k}\cdot\mathbf{R}_\alpha} \phi(\mathbf{r} - \mathbf{R}_\alpha) \eta_\sigma c_{k\sigma} \,. \tag{8.74}$$

The magnetization operator is

$$\mathcal{M}(\mathbf{r}) = -\mu_B \psi^\dagger(\mathbf{r}) \boldsymbol{\sigma} \psi(\mathbf{r}) \,. \tag{8.75}$$

Using (8.74) and taking the Fourier transform we obtain, for example,

$$\mathcal{M}_+(\mathbf{q}) = -2\mu_B \sum_k F(\mathbf{q}, \mathbf{k}) c^\dagger_{k-q,\uparrow} c_{k\downarrow} \tag{8.76}$$

where

$$F(\mathbf{q}, \mathbf{k}) \equiv \frac{N}{V} \sum e^{-i\mathbf{k}\cdot\mathbf{l}} \int d\mathbf{r}\, e^{-i\mathbf{q}\cdot\mathbf{r}} \phi^*(\mathbf{r})\phi(\mathbf{r} + \mathbf{l}) \,. \tag{8.77}$$

Here $\mathbf{l}$ is the vector connecting lattice sites, that is, $\mathbf{l} = \mathbf{R}_\alpha - \mathbf{R}_{\alpha'}$. Since the Wannier function $\phi(\mathbf{r} - \mathbf{R}_\alpha)$ is strongly peaked around the site $\mathbf{R}_\alpha$, the dominant contribution to (8.72) arises from the $\mathbf{l} = 0$ term. The function $F(\mathbf{q}, \mathbf{k})$ is then independent of $\mathbf{k}$. By comparing $\mathcal{M}_+(\mathbf{q})$ in this case to the free-electron expression (5.99), we again obtain the relation (8.71).

Spin-Wave Scattering. There are several applications of diffuse scattering which are particularly interesting. One is the very low temperature region where one has *inelastic spin–wave scattering*. As we saw in Chap. 6, the susceptibilities of both localized spin systems and metals contain poles which correspond to spin–wave excitations. Therefore at low temperatures the differential diffuse-scattering cross section is proportional to $|f(\mathbf{q})|^2 \delta(\omega - \omega_q)$. By measuring the energy loss of those neutrons scattered at direction $\mathbf{q}$ with respect to the incident beam we can thus obtain the spin-wave spectrum. The spin–wave spectrum of the antiferromagnet MnF_2 obtained in this way is shown in Fig. 8.8. An example of a metallic spin–wave spectrum which also reveals the presence of phonons is that of the rare earth terbium, shown in Fig. 8.9. We saw in Sect. 7.1 that the exchange interaction between localized moments in rare-earth metals arises through the conduction electrons and therefore depends upon the band structure. Thus from a spin–wave spectrum such as that shown in Fig. 8.9 we can obtain information about the band structure.

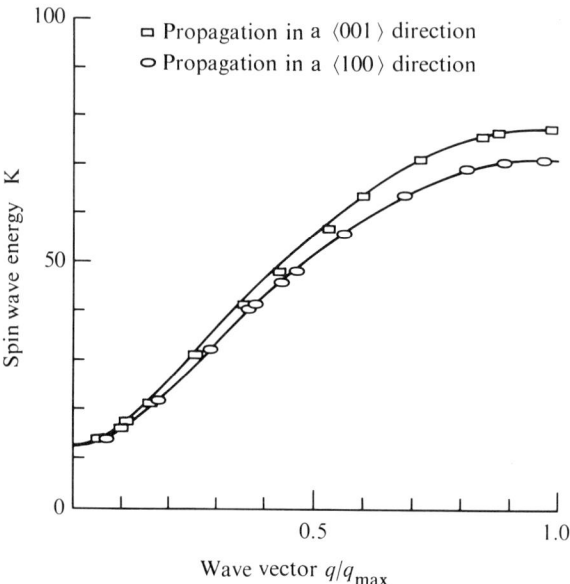

Fig. 8.8. Antiferromagnetic spin-wave dispersion in MnF_2 at 4.2K observed by neutron inelastic scattering. The solid lines are obtained from spin-wave theory, with the parameters $J_1 = 0.32$ K (ferromagnetic), $J_2 = -1.76$ K (antiferromagnetic), $J_3 = 0$ K, and an anisotropy field $H_A = 1.06$ K (at zero wave vector) [8.13]

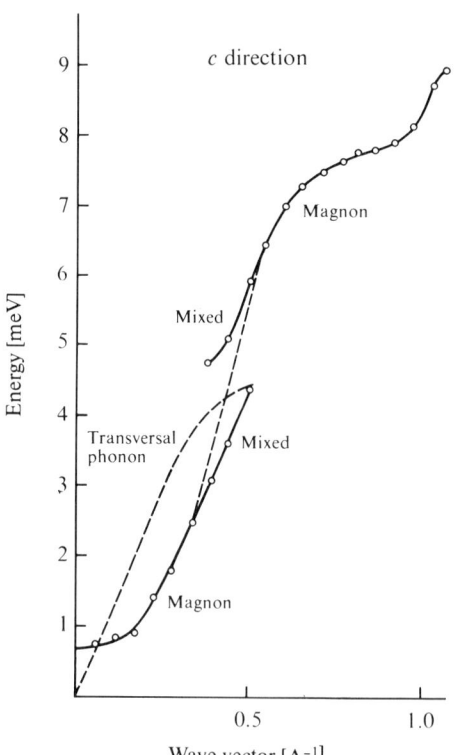

Fig. 8.9. Neutron excitation energies in the c direction of Tb-10% Ho at 110 K. The scan used for these measurements was such that transverse phonons were not observed [8.14]

Critical Scattering. Diffuse scattering is also interesting is in the vicinity of the magnetic phase transition. At the Curie point the Bragg peak transforms to a sharply peaked diffuse intensity, and then when the temperature is increased this diffuse intensity spreads out to approach the paramagnetic limit. The temperature dependence of diffuse scattering at a fixed angle is dramatically illustrated in Fig. 8.10. This is what is referred to as *critical scattering*.

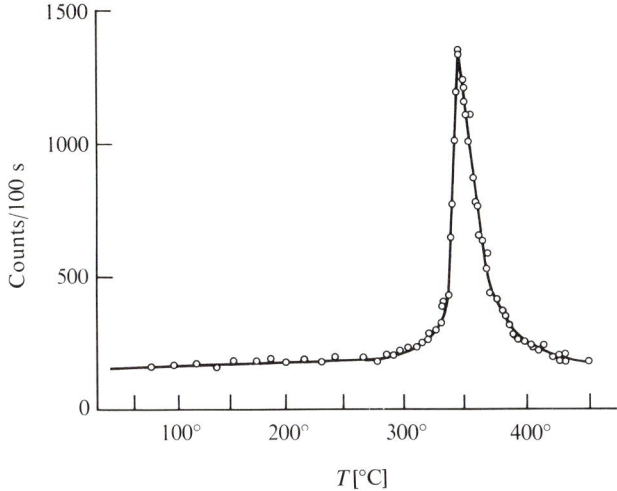

Fig. 8.10. Critical scattering by Ni. Curve gives total intensity of 4.75-Å neutrons scattered at an angle of 2°. [8.15]

The origin of this critical scattering in a localized spin system is contained in our result (4.12). The quantity measured is $d\sigma/d\Omega'$, obtained by integrating (8.70) over ω. If we assume that the major frequency response occurs for frequencies below $k_B T/\hbar$, then $d\sigma/d\Omega'$ is proportional, by the Kramers–Kronig relation (1.64), to the static susceptibility $\chi(\mathbf{q})$. In the vicinity of a Bragg peak momentum transfer is very small. In the limit $q \to 0$ the susceptibility (4.12), and hence the cross section, varies as $(\kappa^2 + q^2)^{-1}$, where κ is given by (4.49). Therefore the critical scattering should increase, in the mean field approximation, as $(T - T_c)^{-1}$. The correlation-range parameter κ^2 is found to diverge more like $(T - T_c)^{-4/3}$, which is in agreement with a more sophisticated theoretical treatment of the correlation function.

In principle we can measure $\chi(\mathbf{q}, \omega)$ for every point in Fourier space and for all temperatures. An example of this is the work of Lowde and coworkers on nickel [8.16]. Such results enable us to study the complete evolution of the magnetic state.

Diffuse Elastic Scattering. Finally, let us mention the use of diffuse scattering to study binary alloys. *Shull* and *Wilkinson* [8.18] were among the first to employ neutrons in this fashion. Consider an alloy containing $N_1 = cN$ atoms of type 1 and $N_2 = (1 - c)N$ atoms of type 2. If we assume that all atoms of type 1 have a moment μ_1, independent of their environment, while all atoms of type 2 have a moment μ_2, then the elastic diffuse cross section becomes

$$\left(\frac{d\sigma}{d\Omega'}\right)_{\text{diffuse}} = \frac{N}{4}\left(\frac{1.91e^2}{mc^2}\right)^2(1-\kappa_z^2)c(1-c)[f_1(\boldsymbol{\kappa})\mu_1 - f_2(\boldsymbol{\kappa})\mu_2] \qquad (8.78)$$

Marshall [8.19] has extended this formula to include environmental effects. From an analysis of the elastic diffuse scattering one can obtain the moments and their form factors. In Ni-Cu alloys, for example, *Medina* and *Cable* [8.20] found that the reduction in the net moment of the alloy is associated with a negative polarization of the Ni sites, and that there is no polarization of the Cu atoms.

References

Chapter 1

1.1 J. H. Van Vleck: *The Theory of Electric and Magnetic Susceptibilities* (Oxford University Press, London 1932) Chaps. 1, 4
1.2 L. D. Landau, E. M. Lifshitz: *Course of Theoretical Physics*, Vol. 8, Electrodynamics of Continuous Media (Pergamon, Oxford 1960) p. 129
1.3 K. Huang; *Statistical Mechanics* (Wiley, New York 1963)
1.4 D. N. Zubarev: Sov. Phys. Usp. **3**, 320 (1960)

Chapter 2

2.1 L. I. Schiff: *Quantum Mechanics*, 3rd ed. (McGraw-Hill, New York 1968) Chap. 8
2.2 J. D. Bjorken, S. D. Drell: *Relativistic Quantum Mechanics* (McGraw-Hill, New York 1965) Chap. 4
2.3 A. Abragam: *The Principles of Nuclear Magnetism* (Oxford University Press, London 1961) p. 165
2.4 C. P. Slichter: *Principles of Magnetic Resonance*, 2nd ed., Springer Series in Solid-State Sciences, Vol. 1 (Springer, Berlin, Heidelberg, New York 1980) Chap. 9
2.5 J. S. Griffith: *The Theory of Transition-Metal Ions* (Cambridge University Press, Cambridge 1961)
2.6 C. J. Ballhausen: *Introduction to Ligand Field Theory* (McGraw-Hill, New York 1962)
2.7 M. T. Hutchings: "Point-Charge Calculations of Energy Levels of Magnetic Ions in Crystalline Electric Fields" in *Solid State Physics*, Vol. 16, ed. by F. Seitz, D. Turnbull (Academic, New York 1964)
2.8 M. Tinkham: *Group Theory and Quantum Mechanics* (McGraw-Hill, New York 1964) Chap. 4
2.9 C. Herring: In *Magnetism*, Vols. IIB, IV, ed. by G. T. Rado, H. Suhl (Academic, New York 1966)
2.10 W. Heitler, F. London: Z. Phys. **44**, 455 (1927)
2.11 J. H. Van Vleck: Rev. Univ. Nac. Tucuman Ser. A**14**, 189 (1962)
2.12 J. B. Goodenough: *Magnetism and the Chemical Bond* (Interscience, New York 1963) Chap. 3
2.13 J. Kanamori: Phys. Chem. Solids **10**, 87 (1959)

2.14 P. W. Anderson: In *Magnetism*, Vol. I, ed. by G. T. Rado, H. Suhl (Academic, New York 1963)
2.15 P. J. Hay, J. C. Thibeault, R. Hoffmann: J. Am. Chem. Soc. **97**, 4884 (1975)
2.16 R. D. Willett, C. P. Landee: J. Appl. Phys. **52**, 2004 (1981)
2.17 H. G. Bohn, W. Zinn, B. Dorner, A. Kollmar: J. Appl. Phys. **52**, 2228 (1981)
2.18 J. W. Corbett, G. D. Watkins: Phys. Rev. Lett. **1**, 314 (1961)
2.19 D. K. Biegelsen: *Nuclear and Electron Resonance Spectroscopies Applied to Materials Science,* Materials Research Society Symposia Proceedings, Vol. 3 (Elsevier, North-Holland, Amsterdam 1981) p. 85
2.20 M. Tachiki: J. Phys. Soc. Jpn. **25**, 686 (1968)
2.21 E. O. Kane: *Semiconductors and Semimetals,* Physics of III–V Compounds, Vol. 1 (Academic, New York 1966) pp. 75–100
2.22 L. M. Roth, B. Lax, S. Zwerdling: Phys. Rev. **114**, 90 (1959)
2.23 Y. Yafet: In *Solid State Physics,* Vol. 14, ed. by F. Seitz, D. Turnbull (Academic, New York, 1943) p. 93

Chapter 3

3. 1 J. H. Van Vleck: *The Theory of Electric and Magnetic Susceptibilities* (Oxford University Press, London 1932) p. 276
3. 2 J. S. Griffith: *The Theory of Transition-Metal Ions* (Cambridge University Press, Cambridge 1961) pp. 437–439
3. 3 M. Tinkham: *Group Theory and Quantum Mechanics* (McGraw-Hill, New York 1964) p. 132, Eq. (5.69)
3. 4 L. D. Landau: Z. Phys. **64**, 629 (1930) [English transl.: In *Collected Papers of L. D. Landau,* ed. by D. ter Haar (Gordon and Breach, New York 1965)]
3. 5 R. Peierls: Z. Phys. **80**, 763 (1933)
3. 6 A. H. Wilson: Proc. Cambridge Philos. Soc. **49**, 292 (1953)
3. 7 E. H. Sondheimer, A. H. Wilson: Proc. R. Soc. London **A210**, 173 (1951)
3. 8 J. Bardeen: In *Handbuch der Physik,* Vol. 15, Low Temperature Physics II, ed. by S. Flügge (Springer, Berlin, Göttingen, Heidelberg 1956) p. 305
3. 9 F. Seitz: *Modern Theory of Solids* (McGraw-Hill, New York 1940) p. 590
3.10 H. Fukuyama: Phys. Lett. **A32**, 111 (1970)
3.11 R. Kubo, Y. Obata: J. Phys. Soc. Jpn. **11**, 547 (1956)
3.12 W. J. de Haas, P. M. van Alphen: Proc. Amsterdam Acad. **33**, 1106 (1936)
3.13 J. M. Ziman: *Principles of the Theory of Solids* (Cambridge University Press, Cambridge 1964) p. 275
3.14 K. v. Klitzing, G. Dorda, M. Pepper: Phys. Rev. Lett. **45**, 494 (1981)
3.15 D. C. Tsui, A. C. Gossard: Appl. Phys. Lett. **38**, 550 (1981)
3.16 R. E. Prange: Phys. Rev. **B23**, 4802 (1981)
3.17 D. C. Tsui, S. J. Allen: Phys. Rev. **B24**, 4082 (1981)
3.18 R. B. Laughlin: Phys. Rev. **B23**, 5632 (1981)
3.19 *Landolt-Börnstein Zahlenwerte und Funktionen aus Physik, Chemie, Astro-*

nomie, *Geophysik und Technik*, 6th ed., Vol. 2, Part 9 Magnetic Properties I (Springer, *Geophysik und Technik*, 6th ed., Vol. 2, Part 9 Magnetic Properties I (Springer, Berlin, Heidelberg, New York 1962)

3.20 C. J. Krissman, H. B. Callen: Phys. Rev. **94**, 837 (1954)
3.21 R. M. Bozorth, H. J. W. Williams, D. E. Walsh: Phys. Rev. **103**, 572 (1956)
3.22 S. Foner: Rev. Sci. Instrum. **30**, 548 (1959);
S. Foner, E. J. McNiff, Jr.: Rev. Sci. Instrum. **39**, 171 (1968)
3.23 F. R. McKim, W. P. Wolf: J. Sci. Instrum. **34**, 64 (1957)
3.24 R. T. Schumacher, C. P. Slichter: Phys. Rev. **101**, 58 (1956)
3.25 A. M. Clogston, V. Jaccarino, Y. Yafet: Phys. Rev. **A134**, 1650 (1964)

Chapter 4

4. 1 J. S. Smart: *Effective Field Theories of Magnetism* (Saunders, Philadelphia 1966)
4. 2 L. D. Landau, E. M. Lifshitz: *Statistical Physics* (Pergamon, Oxford 1958) Sect. 149
4. 3 R. J. Elliott (ed.): *Magnetic Properties of Rare-Earth Metals* (Plenum, New York 1972)
4. 4 A. Yoshimori: J. Phys. Soc. Jpn. **14**, 807 (1959)
4. 5 R. B. Griffiths: Phys. Rev. Lett. **24**, 715 (1970)
4. 6 K. Huang: *Statistical Mechanics* (Wiley, New York 1963)
4. 7 M. Fisher: Rev. Mod. Phys. **46**, 597 (1974)
4. 8 P. Weiss, R. Forrer: Ann. Phys. **5**, 153 (1926)
4. 9 N. D. Mermin, H. Wagner: Phys. Rev. Lett. **17**, 1133 (1966)
4.10 V. L. Berezinskii: Sov. Phys. JETP **32**, 493 (1970)
4.11 J. M. Kosterlitz, D. J. Thouless: J. Phys. **C6**, 1181 (1973)
4.12 H. Kanazawa, N. Matsudaira: Prog. Theor. Phys. **23**, 433 (1960)
4.13 L. D. Landau: Sov. Phys. JETP **2**, 920 (1956); **5**, 101 (1957)
4.14 V. P. Silin: Sov. Phys. JETP **6**, 945 (1958)
4.15 C. Herring: In *Magnetism*, Vol. 4, ed. by G. T. Rado, H. Suhl; Academic, New York 1966)
4.16 J. C. Slater: Phys. Rev. **49**, 537 (1936)
4.17 E. C. Stoner: Proc. R. Soc. London **A165**, 372 (1938)
4.18 P. A. Wolff: Phys. Rev. **120**, 814 (1960)
4.19 A. W. Overhauser: Phys. Rev. **128**, 1437 (1962)
4.20 D. Wohlleben: Phys. Rev. Lett. **21**, 1343 (1968)
4.21 J. C. Slater: Phys. Rev. **81**, 385 (1951).
4.22 W. Kohn, L. J. Sham: Phys. Rev. A **140**, 1133 (1965)
4.23 T. M. Hattox, J. B. Conklin, Jr., J. C. Slater, S. B. Trickey: J. Phys. Chem. Solids **34**, 1627 (1973)
4.24 P. Hohenberg, W. Kohn: Phys. Rev. **B136**, 864 (1964)
4.25 P. Thiry, D. Chandesris, J. Lecante, C. Guillot, R. Pinchaux, Y. Pétroff: Phys. Rev. Lett. **43**, 82 (1979)

4.26 F. J. Himpsel, P. Heimann, D. E. Eastman: J. Appl. Phys. **52,** 1658 (1981)
4.27 J. Hubbard: J. Appl. Phys. **52,** 1654 (1981)
4.28 M. V. You, V. Heine, A. J. Holden, P. J. Lin-Chung: Phys. Rev. Lett. **44,** 1282 (1980)
4.29 J. Hubbard: Proc. R. Soc. London **A276,** 238 (1963)
4.30 P. Pfeuty, G. Toulouse: *Introduction to the Renormalization Group and to critical Phenomena* (Wiley, New York 1977)

Chapter 5

5. 1 F. Bloch: Phys. Rev. **70,** 460 (1946)
5. 2 J. H. Van Vleck: Phys. Rev. **74,** 1168 (1948)
5. 3 R. Kubo, K. Tomita: J. Phys. Soc. Jpn. **9,** 888 (1954)
5. 4 G. E. Pake: *Paramagnetic Resonance* (Benjamin, New York 1962) p. 142
5. 5 A. Abragam: *The Principles of Nuclear Magnetism* (Oxford University Press, New York 1961) Chap. 10
5. 6 R. E. Dietz, F. R. Merritt, R. Dingle, D. Hone, B. G. Silbernagel, P. M. Richards: Phys. Rev. **26,** 1186 (1971)
5. 7 T. Moriya: Prog. Theor. Phys. **16,** 23, 641 (1956)
5. 8 C. P. Slichter: *Principles of Magnetic Resonance*, 2nd ed., Springer Series in Solid-State Sciences, Vol.1 (Springer, Berlin, Heidelberg, New York 1980)
5. 9 N. Bloembergen, E. M. Purcell, R. V. Pound: Phys. Rev. **73,** 679 (1948)
5.10 S. Doniach: Proc. Phys. Soc. London **91,** 86 (1967)
5.11 N. F. Berk, J. R. Schrieffer: Phys. Rev. Lett. **17,** 433 (1966); S. Doniach, S. Engelsberg: Phys. Rev. Lett. **17,** 750 (1966)
5.12 N. D. Mermin: Phys. Rev. **B1,** 2362 (1970)
5.13 V. P. Silin: Sov. Phys. JETP **8,** 870 (1959)
5.14 P. M. Platzman, P. A. Wolff: Phys. Rev. Lett. **18,** 280 (1967)
5.15 F. J. Dyson: Phys. Rev. **98,** 349 (1955); G. Feher, A. F. Kip: Phys. Rev. **98,** 337 (1955)
5.16 S. Schultz, G. Dunifer: Phys. Rev. Lett. **18,** 283 (1967)
5.17 H. Hasegawa: Prog. Theor. Phys. **21,** 483 (1959)
5.18 S. Schultz, M. R. Shanabarger, P. M. Platzman: Phys. Rev. Lett. **19,** 749 (1967)
5.19 P. S. Pershan: J. Appl. Phys. **38,** 1482 (1967)

Chapter 6

6. 1 F. Keffer: In *Handbuch der Physik*, Vol. 18, Part 2, Ferromagnetism, ed. by S. Flügge (Springer, Berlin, Heidelberg, New York 1966)
6. 2 R. M. White, M. Sparks, I. Ortenberger: Phys. Rev. **A139,** 450 (1965)
6. 3 C. Kittel: Phys. Rev. **73,** 155 (1948)
6. 4 L. R. Walker: J. Appl. Phys. **29,** 318 (1958)
6. 5 R. L. White, I. H. Solt, Jr.: Phys. Rev. **104,** 56 (1956)

6. 6 J. F. Dillon, Jr.: J. Appl. Phys. **31**, 1605 (1960)
6. 7 R. W. Damon, J. R. Eshbach: J. Phys. Chem. Solids **19**, 308 (1981)
6. 8 M. Steiner, K. Kakurai, W. Knop, R. Pynn, J. K. Kjems: Solid State Comm. **41**, 329 (1982)
6. 9 A. P. Ramirez, W. P. Wolf: Phys. Rev. Lett. **49**, 227 (1982)
6.10 F. Bloch: Z. Phys. **61**, 206 (1930)
6.11 G. Low: Proc. Phys. Soc. London **82**, 992 (1963)
6.12 F. J. Dyson: Phys. Rev. **102**, 1217 (1956)
6.13 M. Sparks: *Ferromagnetic Relaxation Theory* (McGraw-Hill, New York 1964)
6.14 N. Bloembergen, R. W. Damon: Phys. Rev. **85**, 699 (1952)
6.15 H. Suhl: Proc. IRE **44**, 1270 (1957)
6.16 S. J. Allen, R. Loudon, P. L. Richards: Phys. Rev. Lett. **16**, 463 (1966)
6.17 D. D. Sell, R. L. Greene, R. M. White: Phys. Rev. **158**, 489 (1967)
6.18 P. A. Fleury, R. Loudon: Phys. Rev. **166**, 514 (1968)
6.19 P. M. Richards: Unpublished
6.20 R. Kubo: J. Phys. Soc. Jpn. **12**, 570 (1957)
6.21 P. M. Richards, P. K. Leichner: Phys. Rev. **B7**, 453 (1973)
6.22 J. C. Burgiel, M. W. P. Strandberg: J. Appl. Phys. **35**, 852 (1964)
6.23 T. Izuyama, E. J. Kim, R. Kubo: J. Phys. Soc. Jpn. **18**, 1025 (1963)
6.24 C. Herring: In *Magnetism*, ed. by G. T. Rado, H. Suhl (Academic, New York 1966) Vcl. IV, Sect. XIV, 3
6.25 G. Barnea, G. Horwitz: J. Phys. **C6**, 738 (1973)
6.26 R. Lowde: Proc. R. Soc. London **A235**, 305 (1956)
6.27 M. H. Seavey, P. E. Tannenwald: Phys. Rev. Lett. **1**, 168 (1958)
6.28 J. R. Sandercock, W. Wettling: IEEE Trans. **M–14**, 442 (1978)

Chapter 7

7. 1 T. Wolfram, J. Callaway: Phys. Rev **130**, 2207 (1963)
7. 2 D. Hone, H. Callen, L. Walker: Phys. Rev. **144**, 283 (1966)
7. 3 W. J. L. Buyers, R. A. Cowley, T. M. Holden, R. W. Stevenson: J. Appl. Phys. **39**, 1118 (1968)
7. 4 A. Osheroff, P. S. Pershan: Phys. Rev. Lett. **21**, 1593 (1968); P. Moch, G. Parisot, R. E. Dietz, H. J. Guggenheim: Phys. Rev. Lett. **21**, 1596 (1968)
7. 5 R. Weber: Phys. Rev. Lett. **21**, 1260 (1968)
7. 6 S. M. Rezende, C. B. de Araujo, E. Montarroyos, V. Jaccarino: Solid State Commun. **35**, 627 (1980)
7. 7 J. Friedel: Nuovo Cimento Suppl. **2**, 287 (1958)
7. 8 A. Clogston, B. Matthias, M. Peter, H. Williams, E. Corenzwit, R. Sherwood: Phys. Rev. **125**, 541 (1962)
7. 9 P. W. Anderson: Phys. Rev. **124**, 41 (1961)
7.10 D. C. Abbas, T. J. Aton, C. P. Slichter: Phys. Rev. Lett. **41**, 719 (1978)
7.11 G. T. Rado, H. Suhl (eds.): *Magnetism*, Vol. V (Academic, New York 1973)

7.12 G. Grüner, A. Zawadowski: Rep. Prog. Phys. **37**, 1497 (1974); Prog. Low Temp. Phys. **B7**, 591 (1978)
7.13 J. R. Schrieffer, P. A. Wolff: Phys. Rev. **149**, 491 (1966)
7.14 J. Kondo: Prog. Theor. Phys. **32**, 37 (1964)
7.15 N. Andrei: Phys. Rev. Lett. **45**, 379 (1980)
7.16 P. B. Wiegmann: J. Phys. **C14**, 1463 (1981)
7.17 K. Yosida, A. Yoshimori: In *Magnetism*, Vol. V, ed. by G. T. Rado, H. Suhl (Academic, New York 1973)
7.18 P. W. Anderson: Phys. Rev. **164**, 352 (1967)
7.19 C. Lacroix: J. Phys. **F11**, 2389 (1981)
7.20 P. Nozières: J. Low Temp. Phys. **17**, 31 (1974)
7.21 K. G. Wilson: Nobel Symp. **24**, 68 (1973)
7.22 H. Fröhlich, F. R. N. Nabarro: Proc. R. Soc. London **A175**, 382 (1940)
7.23 M. A. Ruderman, C. Kittel: Phys. Rev. **96**, 99 (1954)
7.24 C. Zener: Phys. Rev. **81**, 440 (1951)
7.25 T. Kasuya: Prog. Theor. Phys. **16**, 45, 58 (1956)
7.26 K. Yosida: Phys. Rev. **106**, 893 (1957)
7.27 H. Levine: Personal communication
7.28 J. B. Boyce, C. P. Slichter: Phys. Rev. **B13**, 379 (1976)
7.29 S. F. Edwards, P. W. Anderson: J. Phys. F **5**, 965 (1975)
7.30 D. Sherrington, S. Kirkpatrick: Phys. Rev. Lett. **35**, 1792 (1975)
7.31 J. R. L. de Almeida, E. J. Thouless: J. Phys. **A11**, 983 (1978)
7.32 H. Sompolinsky, A. Zipplius: Phys. Rev. B **25**, 6860 (1982)
7.33 S. Chakravorty, P. Panigrahy, P. A. Beck: J. Appl. Phys. **42**, 1698 (1971)
7.34 C. P. Bean: J. Appl. Phys. **26**, 1381 (1955)
7.35 C. Schwink, K. Emmerich, U. Schulze: Z. Phys. **B31**, 385 (1978)
7.36 S. Crane, H. Claus: Phys. Rev. Lett. **46**, 1693 (1981)
7.37 S. Fishman, A. Aharony: J. Phys. C **12**, L 729 (1979)

Chapter 8

8.1 P. A. Egelstaff (ed.): *Thermal Neutron Scattering* (Academic, New York 1965)
8.2 W. Marshall, S. W. Lovesey: *Theory of Thermal Neutron Scattering* (Oxford University Press, Oxford 1971)
8.3 S. W. Lovesey, T. Springer (eds.): *Dynamics of Solids and Liquids by Neutron Scattering*, Topics in Current Physics, Vol. 3 (Springer, Berlin, Heidelberg, New York 1977)
8.4 R. E. Rundle, C. G. Shull, E. O. Wollen: Acta Crystallogr. **5**, 22 (1952)
8.5 N. W. Ashcroft, N. D. Mermin: *Solid State Physics* (Holt, Rinehort, and Winston, New York 1974) Appendix L
8.6 R. M. Moon, T. Riste, W. C. Koehler: Phys. Rev. **181**, 920 (1969)
8.7 R. A. Erickson: Phys. Rev. **90**, 779 (1953)
8.8 T. Izuyama, D. Kim, R. Kubo: J. Phys. Soc. Jpn. **18**, 1025 (1963)

8.9 C. C. Becerra, Y. Shapira, N. F. Oliveira, T. S. Chang: Phys. Rev. Lett. **44**, 1692 (1980)
8.10 R. M. Moon, J. W. Cable, Y. Shapira: J. Appl. Phys. **52**, 2025 (1981)
8.11 C. G. Shull, Y. Yamada: J. Phys. Soc. Jpn. **17**, Suppl. B-III, 1 (1962)
8.12 C. G. Shull, H. A. Mook: Phys. Rev. Lett. **16**, 184 (1966)
8.13 G. C. Low, A. Okazaki, R. W. H. Stevenson, K. C. Tuberfield: J. Appl. Phys. **35**, 998 (1964)
8.14 H. Möller, J. Houmann, A. Mackintosh: J. Appl. Phys. **39**, 807 (1968)
8.15 D. Cribier, T. Jacrot, G. Parette: J. Phys. Soc. Jpn. **17**, Suppl. B-III, 67 (1962)
8.16 R. D. Lowde, C. G. Windsor: Adv. Phys. **19**, 813 (1970)
8.17 P. G. DeGennes: J. Phys. Chem. Solids **4**, 223 (1958)
8.18 C. G. Shull, M. K. Wilkinson Phys. Rev. **97**, 304 (1955)
8.19 W. Marshall J. Phys. C1, 88 (1968)
8.20 R. A. Medina, J. W. Cable: Phys. Rev. B **15**, 1539 (1977)

Subject Index

Absorption
    infrared 203
    microwave 147, 175–176, 195
    optical 203
Amorphous magnets 111
Anderson Hamiltonian 228
Angular momentum
    and rotation of coordinates 33
    orbital 4
Anisotropy 60, 112
Antiferromagnetism 105

Bitter magnet 99
Bloch equations 147–148, 180
Boltzmann distribution function 7, 249
Boltzmann equation 170
Born approximation 248, 257
Bottleneck condition 180
BPP theory 165
Bragg scattering
    magnetic 252
    nuclear 261–266
Brillouin functions 75
Brillouin scattering 206
Broken symmetry 119, 183–184

Causality 13, 146
Chemical potential 84, 125
Compensation temperature 111
Conduction electron spin resonance 173–177
Core polarization 98
Correlation 130
Correlation function 18, 116, 151
Correlation length 117
Critical exponents 120–121
Critical neutron scattering 269
Cross section
    electrons 226
    neutrons 249
Crystal fields 38–44
    cubic 39
    tetragonal 62
Curie constant 24
Curie temperature 104
Curie's law 74
Curie–Weiss law 95, 105

Debye–Waller factor 254
deHaas–van Alphen effect 83–87
Demagnetization factor 190
Density matrix
    definition 10
    static 70, 84
    time-dependent 16, 143
Diamagnetism 72–73, 78–81
    in bismuth 82
Diffuse neutron scattering 266–270
Diffusion 158, 175
Dipole–dipole interaction 44, 153, 189
Dirac delta function 12
    integral representation 146
    Lorentzian representation 149
Dirac equation 29

Exchange
    anisotropic 61
    antisymmetric 61
    direct 45–52
    Heisenberg 52
    macroscopic 119
    RKKY 239–242
    sd 233
    Slater 135
    superexchange 52–55
Exchange enhancement 127
Exchange narrowing 154, 200, 208–209

Exchange splitting 134
Exchange stiffness 120, 186

Faraday balance 96–97
Faraday effect 181–182
Fermi liquid theory 124–129, 169–174, 209
Ferrimagnetism 107
Ferromagnetism 104
Field operators 24
    for Kondo problem 228
    free electron 26
    Wannier functions 50
Fluctuation–dissipation theorem 15–19, 116, 151, 166, 207, 223
Form factor, magnetic 260, 266
Friedel sum rule 230
Frustration 243

Gaussian line 150
Giant moments 133
Gouy balance 97
Green's function
    and density of states 221
    and susceptibility 146, 223
    double-time 17, 146
    for Anderson Hamiltonian 229
    lattice 119
    of differential operator 145–146, 148
    of wave equation 205
    operator 216, 221
Group theory 41–43
g-value 58
    III–V semiconductors 66–68
    Landé 64
    Mn in Cu 180–181
    silicon 59
Gyromagnetic ratio 185

Hartree–Fock approximation 129–131, 133, 136, 141, 228
Heitler–London model 48–49
Helimagnetism 106
Holstein–Primakoff approach 187
Hubbard model 139–142
Hyperfine field
    contact 37, 98, 241
    dipolar 37

    orbital 36

Inhomogeneous broadening 150
Ising model 63

Knight shift 98
Kondo effect 232–239
Kondo temperature 236
Korringa relaxation 179
Kramers doublet 65
Kramers–Kronig relations 13–14, 97, 116, 147, 166, 269

Landau susceptibility 81
Landé g-value 64
Langevin susceptibility 75
Lifshitz point 263–264
Ligand fields *see* Crystal fields
Light scattering 204–206
Linewidth 152, 154, 210
Local modes 214–224
Long-range order 111, 121
Lorentz field 190
Lorentzian line 150

Magnetic energy 5
Magnetic moment 3, 5
    and neutron cross-section 258
    effective 75, 77
    nuclear 36
    operator 9
    spin 9
Magnetization 4, 7, 11
Magnetometer 96
Magnetostatic modes 193–195
Magnetron
    Bohr 32
    nuclear 36
Magnons 185
Maxwell's equations
    macroscopic 3, 174, 193
    microscopic 2
Mean field theory 102, 142
Metamagnetism 114
Mictomagnetism 245
Miss van Leeuwen's theorem 8, 79
Molecular field *see* Mean field theory
Moments
    fourth 152

Moments
    method of 150–154
    second 152, 207
Multipole expansion 32

Néel temperature 105
Nonlinear effects 11
    deHaas–van Alphen effect 83–86
    resonance 158
    saturation 75
    solitons 196
    spin echoes 159
    spin-wave 201
Nuclear-spin relaxation 162–163

Onsager reaction field 117
Onsager relation 19–20
Operator equivalents 33–35, 39
Orbital angular momentum 4
Order parameter 119

Parallel pumping 202
Paramagnetic resonance 147
Paramagnetism 73–78, 90–95
Paramagnons 167–168
Partition function 7, 70, 83, 84
Pauli susceptibility 91
    band structure corrections 95
    many-body corrections 129
Pendulum galvanometer 96
Phase transition 119
Phonons 253
Photoemission 137

Quadrupole Hamiltonian 32–35
Quantized Hall effect 86–90
Quenching 43

Raman scattering 206, 224
Random fields 224
Random phase approximation 102
Rare earths
    ionic configurations 64
    metallic magnetic state 108
Relaxation
    electron distribution 173
    spin 148
    spin waves 200
Relaxation function 157

Resistivity
    aluminum 224
    copper 231
Resonance
    conduction electron 173
    ferromagnetic 192
Resonant mode 223
Response function 17
RKKY interaction 239–242

Scattering theory 224–226
sd interaction 234, 240
sd mixing 228
Second quantization 20–28
    Coulomb interaction 26–27, 130
    momentum 81
    Zeeman interaction 28, 90
Short range order 116
Slater determinant 21, 45
Slater exchange 135
Solitons 196
Specific heat
    Fermi liquid 127
    Kondo system 232
    near T_c 118
    spin waves 199
Spin density functional 136–137
Spin density wave 133
Spin echoes 159–160
Spin-flop 114
Spin fluctuation 167
Spin glass 243
Spin Hamiltonian 55–66, 74
Spin-lattice relaxation 162
Spin-orbit interaction 30, 58
Spin waves
    above T_c 213
    in ferromagnetic insulators 184–193
    in ferromagnetic metals 211–212
    in sodium 177–178
    on surfaces 196
Spin-wave sidebands 203
Spin-wave theory 186–206
Stoner excitations 211
Stoner model 130–139
Structure factor
    magnetic 261
    nuclear 252

Suhl instability 202
Superconductors
  diamagnetism 83
  magnets 99
  paramagnetism 92
Superexchange 52
Superparamagnetism 246
Susceptibility
  and Faraday effect 181
  and neutron scattering 266
  antiferromagnetic 106
  complex 146–149, 185
  diamagnetic 73
  ferrimagnetic 110
  ferromagnetic 105
  generalized 12
  Kubo–Obata 82
  Landau 81, 123
  Landau–Peierls 82
  Langevin 74, 98
  measurements of 96–98
  nonlocal 12, 165
  Pauli 90–95
  superconductor 83, 92
  Van Vleck 60, 74

T_1
  calculation of 161–162
  definition 148
  measurement of 158–160, 176
T_2
  definition 148
  measurement of 160
Ten-thirds effect 207
Time reversal 19, 43–44, 65
TMMC 158
Topological order 121
Transition metals
  effective moments 75
  ionic configurations 56
  Kondo temperatures in Cu 233
  susceptibilities 94
Tricritical point 114
Triple-axis spectrometer 256

Van Vleck susceptibility 74
Vibrating sample magnetometer 96–97
Virtual bound state 225

Walker modes 195
Wannier functions 50
Wigner-Eckart theorem 34, 58, 70, 260

XY model 63, 122

Zeeman interaction 31, 58, 76

Electron Correlation and Magnetism in Narrow-Band Systems

Proceedings of the Third Taniguchi International Symposium, Mount Fuji, Japan, November 1–5, 1980
Editor: **T. Moriya**
1981. 99 figures. XIV, 257 pages
(Springer Series in Solid-State Sciences, Volume 29). ISBN 3-540-10767-3

Contents: d Metals and Compounds. – Finite Temperature Properties. – d Metals and Compounds – Ground-State Properties. – Some Theoretical Aspects of Narrow-Band Problems. – Mixed Valence in 4f Compounds. – Index of Contributors.

R. P. Huebener

Magnetic Flux Structures in Superconductors

1979. 99 figures, 5 tables. XI, 259 pages
(Springer Series in Solid-State Sciences, Volume 6). ISBN 3-540-09213-7

"...The emphasis is on exposing the reader to the basic physical phenomena and the diverse experimental techniques which are used to investigate them. Many references (over 500) are given throughout the text to enable those so inclined to follow up any of the topics... This book will be useful to researchers and to people involved in engineering aspects of superconductivity, particularly of small low field devices."
Nature

D. C. Mattis

The Theory of Magnetism I

Statics and Dynamics
1981. 58 figures. XV, 300 pages
(Springer Series in Solid-State Sciences, Volume 17). ISBN 3-540-10611-1

"...Mattis's relentless approach will not be to everyone's taste, but his book will be a useful addition to the library of anyone deeply interested in the origins of magnetism and the careful study of mathematical models. The statistical mechanician, or the particle theorist looking for hints on how to solve the lattice gauge theory problem, may, however, prefer to wait for the second volume which will cover thermodynamics and statistical mechanics. Finally, praise must be given for the introductory chapter, 38 pages long, which spells out the history of magnetism from the earliest days to the present, places it in the perspective of the general evolution of physics and the development of Western thought, and is backed up by marvellous quotations and an impressive bibliography. This chapter can be strongly recommended to anyone interested in the history of science and, almost alone, would justify purchase of the book."
Nature

Physics in High Magnetic Fields

Proceedings of the Oji International Seminar Hakone, Japan, September 10–13, 1980
Editors: **S. Chikazumi, N. Miura**
1981. 257 figures. X, 358 pages
(Springer Series in Solid-State Sciences, Volume 24). ISBN 3-540-10587-5

Contents: Recent Progress in the Steady High Field System. – Physics in Pulsed High Magnetic Fields. – Cyclotron Resonance and Laser Spectroscopy of Semiconductors. – Impurity States in High Magnetic Fields. – Magneto-Transport Phenomena. – Excitons and Magneto-Optics. – Electron-Hole Drops and Semimetals. – Narrow Gap Semiconductors. – Space Charge Layer and Superlattice. – Layered Materials and Intercalation. – Magnetism and Magnetic Semiconductors. – Photograph of the Participants of the Seminar. – List of Persons in the Photograph. – Index of Contributors.

Springer-Verlag
Berlin
Heidelberg
New York

C. P. Slichter

Principles of Magnetic Resonance

Corrected 2nd printing of the 2nd revised and expanded edition. 1980. 115 figures, 9 tables. XII, 397 pages
(Springer Series in Solid-State Sciences, Volume 1). ISBN 3-540-08476-2

"...an improved introductory book... Slichter, a professor of physics, and his graduate students... are responsible for numerous original and significant research contributions that appear in the book. The clarity and style in which the book is written reveals Slichter's research expertise and talent as an excellent teacher and expositor... The referencing is so good that certain new priorities in research contributions to nmr appear that were not previously obvious..."

Physics Today

Superconductivity in Ternary Compounds I

Structural, Electronic, and Lattice Properties
Editors: Ø. Fischer, M. B. Maple
With contributions by numerous experts
1982. 152 figures. XIV, 283 pages
(Topics in Current Physics, Volume 32)
ISBN 3-540-11670-2

Superconductivity in Ternary Compounds II

Superconductivity and Magnetism
Editors: M. B. Maple, Ø. Fischer
With contributions by numerous experts
1982. 136 figures. XVI, 306 pages
(Topics in Current Physics, Volume 34)
ISBN 3-540-11814-4

This two-volume treatment deals with the extraordinary developments that have take place in the domain of ternary superconductors over the last couple of years. They contain comprehensive reviews written by experts in the field covering a large variety of the striking properties of ternary superconductors. The presentation is such that the books will be useful both to students and senior researchers seeking an introduction to the subject, as well as to specialists with its compilation of the results obtained so far in this field.

S. V. Vonsovsky, Y. A. Izyumov, E. Z. Kurmaev

Superconductivity of Transition Metals

Their Alloys and Compounds
Translated from the Russian by E. H. Brandt, A. P. Zavarnitsyn
1982. 182 figures. XIII, 512 pages
(Springer Series in Solid-State Sciences, Volume 27). ISBN 3-540-11382-7

Contents: Introduction. – The Theory of Strong Coupling Superconductors. – Superconductivity and Magnetism. – Superconductivity in Transition Metals. – Superconductivity in Transition-Metal Alloys. – Compounds with A-15 Structure. – Other Compounds Based on Transition Metals. – High-Temperature Superconductors and Lattice Instability of Compounds. – Radiation Effects on Superconductors. – References. – Subject-Index.

Springer-Verlag
Berlin
Heidelberg
New York